Springer Hydrogeology

Series Editor

Juan Carlos Santamarta Cerezal, San Cristóbal de la Laguna, Sta. Cruz Tenerife, Spain

The *Springer Hydrogeology* series seeks to publish a broad portfolio of scientific books, aiming at researchers, students, and everyone interested in hydrogeology. The series includes peer-reviewed monographs, edited volumes, textbooks, and conference proceedings. It covers the entire area of hydrogeology including, but not limited to, isotope hydrology, groundwater models, water resources and systems, and related subjects.

More information about this series at http://www.springer.com/series/10174

Robert Maliva

Climate Change and Groundwater: Planning and Adaptations for a Changing and Uncertain Future

WSP Methods in Water Resources Evaluation Series No. 6

Robert Maliva
WSP and Florida Gulf Coast University
Fort Myers, FL, USA

ISSN 2364-6454 ISSN 2364-6462 (electronic)
Springer Hydrogeology
ISBN 978-3-030-66815-0 ISBN 978-3-030-66813-6 (eBook)
https://doi.org/10.1007/978-3-030-66813-6

This Springer imprint is published by the registered company Springer Nature Switzerland AG
The registered company address is: Gewerbestrasse 11, 6330 Cham, Switzerland

Preface

Fresh groundwater is a vital resource for global potable, irrigation, and industrial water supply. Groundwater has the great advantages as a water supply of often enormous volumes in storage, high-quality (and thus low treatment costs), and widespread and ready access across groundwater basins, which allows for convenient decentralized use. The advantages of groundwater use have led to overexploitation in many areas of the world as manifested by declining groundwater levels and associated adverse impacts, such as salinization of aquifers, land subsidence, declining spring flows, and drying of wetlands. Global groundwater use, and thus aquifer depletion, has increased over time due to population growth, economic development, climate variation, and other factors.

Global climate change will result in increased temperatures, which will likely increase water demands, and modifications of global precipitation rates and patterns, and thus aquifer recharge. There is still considerable uncertainty as to local directions and magnitudes of changes in precipitation. Some areas, such as southwestern North America and Mediterranean region, are anticipated to experience drier conditions, and droughts may become more frequent and intense. Groundwater will increasingly be needed to perform a stabilization role in mitigating fluctuations in the supply of surface waters, serving as a buffer against droughts.

Climate change has become a frequent subject in the mass media, and the academic literature on the subject is now enormous with many thousands of papers and numerous dedicated journals, books, organizations, conferences, conference sections, and Internet sites. The Intergovernmental Panel on Climate Change (IPCC) was established by the United Nations to provide "policymakers with regular scientific assessments on climate change, its implications and potential future risks, as well as to put forward adaptation and mitigation options." Virtually, every aspect of climate change has been the subject of multiple technical papers or dissertations. The challenge is not that the climate change literature is incomplete but rather that it is overwhelming to people outside of the climate change research community.

A number of workers have observed with respect to climate change and water that a chasm exists between the climate change research community and the decision-making community. Decision makers are water users and the staff of water utilities and regional water suppliers who are actually responsible for developing water supplies to meet their own needs and those of their customers. Researchers are frustrated that the result of their climate change modeling and projections are not adequately being considered in water supply planning and implementation. Decision makers often find the output of the climate change research community to not be directly actionable, that is, capable of being acted upon in their water supply planning processes. High degrees of uncertainty in climate change projections limit their usefulness in the traditional water supply planning process.

The "Methods in Water Resources Evaluation" series approaches hydrogeology and water resources evaluation and development from an applied perspective. Key considerations are what is technically and economically practicable for those responsible for developing and managing water resources, which may not necessarily be the current technical state of the art in academia. Water supply investigations have finite budgets and time constraints. *Climate Change and Groundwater: Planning and Adaptations for a Changing and Uncertain Future* attempts to bridge the chasm between the climate change research and decision-making communities with respect to the impacts of climate change on groundwater. This book is a review and numerous references are provided to key papers and data sources that can provide additional information and further guidance on climate change adaptation with respect to groundwater.

An overview is provided of the climate change modeling process and avenues for decision makers to access and efficiently use the results of the modeling. A summary is also provided of the actual decision-making processes in the water field. For climate change research to have impact, it needs to reach the actual people who make water supply decisions in a form that they can use. Issues explored are who actually makes water supply decisions, how are those decisions made, and what is the water supply planning horizon. Despite calls for climate change adaptation decisions to be broadly based with contributions from numerous stakeholders within society, water supply planning is still largely siloed to a small number of technical experts.

Climate change will impact groundwater through changes in water demands, impacts to groundwater recharge (and thus the sustainable supply of water), and through sea level rise and its impacts on the salinity of coastal aquifers and groundwater levels. Rising groundwater levels induced by sea level rise and increases in precipitation can cause flooding of low-lying areas and interfere with the operation of or damage underground infrastructure. The impacts of climate change on water resources (and society in general) are insidious in that they tend to occur so slowly as to not be noticeable by the casual observer on a year-to-year basis, which leads to complacency. However, progressively increasingly impacts from accelerating climate change may eventually profoundly impact vulnerable individuals, communities, cities, and countries.

Although climate change will impact groundwater resources in multiple manners, the impacts of population growth (and associated increases in water demands for domestic use and food supply) and economic development will tend to have a much greater impact on groundwater resources. Climate change may exacerbate an already deteriorating situation in arid and semiarid regions. It is also important to recognize that some areas may experience net benefits from climate change. Vast disparities exist between countries and regions in their ability to adapt to climate change (i.e., their adaptive capacity).

Climate change vulnerability or risk assessments involve systematic consideration of how various elements of water supply systems and other infrastructure could be impacted under different climate change scenarios, the probability of the damaging scenarios, when in the future the damage might occur, and the magnitude of the harm. Once risks are assessed, the next step is evaluation of adaptation options to ameliorate the risks. Most adaptation options for climate change are essentially the same as those for water scarcity in general. In areas experiencing drier conditions or growing water demands that exceed their sustainable supplies of groundwater and local surface water, water supplies and demands will have to be brought into balance by some combination of demand reduction, reallocation of water to higher value uses (i.e., from agricultural to domestic and commercial/ industrial uses), development of alternative water sources (e.g., desalination and wastewater reuse), and optimization of the use of existing resources through conjugate use and managed aquifer recharge. The impacts of climate change on groundwater will be gradual. Rather than requiring immediate adaptive actions, climate change is an additional factor that needs to be incorporated (i.e., mainstreamed) into existing water supply and land-use planning processes.

Fort Myers, USA Robert Maliva

Contents

About the Author

Dr. Robert Maliva has been a consulting hydrogeologist since 1992 and is currently a Principal Hydrogeologist with WSP USA Inc. and a Courtesy Faculty Member at the U.A. Whitaker College of Engineering, Florida Gulf Coast University. He is currently based in Fort Myers, Florida. Dr. Maliva specializes in groundwater resources development including alternative water supply, managed aquifer recharge, and desalination projects.

Robert Maliva completed his Ph.D. in geology at Harvard University in 1988. He also has a Masters degree from Indiana University (Bloomington) and a BA in geological sciences and biological sciences from the State University of New York at Binghamton. Upon completion of his doctorate degree, Dr. Maliva has held research positions in the Department of Earth Sciences at the University of Cambridge, England, and the Rosenstiel School of Marine and Atmospheric Science of the University of Miami, Florida. He grew up in New York City and attended Stuyvesant High School in Manhattan.

Dr. Maliva has also managed or performed numerous other types of water resources and hydrologic investigations including contamination assessments, environmental site assessments, water supply investigations, wellfield designs, and alternative water supply investigations. He has maintained his research interests and completed studies on such diverse topics Precambrian silica diagenesis, aquifer heterogeneity, precipitates in landfill leachate systems, carbonate diagenesis, and various aspects of the geology of Florida. Dr. Maliva gives frequent technical presentations and has numerous peer-reviewed papers and conference proceedings publications on ASR and injection well and water supply issues, hydrogeology, and carbonate geology and diagenesis. He authored or coauthored the books, *Aquifer Storage and Recovery and Managed Aquifer Recharge Using Wells, Arid Lands Water Evaluation and Management, Aquifer Characterization, and Anthropogenic Aquifer Recharge.*

Chapter 1
Introduction to Climate Change and Groundwater

1.1 Introduction

There is now little doubt in the scientific community that the Earth's climate is changing at an accelerating rate and that rising global temperatures are due largely to increasing atmospheric concentrations of greenhouse gases (GHGs). Climate change is already impacting ecosystems and will have growing disruptive impacts on human health and many important sectors of societies, including housing, healthcare, transportation, energy, food, basic materials, and water supplies. Climate change has become the subject of overwhelming attention in academia with an enormous volume of papers, dissertations, and books published, and dedicated journals, website sites, and conference and conference sessions on all aspects of the subject. Despite all the attention that climate change is receiving in academia and the popular press, there is widespread frustration that not enough is being done to mitigate and adapt to climate change.

Climate change mitigation, which is primarily the reduction of GHG emissions, ultimately requires international agreement and national government-level commitments and actions. Adaptation actions, on the contrary, are largely undertaken by the impacted parties, although higher level (i.e., international, national, or state) technical and financial support may be required. Activities taken in response to anthropogenic climate change contribute to the sustainable development goal of making societies more resilient to changes in general (Pielke et al. 2007). "Virtually every climate impact projected to result from increasing greenhouse-gas concentrations—from rising storm damage to declining biodiversity—already exists as a major concern. As long as adaptation is discussed in terms of its marginal effects on anthropogenic climate change, its real importance for society is obscured" (Pielke et al. 2007, p. 598).

Changes in temperature and precipitation will potentially affect local water demands and supplies, with the degrees of impacts depending on the direction, magnitude, and type of climate change and local circumstances, including existing

R. Maliva, *Climate Change and Groundwater: Planning and Adaptations for a Changing and Uncertain Future*, Springer Hydrogeology,
https://doi.org/10.1007/978-3-030-66813-6_1

climate conditions, the amount and types of water use (e.g., domestic, agricultural, and industrial), and the water sources exploited. The impacts of climate change in many areas will be superimposed on increasing water scarcity associated with population growth and economic development. Groundwater was estimated to globally provide 50% of domestic water supplies, 40% of self-provided industrial supplies, and 20% of irrigation supplies (Zekster and Everett 2004). More recent estimates by Siebert et al. (2010) indicate that the total global consumptive use of groundwater for irrigation is about 545 km^3/yr, which corresponds to 43% of the total consumptive irrigation water use of 1277 km^3/yr. The countries with the largest areas equipped for irrigation with groundwater, in absolute terms, are India (39 million ha), China (19 million ha) and the USA (17 million ha; Siebert et al. 2010). Irrigation is critical for meeting global food demands but expanding groundwater use for irrigation (both in absolute terms and as a percentage of total irrigation) is resulting in increasing aquifer overdrafts and declining aquifer water levels (Seibert et al. 2010).

Groundwater resources will be under even greater pressure if declines in surface water availability are compensated for by increased groundwater use. A number of studies reviewed the impacts of climate change on groundwater (e.g., Arnell 1999; Vörösmarty et al. 2000; Ranjan et al. 2006; Bates et al. 2008; Dragoni and Sukhija 2008; Kundzewicz et al. 2008; Earman and Dettinger 2011; Green et al. 2011; Taylor et al. 2013; Green 2016). Groundwater will be needed to play a key role in the adaptation of water supplies to climate change by serving as a buffer against variations in surface water supplies (Alley 2001; Green 2016). A basic conjugative use strategy is that surface water is used when available to meet immediate demands and to recharge aquifers. Groundwater is reserved for times when surface water supplies are inadequate to meet demands.

The option of using groundwater as a long-term buffer against surface-water shortages requires that groundwater use be sustainable. Groundwater will likely not be able to ease freshwater stress in those areas where climate change is projected to decrease groundwater recharge or where groundwater use is unsustainable under current climate conditions (Kundzewicz and Döll 2009). The unfortunate reality is that most of the major aquifers in the world's arid and semiarid zones are already experiencing rapid rates of groundwater depletion (Döll et al. 2014; Famiglietti 2014).

A major impediment to climate change adaptation, in general, is poor communication between the climate change research community and decision makers. With respect to water supply, decision makers are normally individual water users (in the case of self-supply) and the technical staff and management of local water utilities and regional water suppliers. Regional water suppliers provide wholesale raw or treated water to municipal water utilities and other large customers. Examples, of large regional water suppliers in the United States are Tampa Bay Water, Peace River Manasota Regional Water Supply Authority (Florida), Southern Nevada Water Authority, Tarrant Regional Water District (Texas), and Metropolitan Water District of Southern California. In the author's home in Southwest Florida, for example, water supply decisions are made by a local public water utility (Lee County Utilities), whose choices of supply (e.g., amount of fresh groundwater that can be pumped) are constrained by a regional governmental water management agency (South Florida Water Management District). Fresh groundwater is the preferred (least expensive) source of water, but its permitted use is capped because of environmental concerns, particularly impacts to wetlands.

The Intergovernmental Panel on Climate Change (IPCC) observed that

> Previous assessment methods and policy advice have been framed by the assumption that better science will lead to better decisions. Extensive evidence from the decision sciences shows that while good scientific and technical information is necessary, it is not sufficient, and decisions require context-appropriate decision-support processes and tools. (Jones et al. 2014, p. 198)

Much of the academic climate change research literature is written for other members of the research community and published in obscure (to the lay public) places. Decision makers inherently need "actionable" information that they can use in their water supply planning activities.

Adger et al. (2005, p. 79) made the key observations that "Adapting to climate change involves cascading decisions across a landscape made up of agents from individuals, firms and civil society, to public bodies and governments at local, regional and national scales, and international agencies" and that "a broad distinction can be drawn between action that often involves creating policies or regulations to build adaptive capacity and action that implements operational adaptation decisions." With respect to water supply planning, and thus climate change adaptation decision making, scientific and technical information on climate change needs to flow to water users and suppliers in a form that is understandable and usable to them, which includes projections of the probability and likely timing and magnitude of changes.

The objective of this book is to explore the impacts of climate change on groundwater, and adaptation and resiliency enhancement options from an applied perspective. Overviews are provided of historical anthropogenic (human-caused) climate change, climate modeling, the impacts of climate change on groundwater, vulnerability assessments, adaptation options, and the data and methods available to assess potential local impacts. A key focus is on available data sources and methodologies that can provide required information or guidance and are within the technical and financial resources of water users, utilities, and other water suppliers.

1.2 Climate Change and Groundwater

Sunlight (relatively shortwave energy) passes through the Earth's atmosphere and is reradiated back into space as longer wavelength infrared (thermal) energy (Fig. 1.1). GHGs are chemical compounds present in the Earth's atmosphere that allow sunlight to pass through but allow less heat to be reradiated back into space, causing a heating of the lower atmosphere. GHGs include compounds that are naturally present in the atmosphere (e.g., carbon dioxide, methane, water vapor, and

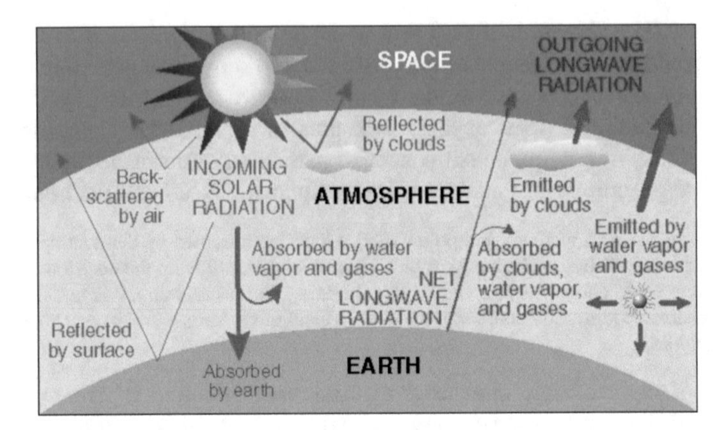

Fig. 1.1 Diagram of the greenhouse effect. Incoming solar radiation in absorbed by the Earth's surface and atmosphere and converted to long-wave thermal radiation, which is mostly radiated back into space. Increasing GHG concentrations result in the retention of more heat in the lower atmosphere. *Source* Markewich et al. (1997)

nitrous oxide) and synthetic compounds (e.g., chlorofluorocarbons, hydrofluorocarbons, perfluorocarbons, and sulfur hexafluoride; NOAA n.d.). Rising GHG concentrations increase the amount of thermal energy trapped in the lower atmosphere. The contributions of the various GHGs to global warming depend on their concentration, absorption strength, atmospheric residence time, and molecular mass, and the time period over which climate effects are of concern (Loaiciga et al. 1996).

The greenhouse effect resulting from increasing atmospheric concentrations of GHGs is causing a progressive rise in the temperature of the Earth's surface and lower atmosphere. Increased atmospheric temperatures are causing sea levels to rise mainly by the thermal expansion of the oceans and the melting of glaciers and polar ice caps. Increasing temperatures result in greater evaporation rates and, in turn, greater precipitation rates, which has been referred to as an intensification or acceleration of the global hydrologic cycle (Loaiciga et al. 1996).

An important distinction, often not understood by some of the general public, is between weather and climate. Spells of cold weather have been cited by climate change skeptics as evidence against global warming. Weather is short-term changes in the state of the atmosphere (temperature, humidity, cloudiness, precipitation) in a region, whereas climate is long-term average conditions. Global temperature projections indicate increasing variation around a progressively rising mean. The impacts of climate change are also projected to have considerable geographic variability. Whereas mean annual temperature is expected to increase essentially everywhere (but to varying degrees), some regions will experience either lesser or greater annual precipitation. The timing, form, and intensity of precipitation will also change with expected lesser snowfall and a greater intensity of rainfall events. Changes in temperature and precipitation will impact both the supply of and demand for water.

Climate has considerable natural variation on different time scales. Regional temperature and precipitation patterns (droughts and wet periods) are impacted by natural oceanic cyclicity, particularly the El Niño/Southern Oscillation (ENSO), Pacific Decadal Oscillation (PDO), and Atlantic Multidecadal Oscillation (AMO; McCabe et al. 2004; Hu and Huang 2009; Kuss and Gurdak 2014; Abiy et al. 2019). In evaluations of current climate changes in response to historical increases in GHG concentrations, an important consideration is the signal-to-noise ratio (Loaiciga et al. 1996). Over short periods of time, the climate change signal may be significantly smaller than the natural variability. However, in recent years unusually warm periods have become more frequent than can be explained as random events that are part of the natural climate variation. Hansen et al. (2012) described the tendency toward more common extreme events as the "climate dice" having become more and more "loaded" over the past 30 years, coincident with rapid global warming.

Groundwater is a critical source of water for both potable and irrigation uses and, to a lesser extent, industrial uses, and supports groundwater dependent ecosystems. Groundwater is generally less sensitive to short-term climatic variations than surface water supplies because of the very large volumes of water in underground storage in aquifers. Groundwater use is already at or above sustainable levels in much of the drier parts of the world and demands on groundwater are increasing due to population growth and economic development. In areas that are experiencing drier conditions due to climate change, there will likely be increasing pressure on groundwater resources to compensate for shortages in surface water supplies and decreasing soil moisture.

The literature on climate change focuses on two plausible strategic responses: mitigation and the related concepts of adaptation, coping, and resilience. Mitigation involves primarily policies to reduce greenhouse gas (GHG) emissions and thus decrease the rate and magnitude of climate changes. Adaptation in human systems was defined by the IPCC (2012) as "the process of adjustment to actual or expected climate and its effects, in order to moderate harm or exploit beneficial opportunities." Coping is defined as the "use of available skills, resources, and opportunities to address, manage, and overcome adverse conditions, with the aim of achieving basic functioning in the short to medium term" (IPCC 2012). Resilience is defined as "the ability of a system and its component parts to anticipate, absorb, accommodate, or recover from the effects of a hazardous event in a timely and efficient manner, including through ensuring the preservation, restoration, or improvement of its essential basic structures and functions" (IPCC 2012).

It is clearly desirable to mitigate against climate change. However, even with the implementation of drastic mitigation actions, climate change will continue to occur because of historical GHG emissions. Therefore, humans will be forced out of necessity to adapt to climate change with the preferred option being to increase the resilience of systems to accommodate a wide range of possible future climates. Human societies have varying abilities to adapt to local climate change, which is referred to as their adaptive capacity. Wealthier countries have the technical and financial resources to increase their resilience to climate change, whereas poorer developing countries have more limited options to cope with and adapt to climate change.

Adaptation to climate change continues to be the subject of considerable academic research, but for that research to be useful, the gap between climate change researchers and decision makers who need to use the results of the research must be bridged (Ferguson et al. 2014; Hewitt et al. 2017). With respect to groundwater, the term "decision maker" broadly includes all managerial, technical, and regulatory staff who are actively involved in the water supply planning process. Evaluating the potential impacts of climate change on local groundwater resources and identifying and applying options to optimize the management of groundwater under changing climate conditions is squarely in the realm of applied hydrogeology. The scientific community is needed to provide decision makers with information that is of practical use in the water supply planning process (i.e., is actionable).

Although this volume is focused on groundwater, effective management of groundwater resources requires a holistic approach in which other water resources are considered along with factors controlling water demand, and water governance rules and processes. The basic information requirements for adaptation of groundwater use to climate change and developing more resilient water supplies start with an understanding of the potential local direction and rates of changes of climate variables (particularly temperature, precipitation, and local seal level) and how local groundwater resources will respond to changes in those variables.

The initial step in adaptation planning is an evaluation of how climate change might locally impact groundwater quantity and quality, which, in additional to evaluating the direct effects of changes in temperature, precipitation, and sea level, also involves consideration of potential changes in the water demands of anthropogenic and natural systems, and potential changes in land uses and land cover. Climate change vulnerability assessments commence with a qualitative screening of potential risks from climate change. Qualitative screenings are followed by quantitative assessments that consider the probability, potential magnitude, and timing of identified risks. Quantitative assessments of impacts to groundwater systems often involve some type of numerical modeling. Vulnerability assessments, by their nature, involve consideration of adverse changes, but it is important to also recognize that climate change can also provide some benefits. For example, increased precipitation may locally reduce irrigation demands and increase aquifer recharge.

Adaptation planning is an exercise in management under uncertainty because there is a broad range of possible future climate scenarios. Given identified vulnerabilities, various actions are identified that can be employed to improve the resilience of groundwater resources, vulnerable infrastructure, and water supplies in general. It is now widely appreciated in the adaptation literature that robust ("no regrets" or "win-win") interventions that would provide benefits under current and a wide range of future climate conditions are preferred (Hallegatte 2009; Heltberg et al. 2009). Adaptation options should be avoided that would perform well under some future climate scenarios but may be maladapted to other plausible scenarios that might actually come to pass.

Where groundwater levels are expected to decline as a result of decreased recharge, managed aquifer recharge (MAR) technologies are available to increase groundwater recharge. Managed aquifer recharge (MAR), which is defined as the "purposeful recharge of water to aquifers for subsequent recovery or environmental benefit" (Dillon 2009), includes a broad suite of technologies to store and treat water underground.

The implementation of climate change adaptation actions can be either prompted or constrained by local groundwater governance regimes. Governmental agencies that regulate water use through restrictions on groundwater pumping and surface water withdrawals may create incentives for investments in MAR and alternative water supply systems (e.g., brackish groundwater desalination). Research performed or financially supported by governmental agencies provides the scientific foundation for the decision-making process. Governmental financial assistance may be required to support local implementation of measures to increase resilience. International support will needed for poorer developing countries that have neither the technical nor financial resources to adapt to climate changes for which they have had a minimal contribution. Conversely, environmental protection rules and policies for the quality of recharged water, and thus pretreatment requirements, can make some options, such as MAR using reclaimed wastewater, economically unviable.

The water planning horizons for water users and regulatory agencies influence adaptation responses. If expected local climate changes and sea level rise are small over planning horizons, then water users and suppliers may not place a high priority now on implementing adaptation actions against climate change.

1.3 Climate Change Modeling

The foundation of climate change predictions is general circulation models (GCMs), which are also referred to as global climate models. GCMs are three-dimensional models that include the entire planet. Coupled atmosphere-ocean general circulation models (AOGCMs) simulate all major processes that impact climate, including sea ice and evapotranspiration over land. Even more complex and comprehensive are earth system models (ESMs) that include representations of various biogeochemical cycles.

A key input to GCMs is the future atmospheric concentrations of GHGs (and other substances impacting atmospheric warming, e.g., aerosols). Future GHG concentrations are inherently unknowable as they depend upon future economic growth, technological developments, energy supply choices, and mitigation actions. To allow for direct comparison of the results from different GCMs, model runs have been performed using the same set of future GHG concentration scenarios. The IPCC for its Fifth Assessment Report (AR5) in 2014 adopted four Representative Concentration Pathways (RCPs) that represent a range of long-term concentration levels and trajectories taken over time to reach them. The results from an ensemble of runs of multiple GCMs using the RCPs, and earlier SRES (Special Report on Emissions Report) scenarios, provide a broad range of plausible future climate conditions. The Coupled Model Intercomparison Project Phase 5 (CMIP5) is a collaborative effort involving more than 20 climate modeling groups from around the world. A standard set of model simulations were run (using the RCPs) in order to evaluate how realistic the models are in simulating the recent past and to understand some of the factors responsible for differences in model projections (PCMDI n.d.). The IPCC AR5 relied heavily on the CMIP5 results, which have also been used in many other climate change studies.

The typical horizontal resolution (horizontal grid spacing) of current AOGCMs and ESMs is approximately 1°–2° for the atmospheric component and around 1° for the ocean and land (Flato et al. 2013). GCMs are thus suitable for simulating only large spatial-scale climate changes rather than more local changes that are pertinent for water supply planning. For more accurate local projections of climate change, GCM data need to be downscaled, which is most commonly performed using dynamic and statistical methods. Dynamic downscaling involves the creation of a finer-grid numerical model (i.e., a regional climate model; RCM) that is forced by (obtains boundary conditions from) a GCM. Statistical downscaling involves the development of fine-scale maps of climate change through the statistical relationships between local climate variables (e.g., temperature and precipitation) and the large-scale outputs of GCMs (Wilby et al. 2004).

Climate modeling is a very complex and specialized discipline with only a small number of research groups capable of developing GCMs and RCMs. The CMIP5 simulations results have been archived and are available to external users. The CMIP5 results are summarized in the IPCC AR5 (IPCC 2013) and are accessible using some interactive viewer programs such as the U.S. Geological Survey CMIP5 Global Climate Change Viewer (Alder et al. 2013; USGS n.d.) The Climate Wizard tool accesses some of the earlier CMIP4 modeling results that supported the IPCC Fourth Assessment Report (AR4; CGIAR n.d.)

1.4 Projected Global Climate Changes

Numerical modeling, whether it be of climate, groundwater, or surface water, is not an exact science and there is inherently considerable uncertainty in the results. Climate modeling predictions are subject to a "cascade of uncertainties," which flows from uncertainties in emissions scenarios (i.e., future GHG concentrations), to uncertainties in the GCM simulations of the climatic response to a given GHG scenario, and then to the downscaling of the GCM data to the regional and local scales, and subsequent hydrologic modeling (e.g., Foley 2010; Mitchell and Hulme 1999; Maslin 2013; Vaghefi et al. 2019; Falloon et al. 2014). The uncertainties in each step are compounded to result in large uncertainties in final projections. Nevertheless, both historical trends and a consensus of the modeling results unequivocally indicate that the Earth's climate has been warming over the past century and will continue to do so at an accelerating rate into the future.

Depending on which RCP most closely comes to pass, the increase in global mean surface temperature by the end of the 21st century (2081–2100) relative to 1986–2005 is projected to likely range from 0.3 to 1.7 °C under the lowest emissions RCP2.6 to between 2.6 and 4.8 °C under the highest emission RCP8.5 (IPCC 2014). Land areas will warm more than the oceans and the greatest temperature increases will occur at high latitudes in the northern hemisphere (IPCC 2014). In additional to an increase in mean temperatures, extreme high temperature events (heat waves) are projected to become more common.

Changes in precipitation will not be uniform. As a broad generalization, many mid-latitude and subtropical dry regions will likely experience decreasing precipitation, while many mid-latitude wet regions will likely become wetter (IPCC 2014). High latitude areas and the equatorial Pacific are likely to experience an increase in annual mean precipitation under the RCP8.5 scenario (IPCC 2014). Extreme precipitation events and droughts are generally expected to become more common.

Global warming is expected to alter the hydrologic cycle in snowmelt-dominated regions by decreasing the amount of precipitation that falls in mountainous areas as snow, and thus the thickness of the winter snowpack, and cause an earlier melting of the snowpack. Changes in hydrologic conditions are expected to increase flood risk in the winter and early spring and decrease water availability in the summer. The changes in hydrologic conditions in snowmelt-dominated regions may impact groundwater recharge in mountainous areas although the direction and magnitude of changes in recharge are generally uncertain (Green 2016).

Modeling results indicate that sea level will continue its historical rise and that the rate of rise will very likely increase as the atmosphere and oceans warm. Local sea level rise may significantly vary from the global mean rate depending on such factors as changes in land elevation (e.g., from subsidence and isostatic rebound), winds, and ocean circulation. The latest IPCC (2019) projections are that the global mean sea level (GMSL) rise with respect to 1986–2005 under the low emissions RCP2.6 scenario will be 0.39 m (0.26–0.53 m, *likely range*) for the period 2081–2100, and 0.43 m (0.29–0.59 m, *likely range*) in 2100. For the higher emissions RCP8.5 scenario, the corresponding projected GMSL rise is 0.71 m (0.51–0.92 m, *likely range*) for 2081–2100 and 0.84 m (0.61–1.10 m, *likely range*) in 2100 (IPCC 2019). The IPCC (2019) also projected that there is a high confidence that extreme sea level events that are historically rare (once per century in the recent past) will occur frequently (at least once per year) at many locations by 2050 in all RCP scenarios, especially in tropical regions. The wild card with respect to sea level rise is the rate of melting of the polar ice caps. If the rate of melting is faster than expected, then sea level rise will be substantially more rapid than historical and projected rates.

1.5 Climate Change and Groundwater Recharge and Use

Climate change will impact groundwater resources mainly through changes in the amount of recharge and the demand for groundwater, and through impacts to water quality. Recharge is broadly defined as the flow of water from surface water bodies and land surfaces to underlying aquifers. Changes in annual average precipitation impact the amount of water that is available for recharge. However, the relationship between precipitation and recharge is not linear. Recharge rates also depend on the intensity and duration of precipitation events, its form (rain versus snow), and its seasonality and temporal distribution. Changes in land cover and land uses can impact recharge rates through changes in vegetation evapotranspiration (ET) and infiltration rates. The expected trend of precipitation occurring in more intense events, depending on local hydrological conditions, could result in either less recharge as more water runs off, or greater recharge if it allows sufficient water to accumulate to overcome soil-moisture deficits, permitting water to percolate to the water table. Green (2016) observed that the data necessary for confident prediction of recharge responses to future climate conditions (e.g., long-term continuous monitoring of recharge processes) are not available in most areas. Therefore, in many regions of the world, it is unknown whether (and how much) recharge will increase or decrease under projected future climates (Green 2016).

Locally decreasing precipitation may result in increased water demands for irrigation, which could be offset by a local abandonment of agriculture, a change to less water intensive crops or cultivars, and/or increasing water use efficiency. Rising temperatures can prompt increases in agricultural irrigation through increased ET rates and, in some areas, a longer crop growing season. The transpiration rates of some plants have been shown to decrease with increasing atmospheric carbon dioxide concentrations as the result of partial stomatal closure, which could offset the temperature effect on transpiration. Domestic water use may directly increase through greater residential lawn and landscaping irrigation and indirectly through increased water consumption associated with additional energy use for cooling. The most vulnerable regions will be areas where groundwater resources are already being exploited unsustainably and a drying climate will increase water demands and reduce supplies of fresh surface water (e.g., southwestern United States, parts of India).

Groundwater quality can be impacted through saline-water intrusion induced by sea level rise or, more importantly, excessive pumping of coastal aquifers. Changes in precipitation and infiltration rates can impact the quality of recharged water and thus shallow aquifers. Decreased precipitation rates and warmer temperatures could result in an increase in the total dissolved solids concentration of recharged water (Dragoni and Sukhija 2008; Green 2016). Increased infiltration could result in either a freshening of shallow groundwater (and thus improve its quality) or a deterioration of groundwater quality if it causes the leaching of salts that accumulated in the vadose (unsaturated) zone.

1.6 Sea Level Rise and Groundwater

Sea level rise can adversely impact the quality of groundwater through the salinization of coastal aquifers. Rising sea level will most directly impact coastal aquifers by the permanent inundation of low-lying areas, coastal erosion, and infiltration of sea water during flooding from storm and extreme tidal events. Freshwater lens aquifers on small, low-elevation islands are highly vulnerable to sea level rise.

Rising sea levels can induce lateral saltwater intrusion (i.e., the landward migration of the fresh-saline groundwater interface), but the extent of intrusion will depend on whether the impacted coastal aquifer is flux or head controlled (Werner and Simmons 2009). In a flux-controlled system, the rate of ground water discharge to the sea is persistent despite changes in sea level and, as a result, saline water intrusion tends to be limited. Rising sea levels will tend to cause the water table to rise, maintaining the seaward hydraulic gradient. Where the rise of the water table is limited by drainage, head-controlled conditions will occur, and the rate of saline-water intrusion will tend to be much greater (Werner and Simmons 2009).

A rising water table can cause local flooding of low-lying areas, which is referred to as groundwater inundation. The impacts of sea level rise on coastal and island aquifers are now typically evaluated by numerical modeling using a density-dependent solute transport modeling code (e.g., SEAWAT, Guo and Langevin 2002). Vulnerability to direct inundation and groundwater inundation can be assessed using high-resolution topographic maps (digital elevation models; DEMs) and projected local sea level and groundwater rise, and by hydrodynamic modeling.

1.7 Evaluating Climate Change Impacts on Groundwater Storage

Numerical groundwater modeling incorporating GCM projections is the state-of-the-art for evaluating the impacts of climate change on groundwater resources. The technical challenge is developing practicable workflows for extracting large-scale climate change data from GCMs, downscaling the data to a local scale relevant for groundwater recharge, simulation of the surficial hydrological processes that partition precipitation into runoff, ET, and recharge, and then developing a groundwater flow model calibrated to available historical observation data (e.g., water levels). GCM projections for the different RCPs can be used to evaluate the groundwater responses to a range of future emissions scenarios. However, the cascade of the uncertainties in each modeling step can result in large uncertainties in final projections. Uncertainties related to GCMs and RCMs are typically assessed by an ensemble approach where multiple climate models run for different RCPs are used for making semi-probabilistic projections. Published studies illustrate a wide range of approaches for simulating the impact of climate change on groundwater recharge, which vary in the technical expertise and effort (and thus time and cost) required. Irrespective of the modeling procedures employed, the results of a rigorous modeling program will at best define a "cone of uncertainty" of future groundwater conditions.

An alternative to the "top-down" approach of starting with GCM projections, is the "bottom-up" scenario-based approach that starts with consideration of various plausible hydrological changes (Brown 2011). For example, groundwater flow simulations could be run with hypothetical changes in recharge rates (e.g., 10% and 20% decreases) to evaluate their impacts. If the simulation results indicate that some specific changes in recharge rates (or other variables) could materially impact groundwater levels (or other aspects of hydrologic systems of concern, such as spring flows and water levels in groundwater dependent ecosystems), then the probability of such changes is evaluated using climate modeling results.

1.8 Adaptation Options

Adaptation options for climate change are essentially the same as those available for dealing with water scarcity in general. Changes in the local supply of water can be addressed through either management of demands or changes in the water sources utilized. Groundwater depletion can be reduced through decreasing extractions by either conservation (curtailing some water uses or increasing water use efficiency) or adopting alternative water sources (e.g., reclaimed water and desalination). Unsustainable groundwater use (and thus aquifer depletion) may alternatively be allowed to continue either in a planned or unplanned manner. Indeed, production of non-renewable (fossil) groundwater is inherently unsustainable, but it may support societal development goals if done for a limited period of time to transition to more sustainable water use.

Adaptation options include conjunctive use of groundwater and surface water, seawater and brackish groundwater desalination, and wastewater reuse. Desalination is increasingly being adopted as an alternative water source, but it is expensive and energy intensive, and thus is contrary to climate change mitigation goals (unless alternate, non-fossil fuel energy sources are used). Groundwater depletion resulting from climate change can also be addressed by MAR. A deep toolbox of MAR technologies is available, such as infiltration basins and vadose and phreatic injection wells, that can be employed to recharge aquifers using excess surface water and reclaimed water supplies.

Adaptations to saline-water intrusion associated with sea level rise include reducing near coastal pumping (i.e., relocating production wells inland) and MAR techniques to restore a seaward hydraulic gradient at the fresh-saline groundwater interface. Groundwater inundation can be addressed by drainage, or elevation or relocation of high-value infrastructure.

1.9 Water Planning and Governance

"Adaptation encompasses both national and regional strategies as well as practical measures taken at all political levels and by individuals" (Green 2016, p. 124). There are great differences in the degree of governmental involvement in water management within and between countries. In the United States and many other countries, water use is governed (regulated) on the national, state, or intrastate level. The regulatory process constrains what water users may do. Permitting systems normally endeavor to limit extractions to what are considered to be the safe yield of the source, which may consider the physical availability of water, and impacts to existing users, the environment, and the aquifer (e.g., declining of water levels and salinization of aquifers). National and state governments may perform large-scale planning, support or self-perform research, and construct or subsidize water projects. However, water supply decisions are typically made by individual water users, water utilities, and regional water suppliers, which may be either governmental agencies, not-for-profit private entities, or for-profit corporations.

Climate change adaptation planning and implementation occur in the context of existing water utility and regional water supplier organizational structures and regulatory governance. Key issues that have received relatively little attention in the climate change research literature are who actually makes water supply decisions, how are decisions made, and what are the planning time horizons. Adaptation to climate change has societal impacts and it has often been advocated in the climate change literature that input from all stakeholders and a broad suite of issues (e.g., social equity) should be considered in the decision-making process (e.g., Green 2016). However, in practice, water planning decisions tend to be largely siloed to technical experts either internal to or contracted by the organizations responsible for water supplies.

Public water utilities are typically run by directors or heads who are appointed by elected officials, such as county commissions and town and city councils. Planning is performed either directly by water utility staff or in conjunction with contracted external engineering, hydrogeology, and hydrology consultants. Regional and local water supply authorities and districts are governed by a board of directors who may either be appointed by their member governments or elected by landowners or the residents of a district. Minimum planning horizons may be regulatorily prescribed. Water supply planning in the United States tends to commonly be on 20-year (in some instances 50-year) horizons. Longer-term planning horizons are employed for expensive infrastructure projects with long operational lives.

Climate change is an insidious issue in that the impacts from year to year are not so severe or even noticeable so as to prompt immediate response. Changes in precipitation related to anthropogenic climate change over a 20-year planning period may not fall outside of the range of natural variation. The principal threat of climate change on groundwater (and water in general) supply and demand may not be gradual changes in mean precipitation rates but rather changes in the frequency and intensity of extreme events. The nightmare situation for water suppliers is droughts more severe than those in the modern climate record that cause water supplies to become exhausted. Tree ring data are being used in the western United States for climate reconstruction to identify past extreme droughts that could be a harbinger of future conditions in a drying climate (e.g., Bekker et al. 2014; Williams et al. 2020).

Groundwater governance (i.e., regulatory environment) controls the manner and extent to which water use is effectively regulated. Water governance addresses who has access to water and how water is allocated during times of shortage. Where water governance in strong, opportunities exist to control groundwater use to sustainable levels. Weak governance can lead to a "tragedy of the commons" situation resulting in over exploitation of aquifers to the ultimate harm of all users. Water use and environmental regulations limit the adaptation options that are allowed and impact the costs of their implementation. For example, water quality standards for aquifer recharge, and resulting pretreatment requirements, can make some MAR options economically unviable despite their water supply benefits. Water governance also influences the incentives for implementing some adaptation measures. To encourage MAR, the local regulatory environment should allow the owner and operator of systems to capture the benefits of their recharge and prevent third parties that are not contributing to the systems (i.e., free riders) from stealing the recharged water for their own benefit.

1.10 Climate Change Adaptation Planning Process

Climate change adaption planning starts with vulnerability assessments, which are systematic evaluations of the potential climate changes that are plausible in the area of interest and how the identified changes might impact water and related infrastructure of concern. The assessments usually incorporate the basic risk assessment process of evaluating both the probabilities of a series of adverse impacts and the potential magnitudes of the impacts (degrees of harm). Once vulnerabilities are identified, possible options are explored that could reduce or eliminate each identified risk. Adaptation planning for water supplies involves the investigation of potential responses to the impacts of climate change and how those responses can be best integrated into existing water supply systems. Water supply planning is essentially an exercise in multiple-criteria decision analysis (MCDA) in which overall strategies and specific supply options are evaluated based on the degree to which they achieve each of multiple criteria, of which reliability and cost are usually of greatest importance.

The degree of sophistication of water supply planning varies greatly between water utilities and regional suppliers. Small utilities supplied solely by groundwater may adapt to climate change by implementing conservation measures and perhaps managed aquifer recharge. The choice of practical options is limited, and the decision-making process is relatively simple. A small number of options may be identified that achieve water supply requirements, which are judged by their costs (and other relevant considerations).

Larger utilities and regional water suppliers with multiple water sources and complex supply and distribution infrastructure require more sophisticated planning tools to identify optimal solutions. The current state of the art is decision support systems (DSSs), which incorporate both hydrological simulation and various optimization algorithms. The most sophisticated DSSs also include modules that estimate the impacts of climate change on water demands and other socioeconomic factors (Serrat-Capdevila et al. 2011; Yates and Miller 2011). The economics of various supply options can be evaluated using some type of expected net present value approach, which requires that probabilities be assigned to the various climate change contingencies.

In the face of uncertainty over future climate conditions, common sense dictates constructing robust and resilient water supply and management systems than can handle a wide range of eventualities. Historical climate statistics (e.g., magnitude of the 100-year rainfall event) can no longer be relied upon for the future—stationarity is dead (Milly et al. 2008). Instead of constructing systems that perform optimally under a single or group of future climate conditions, robust decision making (RDM) focuses on identifying a set of choices that perform reasonably well compared to the alternatives across a wide range of plausible future climate scenarios (Grove and Lampert 2007). RDM employs computer simulation models to create large ensembles of possible future states that are used to identify candidate robust strategies and systematically assess their performance (Grove and Lampert 2007). The preferred adaptation strategy is the one that least frequently fails to meet performance goals.

1.11 Case Studies of Adaptation to Climate Change in High Groundwater Use Area

Chapter 13 examines projected climate changes and potential or implemented adaptative actions in some areas that have been identified as climate change hot spots or have a high dependence on groundwater—southwestern North America, High Plains of the western United States, Florida, the Mediterranean (MENA) region, and Africa. Major water suppliers in the United States (and other developed countries) generally have a high awareness of the potential impacts of climate change on their water supplies and there are numerous examples of active interactions between water suppliers and the climate change research community, such as participation in collaborative organizations (e.g., the Water Utility Climate Alliance). However, examples of implementation of actions specifically driven by climate change concerns are uncommon. Water planning tends to be driven by concerns over current climate conditions and water supplies, and the need to develop new supplies to meet the demands of growing populations. Where strong water governance exists, regulatory efforts to achieve or maintain sustainable groundwater use have had the incidental benefit of making water supplies more resilient to climate change. Regulatory limits on additional fresh groundwater withdrawals in Florida, for example, have led water utilities to invest in brackish groundwater desalination plants, a source with a low vulnerability to climate change. Efforts in Texas, Arizona, and Southern California to achieve sustainable groundwater use through reductions in pumping, conjunctive use schemes, and the implementation of MAR will also make their water supplies more resilient to future climate change.

Developed countries have the financial resources to develop more climate resilient water supplies. Areas with weak water governance, on-going aquifer overdraft, a drying climate, and the absence of affordable alternative water supplies may face a bleak future as wells run dry. Small-scale adaptations, such as rainwater harvesting and managed aquifer recharge schemes, that can be operated by local populations offer the best opportunities to provide greater resilience to the water supplies in rural areas of developing countries. Construction of deeper wells can also increase resilience, provided that they don't encourage even greater groundwater use and the depletion of deeper aquifers.

References

Abiy AZ, Melesse AM, Abtew W (2019) Teleconnection of regional drought to ENSO, PDO, and AMO: Southern Florida and the Everglades. Atmosphere 10(6):295

Adger WN, Arnell NW, Tompkins EL (2005) Successful adaptation to climate change across scales. Glob Environ Change 15(2):77–86

Alder JR, Hostetler SW, Williams D (2013) An interactive web application for visualizing climate data. EOS Trans AGU 94:197–198. https://doi.org/10.1002/2013EO220001

Alley WM (2001) Ground water and climate. Ground Water 39(2):161–161

Arnell NW (1999) Climate change and global water resources. Glob Environ Change 9:S31–S49

Bates B, Kundzewicz Z, Wu S, Palutikof JP (eds) (2008) Climate change and water. IPCC Technical paper VI. Intergovernmental Panel on Climate Change. IPCC Secretariat, Geneva

Bekker MF, DeRose RJ, Buckley BM, Kjelgren RK, Gill NS (2014) A 576-year Weber river streamflow reconstruction from tree rings for water resource risk assessment in the Wasatch Front, Utah. J Am Water Resour Assoc 50(5):1338–1348

Brown C (2011) Decision-scaling for robust planning and policy under climate uncertainty. World Resources Report, Washington D.C. https://wriorg.s3.amazonaws.com/s3fs-public/uploads/wrr_brown_uncertainty.pdf. Accessed 14 June 2020

CGIAR (n.d.) Climate change knowledge portal. Climate analysis tool—powered by climate wizard. https://climatewizard.ciat.cgiar.org/. Accessed 26 May 2020

Dillon P (2009) Water recycling via managed aquifer recharge in Australia. Boletín Geológico y Minero 120(2):121–130

Döll P, Schmied HM, Schuh C, Portmann FT, Eicker A (2014) Global-scale assessment of groundwater depletion and related groundwater abstractions: combining hydrological modeling with information from well observations and GRACE satellites. Water Resour Res 50: 5698–5720

Dragoni W, Sukhija BS (2008) Climate change and groundwater: a short review. In: Dragoni W, Sukhija BS (eds) Climate change and groundwater, vol 288. Geological Society of London Special Publications, pp 1–12

Earman S, Dettinger M (2011) Potential impacts of climate change on groundwater resources—a global review. J Water Clim Change 2(4):213–229

Falloon P, Challinor A, Dessai S, Hoang L, Johnson J, Koehler AK (2014) Ensembles and uncertainty in climate change impacts. Front Environ Sci 2(33)

Famiglietti JS (2014) The global groundwater crisis. Nat Clim Change 4:945–948

Ferguson DB, Rice JL, Woodhouse CA (2014) Linking environmental research and practice: lessons from the integration of climate science and water management in the Western United States. Climate assessment for the Southwest, Tucson, AZ. www.climas.arizona.edu/publication/report/linking-environmental-research-and-practice. Accessed 26 May 2020

Flato G, Marotzke J, Abiodun B, Braconnot P, Chou SC, Collins W, Cox P, Driouech F, Emori S, Eyring V, Forest C, Gleckler P, Guilyardi E, Jakob C, Kattsov V, Reason C, Rummukainen M (2013) Evaluation of climate models. In: Stocker TF, Qin D, Plattner G-K, Tignor M, Allen SK, Boschung J, Nauels A, Xia Y, Bex V, Midgley PM (eds) Climate change 2013: the physical science basis. Contribution of working group I to the fifth assessment report of the Intergovernmental Panel on Climate Change. Cambridge University Press, Cambridge UK, pp 741–866

Foley AM (2010) Uncertainty in regional climate modelling: a review. Prog Phys Geogr 34 (5):647–670

Green TR (2016) Linking climate change and groundwater. In: Jakeman AJ, Barreteau O, Hunt RJ, Rinaudo JD, Ross A (eds) Integrated groundwater management. Springer Nature, Cham, pp 97–141

Green TR, Taniguchi M, Kooi H, Gurdak JJ, Allen DM, Hiscock KM, Treidel H, Aureli A (2011) Beneath the surface of global change: impacts of climate change on groundwater. J Hydrol 405 (3–4):532–560

Groves DG, Lempert RJ (2007) A new analytical method for finding policy-relevant scenarios. Glob Environ Change 17:73–85

Guo W, Langevin CD (2002) User's guide to SEAWAT. A computer program for simulation of three-dimensional variable-density ground-water flow. In: U.S. geological survey techniques of water-resources investigations 06-A7

Hallegatte S (2009) Strategies to adapt to an uncertain climate change. Glob Environ Change 19 (2):240–247

Hansen J, Sato M, Ruedy R (2012) Perception of climate change. Proc Natl Acad Sci 109(37): E2415–E2423

Heltberg R, Siegel PB, Jorgensen SL (2009) Addressing human vulnerability to climate change: toward a 'no-regrets' approach. Glob Environ Change 19(1):89–99

Hewitt CD, Stone RC, Tait AB (2017) Improving the use of climate information in decision-making. Nat Clim Change 7(9):614–617

Hu ZZ, Huang B (2009) Interferential impact of ENSO and PDO on dry and wet conditions in the US Great Plains. J Clim 22(22):6047–6065

IPCC (2012). Glossary of terms. In: Field CB, Barros V, Stocker TD, Qin D, Dokken DJ, Ebi KL, Mastrandrea MD, Mach KJ, Plattner G-K, Allen SK, Tignor M, Midgley PM (eds) Managing the risks of extreme events and disasters to advance climate change adaptation. A special report of working groups I and II of the Intergovernmental Panel on Climate Change (IPCC). Cambridge University Press, Cambridge UK, pp 555–564

IPCC (2013) Climate change 2013: the physical science basis. In: Stocker TF, Qin D, Plattner G-K, Tignor M, Allen SK, Boschung J, Nauels A, Xia Y, Bex V, Midgley PM (eds) Contribution of working group I to the fifth assessment report of the Intergovernmental Panel on Climate Change. Cambridge University Press, Cambridge UK

IPCC (2014) Climate change 2014: synthesis report. In: Pachauri RK, Meyer LA (eds) Contribution of working groups I, II and III to the fifth assessment report of the Intergovernmental Panel on Climate Change. IPCC, Geneva

IPCC (2019) Summary for policymakers. In: Pörtner H-O, Roberts DC, Masson-Delmotte V, Zhai P, Tignor M, Poloczanska E, Mintenbeck K, Alegría A, Nicolai M, Okem A, Petzold J, Rama B, Weyer NM (eds) IPCC special report on the ocean and cryosphere in a changing climate. In Press

Jones RN, Patwardhan A, Cohen SJ, Dessai S, Lammel A, Lempert RJ, Mirza MMQ, von Storch H (2014) Foundations for decision making. In: Field CB, Barros VR, Dokken DJ, Mach KJ, Mastrandrea MD, Bilir TE, Chatterjee M, Ebi KL, Estrada YO, Genova RC, Girma B, Kissel ES, Levy AN, MacCracken S, Mastrandrea PR, White LL (eds) Climate change 2014: impacts, adaptation, and vulnerability, part A: global and sectoral aspects. Contribution of working group II to the fifth assessment report of the Intergovernmental Panel on Climate Change. Cambridge University Press, Cambridge UK, pp 195–228

Kundzewicz ZW, Döll P (2009) Will groundwater ease freshwater stress under climate change? Hydrol Sci J 54(4):655–675

Kundzewicz ZW, Mata LJ, Arnell NW, Döll P, Jimenez B, Miller K, Oki T, Şen Z, Shiklomanov I (2008) The implications of projected climate change for freshwater resources and their management. Hydrol Sci J 53(1):3–10

Kuss AJM, Gurdak JJ (2014) Groundwater level response in US principal aquifers to ENSO, NAO, PDO, and AMO. J Hydrol 519:1939–1952

Loaiciga HA, Valdes JB, Vogel R, Garvey J, Schwarz H (1996) Global warming and the hydrologic cycle. J Hydrol 174(1–2):83–127

Markewich HW, Bliss NB, Stallard RF, Sundquist ET (1997) Can the global carbon budget be balanced? U.S. geological survey fact sheet, pp 137–197. https://pubs.usgs.gov/fs/fs137-97/fs137-97.html. Accessed 12 June 2020

Maslin M (2013) Cascading uncertainty in climate change models and its implications for policy. Geogr J 179(3):264–271

McCabe GJ, Palecki MA, Betancourt JL (2004) Pacific and Atlantic Ocean influences on multidecadal drought frequency in the United States. Proc Natl Acad Sci 101(12):4136–4141

Milly PC, Betancourt J, Falkenmark M, Hirsch RM, Kundzewicz ZW, Lettenmaier DP, Stouffer RJ (2008) Stationarity is dead: whither water management? Science 319(5863): 573–574

Mitchell TD, Hulme M (1999) Predicting regional climate change: living with uncertainty. Prog Phys Geogr 23(1):57–78

NOAA (n.d.) Greenhouse gases. https://www.ncdc.noaa.gov/monitoring-references/faq/greenhouse-gases.php. Accessed 26 May 2020

PCMDI (n.d.) CMIP5—Coupled model intercomparison project phase 5—overview. In: Program for climate model diagnosis & intercomparison. https://pcmdi.llnl.gov/mips/cmip5/. Accessed 26 May 2020

Pielke R Jr, Prins G, Rayner S, Sarewitz D (2007) Climate change 2007: lifting the taboo on adaptation. Nature 445(7128):597–598

Ranjan P, Kazama S, Sawamoto M (2006) Effects of climate change on coastal fresh groundwater resources. Glob Environ Change 16(4):388–399

Serrat-Capdevila A, Valdes JB, Gupta HV (2011) Decision support systems in water resources planning and management: stakeholder participation and the sustainable path to science-based decision making. In: Jao C (ed) Efficient decision support systems—practice and challenges from current to future. Intech, Rijeka, Croatia, pp 423–440

Siebert S, Burke J, Faures JM, Frenken K, Hoogeveen J, Döll P, Portmann FT (2010) Groundwater use for irrigation—a global inventory. Hydrol Earth Syst Sci 14(10):1863–1880

Taylor RG, Scanlon B, Döll P, Rodell M, Van Beek R, Wada Y, Longuevergne L, Leblanc M, Famiglietti JS, Edmunds M, Konikow L (2013) Ground water and climate change. Nat Clim Change 3(4):322–329

USGS (n.d.) Regional and climate change. Visualization. http://regclim.coas.oregonstate.edu/visualization/gccv/. Accessed 26 May 2020

Vaghefi SA, Iravani M, Sauchyn D, Andreichuk Y, Goss G, Faramarzi M (2019) Regionalization and parameterization of a hydrologic model significantly affect the cascade of uncertainty in climate-impact projections. Clim Dyn 53(5–6):2861–2886

Vörösmarty CJ, Green P, Salisbury J, Lammers RB (2000) Global water resources: vulnerability from climate change and population growth. Science 289(5477):284–288

Werner AD, Simmons CT (2009) Impact of sea-level rise on sea water intrusion in coastal aquifers. Groundwater 47(2):197–204

Wilby RL, Charles SP, Zorita E, Timbal B, Whetton P, Mearns LO (2004) Guidelines for use of climate scenarios developed from statistical downscaling methods. Supporting material of the Intergovernmental Panel on Climate Change. http://apps.ipcc-data.org/guidelines/dgm_no2_v1_09_2004.pdf. Accessed 26 May 2020

Williams AP, Cook ER, Smerdon JE, Cook BI, Abatzoglou JT, Bolles K, Baek SH, Badger AM, Livneh B (2020) Large contribution from anthropogenic warming to an emerging North American megadrought. Science 368(6488):314–318

Yates D, Miller K (2011) Climate change in water utility planning: decision analytic approaches. Water Research Foundation and University Corporation for Atmospheric Research, Denver

Zekster IS, Everett LG (eds) (2004) Groundwater resources of the world and their use. IHP-VI, Series on groundwater, vol 6. UNESCO (United Nations Educational, Scientific and Cultural Organization), Paris

Chapter 2
Climate and Groundwater Primer

2.1 Aquifer Water Budgets

Climate impacts groundwater quantity mainly through its influences on recharge and the demand for groundwater. Aquifer water budgets can be basically expressed as:

$$\Delta S = \sum \text{Inflows} - \sum \text{Outflows} \tag{2.1}$$

where ΔS is the change in storage. If total outflows exceed total inflows over a given time period, then the amount water stored in an aquifer decreases, which is manifested by a decrease in aquifer water levels or pressures (heads). The aquifer property "storativity" (also called storage coefficient) is defined as the volume of water that is released from a unit area of an aquifer under a unit decline of hydraulic head. In unconfined aquifers, water is released from storage primarily by gravity drainage with an associated lowering of the water table. Water is released from storage in confined aquifers by a combination of the compaction of the aquifer and expansion of water. The storativity of unconfined aquifers (which is approximately equal to their specific yield) is usually orders of magnitude greater than that of confined aquifers. Therefore, a given change in water table elevation results in a much greater change in storage in unconfined aquifers than comparable pressure (head) changes cause in confined aquifers.

Large aquifers have stored water volumes orders of magnitude greater than their annual inflows and outflows. Hence, aquifers tend to be much less immediately impacted by short-term drought and wet periods than surface water bodies. Aquifers are vulnerable to chronic long-term overdraft and changes in hydrologic conditions whose impacts may not raise immediate concerns. Cuthbert et al. (2019) cautioned that adaptation strategies should account for the "hydraulic memory" of groundwater systems, which can buffer climate change impacts on water resources, but

R. Maliva, *Climate Change and Groundwater: Planning and Adaptations for a Changing and Uncertain Future*, Springer Hydrogeology,
https://doi.org/10.1007/978-3-030-66813-6_2

may also lead to a long, but initially unnoticed, legacy of anthropogenic and climatic impacts on aquifer water levels, river flows, and groundwater-dependent ecosystems.

Precipitation either infiltrates into the soil, runs off, or ponds on land surface and is lost to subsequent evaporation. Infiltrated water may either percolate to the water table and become aquifer recharge, be held in the soil under capillary pressure as soil moisture, be lost to evaporation from the soil or transpiration by vegetation (referred to collectively as evapotranspiration; ET), or may flow in the unsaturated zone and later discharge to surface waters or land surface (interflow). Combining all the evaporation and ET terms and assuming a long-term constant (steady-state) soil-moisture profile, the local water budget of a surficial (water table) aquifer can be expressed as

$$\Delta \mathrm{S} = \mathrm{R} - \mathrm{D} - \mathrm{Q} \tag{2.2}$$

and

$$\mathrm{R} = (\mathrm{P} - \mathrm{ET}) + (\mathrm{SW} + \mathrm{IF}) \tag{2.3}$$

where (in units of volume/time)
R = recharge
D = discharge (e.g., to springs and surface water bodies)
Q = groundwater pumping
P = precipitation
ET = evapotranspiration
SW = net surface water inflow
IF = net interflow.

The net surface water inflow term is the difference between any external surface water that flows into the study area and water that flows out. In arid regions, SW is equal to the local transmission losses in streams, corrected for local precipitation into and ET from the stream. Net interflow is the net flow of groundwater in the vadose zone into the study area. The term (P—ET) is referred to as available water and is the amount of water that available for both aquifer recharge and local runoff.

Aquifer-wide water budgets also need to consider net lateral (GH) and vertical (GV) groundwater flows into an aquifer:

$$\Delta \mathrm{S} = R + \mathrm{GH} + \mathrm{GV} - \mathrm{D} - \mathrm{Q} \tag{2.4}$$

2.2 Potential, Reference, and Actual Evaporation

Evapotranspiration is sum of evaporation and plant transpiration. Transpiration rates are related to green leaf area. With respect to agricultural crops, at sowing (when the ground in bare) nearly 100% of ET comes from soil evaporation, whereas more than 90% of ET comes from transpiration at full crop cover (Allen et al. 1998). ET is commonly expressed in units of length of a water column per time (e.g., mm/d, cm/yr, in/yr). ET can also be expressed as the latent heat of vaporization (λ), which is the energy or heat required to vaporize water. Latent heat flux has units of energy divided by the product of area and time (e.g., $Jm^{-2}s^{-1}$ or Wm^{-2}). ET is usually by far the largest outflux of water from land surface and is thus of great importance in water budgets.

There are two main types of ET: actual ET (ET_a) and the similar reference ET (ET_o) and potential ET (PET). Actual ET is the real flux of water from land surface to the atmosphere, whereas ET_o and PET are measures of the evaporative power of the atmosphere. Actual ET is a function of water availability. In arid and semiarid regions, ET_o and PET rates are often very high, whereas ET_a rates can be very low due to the paucity of water.

ET_o and PET are defined as the ET rates that would occur from a reference surface that is not short of water. Penman (1956, p. 43) defined the reference surface for PET as "a short green crop cover, completely shading the ground and never short of water." In an earlier paper, Penman (1948) referred to the surface as just "turf." PET is imprecisely defined because the description of the reference surface as a short green crop could correspond to many types of horticultural and agronomic crops (Irmak and Haman 2003).

The reference surface for ET_o is more precisely defined as "a hypothetical grass reference crop with an assumed crop height of 0.12 m, a fixed surface resistance of 70 s m^{-1} and an albedo of 0.23 (Allen et al. 1998). The reference surface closely resembles an extensive surface of green, well-watered clipped grass of uniform height, actively growing and completely shading the ground (Allen et al. 1998; Task Committee on Standardization of Reference Evapotranspiration 2005). An alternative reference surface for ET_o is a tall crop having an approximate height of 0.50 m, similar to full-cover alfalfa.

Reference ET rate is a climate parameter that is independent of biological and non-weather-related variables, such as crop type, crop development, crop roughness, management practices, ground cover, plant density, and soil moisture (Allen et al. 1998). Crop evapotranspiration under standard conditions (ET_c) refers to the ET of specific crops "that are grown in large fields under optimum soil water, excellent management and environmental conditions, and achieve full production under the given climatic conditions" (Allen et al. 1998). ET_c is related to ET_o by a crop coefficient (K_c):

$$ET_c = ET_o K_c \tag{2.5}$$

As ET_o and ET_c rates are based on fully watered conditions, Eq. 2.5 cannot be used to calculate ET_a rates during periods of less than fully watered conditions when ET_a may be largely controlled by water availability.

ET_o rates can be calculated from meteorological data. Modifications of the Penman-Monteith method (Penman 1948; Monteith 1965), including the Priestley-Taylor (1972) method, are now recommended as the standard methods for the definition and computation of reference evapotranspiration (Allen et al. 1998; Task Committee on Standardization of Reference Evapotranspiration 2005). Penman-Monteith methods require data on radiation, air temperature, air humidity and wind speed. (Allen et al. 1998), whereas the Priestley-Taylor method can be used to estimate ET_o where data on aerodynamic variables (relative humidity and wind speed) are not available.

The evaporative power of a local environment can be estimated from pan evaporation measurements. Pan evaporation rates in the United States are mostly commonly measured using a Class A evaporation pan (Fig. 2.1), which is a steel cylinder with a diameter of 47.5 in (120.7 cm) and a depth of 10 in (25 cm). The pans are placed on open ground upon a level slatted wooden platform. Evaporation is typically measured every 24 h by either the decrease in water level or the amount of water that is added to bring the water level back up to a reference level. Meteorological data, such as temperature (maximum and minimum), rainfall, humidity, and wind speed, are commonly also recorded at evaporation pan stations to assist in the interpretation of the data. ET_o and surface water evaporation rates are related to pan evaporation rates (ET_p) through a pan coefficient (K_p):

Fig. 2.1 Class A evaporation pan, Lake Pleasant Regional Park, Maricopa County, Arizona

$$ET_o = K_p(ET_p) \tag{2.6}$$

Pan coefficients are a function of multiple variables, particularly local meteorological conditions, and pan design, location, and condition (Snyder 1992; Snyder et al. 2005).

The principal weather parameters that affect reference ET (and ET_a) are radiation, air temperature, humidity and wind speed (Allen et al. 1998). Warmer and drier conditions and increased solar radiation will result in higher ET_o rates, and thus greater crop water requirements (and associated irrigation water demands), and potentially lesser aquifer recharge rates.

Actual ET rates tend to be by far the most poorly constrained variable in water budgets because of a sparsity of long-term data in the vast majority of areas and uncertainties associated with the data. Nevertheless, climate modeling data can provide insights into the direction of changes in ET_o and ET_a and allow for quantitative estimates of potential changes in irrigation water demands and aquifer recharge.

2.3 Infiltration

Infiltration is the process by which surface water enters the soil (i.e., the vadose zone). The vadose zone consists of soil and rock located between land surface and the regional water table (as opposed to a water table topping a local perched aquifer). The vadose zone includes the unsaturated zone and the capillary fringe, a usually thin saturated zone located above the water table in which water is held under less than atmospheric pressure (Fig. 2.2). Percolation is the downward

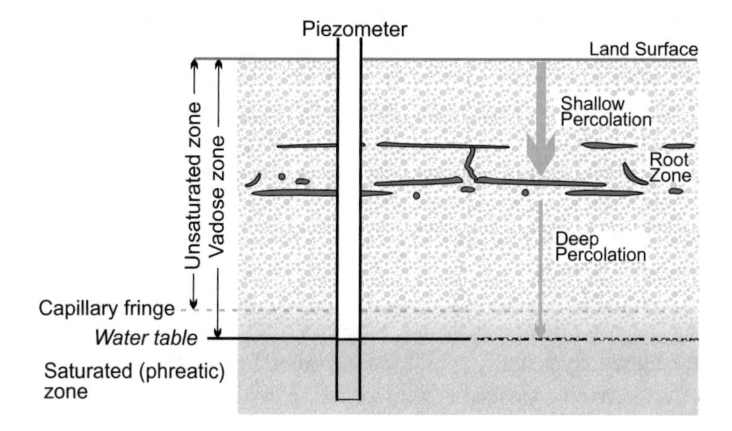

Fig. 2.2 Diagram of the relationship between the vadose zone, unsaturated zone, and water table. The lower part of the capillary fringe may be saturated but is above the water table

movement of water through the vadose zone that occurs after infiltration. A high percentage, and in some cases all, of the water that locally infiltrates into the shallow vadose zone (i.e., the soil) may be subsequently lost to ET (Wheater 2002).

Deep percolation refers to water that percolates past the root zone of plants (Bouwer 1978) after which it becomes much less susceptible to loss via ET and is more likely to recharge the underlying aquifer. The term "net infiltration" has been similarly used to describe the percolation flux that passes below the depth at which the rate of removal by ET becomes insignificant (Flint et al. 2002).

Deep percolation and net infiltration rates are approximately equal to the recharge rate provided that steady-state soil-moisture conditions exist in the deep vadose zone. However, in semiarid and arid lands with deep water tables, it may take an extremely long time for the small volume of water the passes through the root zone to reach the water table and steady-state conditions may not exist.

The vadose zone partitions precipitation (and applied irrigation water) into infiltration, runoff, ET, interflow, and groundwater recharge. Infiltration rate refers to the velocity at which water enters the soil and is typically expressed in units of millimeters or inches per hour. Infiltration rates vary over the duration of an infiltration event. The maximum infiltration rate occurs at the start of an event when the surficial soils are relatively dry and tend to more rapidly draw in water under capillary forces. The term "infiltration capacity" is defined as the maximum rate at which water can be absorbed by a given soil under given conditions (Horton 1933). As the soil becomes saturated, a lower, constant equilibrium rate of infiltration is eventually reached.

Infiltration will be most rapid in dry soils because of their greater capillary action (i.e., more negative soil-water potential). However, the recharge rate in dry sediments will be less than the rate in pre-wetted sediments because infiltrated water is retained in the soil under capillary pressure. Pore spaces must first be largely filled with water before significant percolation to the water table can occur. The terms "field capacity" and "field moisture capacity" refer to the amount of water held in a soil after excess water has drained away and the rate of downward movement has decreased (Horton 1933, 1940). Field moisture capacity, as define by Horton (1933), is roughly equivalent to specific retention. Field moisture capacity usually refers to water held in soils after the rate of drainage slows (which usually takes place about 2 to 3 days after a rain or irrigation event), whereas specific retention refers to water retained after gravity drainage is complete.

Rainfall will infiltrate into the soil, initially at its infiltration capacity, with excess water available for surface runoff. The initially infiltrated water will first make up any soil-moisture deficiency (i.e., soil moisture content below the field moisture capacity). Once the field moisture capacity is exceeded, some infiltrated water will be available for groundwater recharge. Horton (1933) noted that only a fraction of the rainfall excess may actually runoff into streams. Some of the excess rainfall will be captured in transit as surface detention and then either infiltrate into the soil or later evaporate.

Infiltration rates decrease over time in a regular, cyclical manner after each rainfall or irrigation event (Horton 1933; Jury and Horton 2004). Infiltration rates may decrease by the packing of the soil surface, swelling of the soil, infiltration of fine materials into soil-surface openings, and the progressive filling of soil pores with water. The change in infiltration rate over time is expressed by the equation (Horton 1940)

$$f_t = f_c + (f_0 - f_c)e^{-kt} \tag{2.7}$$

where
f_t = infiltration rate at time t (mm/hr)
f_0 = initial or maximum infiltration rate (mm/hr)
f_c = final or equilibrium infiltration rate after the soil has become saturated (mm/hr)
t = time (hr)
k = decay constant specific to the soil (hr^{-1}).

The total volume of infiltration (F_t; units = mm) after time t is calculated as

$$F_t = f_c t + \frac{(f_0 - f_c)}{k}\left(1 - e^{-kt}\right) \tag{2.8}$$

Horton's equations provide a useful conceptual understanding of the infiltration process but has the practical limitation that it is difficult to obtain accurate values for the parameters. A key point to be taken from infiltration theory is that groundwater recharge rates are a complex function of both the intensity and duration of rainfall events, soil moisture content, and land surface and soil properties that control initial and equilibrium infiltration rates. The relationships between mean annual rainfall, available water, and aquifer recharge are not linear.

2.4 Recharge

2.4.1 Recharge Types

Aquifer recharge may be either direct or indirect. Direct recharge, also referred to as diffuse and distributed recharge, is water that enters an aquifer by infiltration and percolation at the general site of precipitation or application. Indirect or focused recharge involves the runoff of water from an area of precipitation and its infiltration and percolation to the water table at another location where it is concentrated. In the case of confined aquifers, the primary recharge area may be located a great distance from groundwater extraction sites.

Indirect recharge sites include depressions and channels into which runoff flows and ponds and where the vertical hydraulic conductivity of the vadose zone is high enough to allow for significant infiltration before the loss of accumulated water to evaporation. Indirect recharge sites include ephemeral stream channels (also referred to as wadis and arroyos), flood plains, alluvial fans, and dry playas (depressions). Direct recharge tends to be of decreasing significance with increasing aridity (Simmers 1990, 1998; De Vries and Simmers 2002). In some desert areas, it has been shown that essentially no recharge occurs outside of ephemeral stream channels (i.e., in interfluve areas) under current climate conditions (e.g., Izbicki et al. 2007; Stonestrom et al. 2007). In less arid areas, both direct and indirect recharge may occur. For example, in the High Plains of the western United States, diffuse recharge occurs across the landscape, but the bulk of the recharge occurs by focused (fast path) recharge in a much smaller fraction of the area (Gurdak et al. 2007).

Recharge is also further subdivided into matrix and macropore recharge. Matrix or interstitial recharge occurs through the intergranular pore spaces of sediment or rock. Macropore recharge, on the contrary, occurs through larger secondary pores zones with enhanced permeability. Macropores includes deep desiccation cracks, animal burrows, root tubes, fissures, and fractures (Stephens 1996; Wood et al. 1997). The higher permeability of macropores results in more rapid infiltration, percolation, and recharge rates compared to matrix recharge. More rapid percolation downward from land surface through macropores decreases ET losses.

Sediment or rock with both matrix porosity and macroporosity are referred to as dual-porosity systems. Multiple generations or types of macroporosity may be locally present. Total recharge is the sum of matrix and macropore recharge. Recharge often occurs simultaneously through both matrix porosity and macroporosity. Water percolating through macropores may be completely absorbed by the surrounding matrix before it reaches the water table (Stephens 1996). Macropores may become clogged over time, which can result in gradual changes the spatial pattern of deep soil water movement over time (Stephens 1996).

Mountain front recharge (MFR) is a broadly encompassing term that is loosely defined to include all water that enters a basin aquifer through the adjoining mountain front and includes water that infiltrates into both fractured rock and alluvium in mountain stream beds (Wilson and Guan 2004). The term "mountain block recharge" (MBR) is used to specifically describe water that enters a groundwater basin from adjacent mountain bedrock via subsurface fractures and fissures. As defined by Wilson and Guan (2004), MFR includes water entering a basin aquifer from both MBR and recharge in alluvial fans, perennial streams, and reaches of ephemeral streams within the mountains and mountain front. Basin floor recharge (BFR) consists of both focused stream recharge and distributed recharge on the basin floor. Greater recharge tends to occur in mountains because of a combination of greater annual precipitation due to orogenic effects, lower temperatures (and thus reduced ET), and thinner soils that can store less water (Wilson and Guan 2004). MFR is a very important in arid and semiarid lands because BFR recharge is often negligible. MFR is the only significant natural source of water for some basin aquifers. Hence, investigations of the impacts of climate change on basin aquifers need to consider the effects of climate changes on recharge in adjoining mountainous areas.

2.4.2 Anthropogenic Aquifer Recharge

The term "anthropogenic aquifer recharge" (AAR) is recommended to broadly describe increases groundwater recharge caused by human activities (Maliva 2019). AAR includes managed aquifer recharge (MAR), which has been described as the "intentional banking and treatment of waters in aquifers" (Dillon 2005). AAR also includes unintentional recharge and unmanaged recharge (UMAR; NRMMC, EPHC & NHMRC 2009). Unintentional recharge includes recharge from unplanned processes (e.g., pipe leaks). UMAR includes intentional activities that have a primary disposal function in which recharge is incidental, such as discharges to septic system leach fields. UMAR and unintentional aquifer recharge include infiltration of stormwater in retention basins, leakage from potable water and sewer mains, discharges to on-site sewage disposal and treatment systems, irrigation return flows, and leakage from canals. UMAR from irrigation return flows and canal leakage is the primary local recharge in some arid areas.

MAR includes technologies that can be employed as part of adaptation strategies to climate change. Unintentional recharge and UMAR may be indirectly affected by climate change. For example, improvements in agricultural water use efficiency and the lining of canals may be employed as water saving adaptation strategies to address declining surface water supplies but could have the adverse impact of locally decreasing aquifer recharge.

2.4.3 Quantification of Recharge

The primary mechanisms and rates of current recharge in a study area need to be understood before the impacts of climate change on recharge can be quantitatively evaluated. Accurate estimation of groundwater recharge rates should proceed from a well-defined conceptualization of the probable flow mechanisms and important features influencing recharge at a given locality (Simmers 1990). The conceptual model should guide in the selection of quantification methods that are most appropriate for the main types of local recharge and local climatic and hydrogeological conditions. Methods used to quantify recharge were reviewed numerous workers (e.g., Allison 1988; Gee and Hillel 1988; Lerner et al. 1990, 1997; Simmers 1990; Allison et al. 1994; Stephens 1996; Scanlon et al. 1997, 2002; Flint et al. 2002; Sophocleous 2004; Herczeg and Leaney 2011; Maliva and Missimer 2012; Maliva 2019).

There are five basic categories of methods used to estimate recharge rates:

- analysis of the rate of water movement downward through the unsaturated zone (i.e., "from above", Allison 1988)
- inferring recharge rates from water table elevation changes (i.e., "from below", Allison 1988)
- geochemical tracers (e.g., chloride mass balance method)

- as the residual in water budget estimates (including numerical modeling investigations)
- from the volumetric rate of flow away from the recharge areas of confined aquifers.

A wide variety of specific methods have been developed in these five categories. However, many of them have had limited application in applied water resources investigations. Methods that analyze the actual downward movement of water through the vadose zone have the great limitation of being point measurements from which regional-scale extrapolation and interpolation of the results introduce great uncertainty.

2.4.3.1 Water Fluctuation Method

The water-table fluctuation (WTF) method is widely used to estimate recharge rates (Healy and Cook 2002). For an unconfined aquifer, the total recharge (R_v) from an individual rainfall event is estimated from the increase in water table elevation (Δh) and the aquifer specific yield (S_y):

$$R_v = S_y \Delta H \quad (2.9)$$

Where recharge is focused (e.g., in an ephemeral stream channel or basin), the increase in water table elevation needs to be integrated over the impacted area of the aquifer, which requires sufficient monitoring data to capture the geometry of the groundwater mound.

Cumulative recharge is the sum of the response to all rain events over a period of time. The WTF method is based on the water table elevation response fully reflecting the water added into storage and that there is no net transport of water from the water table. A considerable delay may occur between a rainfall event and the water table response, depending on the thickness and hydraulic properties of the vadose zone sediment or rock. Extraneous factors that can impact measured water levels include changes in barometric pressure, groundwater flow (in both the saturated zone and as interflow), diurnal fluctuation in evapotranspiration, entrapped air, and groundwater pumping (Healy and Cook 2002).

2.4.3.2 Environmental Tracers

Environmental tracers are used to calculate recharge rates in four basic manners, three of which involve age dating of water:

- age-dating using a time dependent input function (e.g., tritium and chlorofluorocarbons dating)
- age-dating using the radioactive decay of isotopes whose initial concentrations are known (e.g., carbon-14)

- age-dating methods using the increase in daughter isotopes of radioactive decay over time (e.g., helium-3)
- methods using the concentration of a (largely) time-independent input (e.g., chloride mass balance).

Age-dating methods are often used to measure the soil-water flux in the vadose zone, which may significantly exceed the actual recharge rate (i.e., the water that actually reaches the water table, Stephens 1996). Age-dating methods have also been used to estimate flow rates along horizontal flow paths.

The basic principle of the chloride mass balance (CMB) method is that rainfall contains low concentrations of chloride (and other ions) that are concentrated by ET to a degree inversely proportional to the recharge rate (Eriksson and Khunakasem 1969). The total mass of an ion is the product of the volume of rainfall and the ion concentration in the rainfall. If an ion is conservative (i.e., its concentration is not changed over time by reaction with the soil, aquifer solids or other ions in solution), then ET will cause the volume of water to decrease and the concentration of the ion to increase, but the total mass of the ion in the water will remain the same. Chloride is the preferred ion for recharge rate estimation because it usually behaves in a conservative manner and it can be accurately and relatively inexpensively measured.

Recharge rates (R) can be calculated if the precipitation rate (P) and concentrations of chloride in the rainfall (Cl_p) and shallow groundwater (Cl_{GW}) are known:

$$R = \frac{Cl_P}{Cl_{GW}} P \tag{2.10}$$

However, soils typically contain some chloride deposited as dryfall (wind-blown particulate salt), which is mobilized by infiltrating precipitation. The total chloride (Cl_T) is thus the sum of the concentration in rainfall (Cl_p) and the contribution from dryfall (DF) multiplied by the precipitation rate:

$$R = \frac{Cl_P + DF}{Cl_{GW}} P = \frac{Cl_T}{Cl_{GW}} \tag{2.11}$$

Corrections may need to be made for seasonal variations in chloride concentrations and a covariation of precipitation and chloride concentrations (Subyani and Şen 2006; Şen 2008).

2.4.3.3 Water Budget Methods

Recharge is commonly estimated as the residual of water budgets. If all the other terms of an aquifer water budget and the change in storage are known, then recharge rates can be calculated. The main technical challenge of water budget methods is that recharge rates are part of the difference between precipitation and ET rates, with the latter being a large number whose value is poorly constrained. Small uncertainties in ET rates can thus result in large percentage errors in calculated recharge rates.

Transmission losses in streams and the rate of decline of water levels in lakes and reservoirs are used to estimate recharge rates, which are assumed to approximately equal infiltration losses. Estimation of recharge losses along a stream reach requires accurate upstream and downstream gauge data and information on any tributary flows, direct rainfall, and evaporative losses. Estimation of recharge from surface water bodies (e.g., reservoirs) requires detailed topographic data to determine the stage-volume relationship as well as data on all flows into and out of the bodies during the time period in question.

Numerical (computer) modeling has become the most important tool for managing groundwater resources. The model calibration process (which is an inverse modeling exercise) provides an opportunity to refine estimates for hydraulic parameters and recharge rates (Sanford 2002). Varying amounts of recharge are applied to the upper layer of models until a satisfactory match of modeled groundwater levels (and flows) to observed values is achieved. The fundamental limitation of the modeling approach to recharge estimation is that inverse modeling does not yield unique solutions and a well-calibrated model may still be inaccurate. As is the case for water budget estimates in general, large errors in model-determined recharge rates may occur if there are significant errors in the dominant parameters of precipitation, ET, and groundwater extractions. The accuracy of model-derived recharge rates may be increased by inverse modeling using chemical tracers in addition to water levels (Sanford 2011). An advantage of using groundwater modeling to evaluate groundwater recharge is that once a calibrated model has been developed, it can be used as a predictive tool to assess groundwater management options (Scanlon et al. 2006) and potential climate change impacts.

2.4.3.4 Flow-Tube Method

Under steady-state flow conditions, the volumetric groundwater flow rate down gradient from a recharge area is equal to the recharge rate (volume/time), if vadose zone percolation in the recharge area is the only source of water and the aquifer has impermeable boundaries (i.e., minimal vertical leakage; Stephens 1996). As a simplification, an aquifer can be conceptualized as being divided into a series of flow tubes with their long axes oriented parallel to the flow direction from the recharge area to a discharge area (natural or pumping location). For a confined aquifer at steady state, the recharge flux (m^3/d) is equal to the flow rate within a flow tube, which can be calculated from the transmissivity (m^2/d), width (m), and hydraulic gradient (m/m) at one point along the flow tube. Sources of errors include pumping and aquifer leakage between the recharge area and measuring point. Total recharge is equal to the calculated recharge for all flow tubes from the recharge area.

Recharge can also be estimated from measured values of the discharge from an aquifer under steady-state conditions, as recharge must equal discharge (with corrections made for pumping and other sources and sinks of water). Measurement of recharge rates from discharge rates can be quite complicated if there is a large change in aquifer geometry from the recharge to discharge areas, and where there are multiple discharge points (e.g., springs and wells) or the discharged water is quickly lost to ET. For example, it can be very difficult to accurately measure discharge rates in wet playas with high evaporation rates or where discharge occurs primarily by plant transpiration. The errors associated with discharge estimates may be large relative to recharge rates.

Regional aquifers in arid lands may not be under steady-state conditions. For example, the flow volumes estimated from hydraulic gradients in North African regional aquifers were found to be too large based on reasonable estimates of current recharge (Bourdon 1977; Lloyd and Farag 1978). The current hydraulic gradients observed in regional basins, such as in Saudi Arabia and North Africa, may in part be attributed to the on-going decay of recharge mounds created during wetter periods during Pleistocene time (Bourdon 1977; Lloyd and Farag 1978).

2.4.4 Climate Change, Land Use Land Cover Change, and Groundwater Recharge

Most previous investigations of the impacts of climate change on groundwater recharge assume that parameters other than precipitation and temperature remain constant. Holman (2006, p. 645) cautioned that

> However, despite the many uncertainties involved in the use of scenarios, to solely focus on the direct impacts of climate change (arising from temperature and precipitation changes) is to neglect the potentially important role of societal values and economic processes in shaping the landscape above aquifers. There are also changes occurring to soil properties over a range of time scales, so that the soils of the future may not have the same infiltration properties as given in current datasets. These all have implications for the certainty, robustness and confidence of future recharge estimates.

Climate change may impact groundwater recharge through its impacts on land use and land cover (LULC). Global climate change is expected to directly result in rapid shifts in the distribution of terrestrial biological communities (Graham and Grimm 1990; Kelly and Goulden 2008). Indirect anthropogenic impacts include changes in crops and agricultural practices.

Vegetation can greatly affect recharge rates. Rainwater must pass through the root zone before it can recharge an underlying aquifer. Vegetation adapted to water scare conditions (i.e., xerophytes) is highly efficient at extracting soil moisture (Stonestrom and Harrill 2007). A transition from deep-rooted native xerophytic vegetation (trees and shrubs) to shallower-rooted agricultural crops and landscaping may increase groundwater recharge rates as the latter are less efficient at extracting soil moisture (Keese et al. 2005; Scanlon et al. 2005). Recharge rates have increased in some areas by one or two orders of magnitude with a natural or man-caused change in vegetative type (Scanlon et al. 2006).

The impacts of LULC changes on groundwater resources depend on numerous factors including the original vegetation being replaced, the vegetation that is replacing it, whether the change is permanent or temporary, and associated land management practices involving alteration of drainage (Scanlon et al. 2007). Areas with deep-rooted vegetation can have dramatically lower recharge rates compared to ground covered with shallow-rooted vegetation and bare ground (e.g., Gee et al. 1992). Natural forests have greater ET rates than other types of vegetation, and the conversion of forests to cultivated areas can provide more water for groundwater recharge and streamflow (i.e., baseflow). The conversion of native vegetation to rainfed agriculture generally increases groundwater quantity but can decrease water quality through the mobilization of salts and nutrients (e.g., nitrates) that have accumulated in the vadose zone and through soil salinization caused by the rising of the water table to close to land surface (Scanlon et al. 2007)

The wilting point of native arid and semiarid rangeland vegetation (i.e., the minimum matric potential at which plants can take up water) is typically much lower than that of typical agricultural crops (Scanlon et al. 2005). Native rangeland vegetation can draw much more water out of the soil than typical agricultural crops, which is an adaptation to water scarcity. The transition in LULC from rangeland vegetation to cultivated crops can result in an increase in groundwater recharge, in part related to the addition of water through irrigation. Increased recharge associated with dryland agriculture appears to be associated with reduced interception and ET, shallow rooting depths, fallow periods, and increased soil permeability caused by plowing (Scanlon et al. 2005, 2006).

There is evidence that anthropogenic changes in land cover can impact local climates, which, in turn, can impact water resources. For example, rainfall in the peninsular Florida summer wet season occurs predominantly as convective storms associated with sea breeze fronts. A regional-average time series of accumulated convective rainfall for July and August from 1924 to 2000 has a linear trend with a slope of −0.064, with a total decrease of 5 cm or 12% over the period of record (Marshall et al. 2004). Pielke et al. (1999) noted that land cover and use influence the amount of water transpired and evaporated into the atmosphere as well as affecting local wind circulations that focus the cumulonimbus activity. Modeling results support the finding that land cover changes, particularly replacement of native wetlands vegetation, resulted in an increase in summertime maximum temperatures and a decrease in convective rainfall in southern Florida (Pielke et al. 1999; Marshall et al. 2004).

2.4.5 *Effects of Temperature on Recharge*

Temperature can impact groundwater recharge through its effects on ET rates and hydraulic conductivity. Warmer temperatures can result in reduced infiltration rates because of more rapid evaporative losses of standing and shallowly infiltrated water in the upper vadose zone. Plant ET rates also generally increase with increasing temperature. With higher ET rates, soils will tend to have larger soil moisture deficits and larger rainfall events would be required to make up for the deficits and allow water to percolate past the root zone and become recharge. A transition to warmer and drier conditions would result in drier soils that would retain more precipitation.

Infiltration rates are also related to water temperature through the temperature dependency of hydraulic conductivity, which is a function of the kinematic viscosity of water. As temperature increases, the viscosity of water decreases, and hydraulic conductivity increases, which for a given hydraulic gradient will result in more rapid infiltration. Field and modeling studies have demonstrated that large diurnal variations in infiltration rates occurs at some sites that can be attributed to the temperature-dependence of hydraulic conductivity (Jaynes 1990: Constantz et al. 1994; Ronan et al. 1998).

Warmer climatic conditions are expected to decrease the amount of precipitation that falls as snow relative to rain and result in an earlier and more rapid melting of snowpacks. Changes in the seasonal availability of water may impact recharge rates in either direction.

2.4.6 *Climate Change and Recharge*

Changes in average annual precipitation translate to changes in the amount of water that is potentially available for recharge. Infiltration and recharge rates can be influenced by climate changes through the annual precipitation rate, and the volume, rate (intensity), duration, and seasonal timing of precipitation events as follows:

- High intensity rainfall events in excess of soil infiltration capacity result in a greater partitioning of rainfall into runoff. This can be either adverse for recharge, if water is lost to beneficial use and recharge (i.e., flows to tide), or favorable for recharge if more water flows to, and accumulates in, locations more favorable for recharge (i.e., sites of indirect recharge).
- Greater volume (product of duration and rate) rainfall events are needed to overcome soil moisture deficiencies and allow for water to percolate past the root zone of plants and become recharge.
- Most, if not all, rainfall from small volume events may be held in near surface soils and subsequently lost to ET.

- Greater recharge rates can occur from rainfall events that occur outside of peak plant growth (and thus ET) periods.

Climate variability at various time scales is a fundamental control on groundwater resources of some aquifers (Gurdak et al. 2007), and a key issue is how climate change will impact variability in precipitation and temperature rather than just mean values. In semiarid and arid regions, small, temporally isolated rainfall events result in essentially no recharge because the shallowly infiltrated water is retained in the upper part of the vadose zone or is shallowly ponded on impervious surfaces and is quickly lost to ET. Most groundwater recharge in arid and semi-arid regions occurs instead during infrequent wet periods (large rainfall events). A local decrease in mean rainfall with an increased tendency for rainfall to occur in larger events could result in increased recharge rates.

2.5 Climate Change and Water Demand

Climate change can impact the demand side of the aquifer budgets by increasing the demand for water in general or by reducing the amount of available surface water prompting a shift toward more groundwater pumping (Russo and Lall 2017). Groundwater has the great advantage for water supply in that the huge storage volumes of aquifers provide a buffer against variations in recharge and extractions. Aquifer water levels may drop during droughts, but the resource may be far from being depleted. However, water levels may drop below pump intakes or well bottoms disrupting the supplies of some users. Small aquifers (e.g., freshwater lenses on low-lying islands) may have a high vulnerability to droughts.

Much has been written on "groundwater sustainability" and "safe yield" and there are considerable differences in opinion on the specific definitions of the terms. A reasonable, broad definition of groundwater sustainability is (Alley et al. 1999; Alley and Leake 2004, p. 13):

> the development and use of ground water resources in a manner that can be maintained for an indefinite time without causing unacceptable environmental, economic, or social consequences.

For groundwater use to be sustainable, aquifer levels should not drop to a point where long-term use of the resource is compromised or the lower aquifer water levels cause what are considered unacceptable environmental impacts, such as dehydration of wetlands and reductions in spring and stream flows to the detriment of associated groundwater dependent ecosystems.

In some areas where groundwater use is currently unsustainable or at sustainable limits, surface water, a more climate-vulnerable supply of water, is being pursued out of necessity as a means of reducing the current aquifer overdraft and meeting future increases in demands. On the contrary, where surface water supplies may be compromised by drying climate conditions, additional (at least temporary) water supplies may be obtained from more reliable groundwater sources. There are few (if any) areas in world that have large populations, irrigated agriculture, and are facing water scarcity in which local fresh groundwater resources are not already being used at close to or above sustainable levels. Conjugate use schemes involving managed aquifer recharge of temporarily available excess surface water is one tool for increasing the resilience of water supplies.

Climate can impact demands on groundwater, and water in general, through changes in the ET rates of vegetation (natural and crops) and changes in the water demands of domestic, commercial, and industrial users. The latter changes depend on both the direction and magnitude of climate changes and the socioeconomic responses to the changes.

2.5.1 Plant Evapotranspiration Rates and Irrigation Water Demands

Climate change can impact groundwater resources through plant ET rates in several manners. ET of native vegetation and crops, with rare exceptions, is the major water outflow of water budgets. Climate change-induced increases or decreases in ET can, therefore, impact aquifer recharge rates. Decreasing precipitation could result in less water passing through the root zone of vegetation, decreasing groundwater recharge. Higher temperatures can result in increased ET rates and thus decreased soil moistures and increased irrigation water requirements. Clearly, a shift to drier conditions would be expected to result in increased irrigation demands. Climate change in some temperate areas may extend the crop growing (and thus irrigation) season.

Climate change will have the greatest impact on global water use through irrigation because it is the major global water use and crop supplemental water requirements have a high sensitivity to increases in temperature and decreases in precipitation. The global water withdrawal ratio is 69% agricultural, 12% municipal and 19% industrial (FAO n.d.). In the United States, agriculture accounts for approximately 80% of the nation's consumptive groundwater and surface water use, and over 90% in many western States (USDA 2019). The FAO (2009) estimated that feeding a projected world population of 9.1 billion people in 2050 would require raising overall food production by some 70% between 2005/07 and 2050 and that production in developing countries would need to almost double. In additional to increases in irrigation water demand associated with the need to feed a growing global population, climate change associated increases in water demand may cause an additional severe stress on local water resources.

Increasing temperatures will tend to result in higher ET rates and thus greater plant water requirements. However, numerous experiments performed since the 1800s have shown that atmospheres enriched in CO_2 compared to ambient air result in better plant growth (Kimball 1983; Bazzaz 1990). The effects of CO_2 on plant growth have been investigated in laboratory (cabinet), greenhouse, and free-air carbon dioxide enrichment (FACE) experiments. The latter are field experiments in which CO_2 is emitted and sensors are used to maintain a target elevated CO_2 concentration, which is compared to a nearby non-enriched area. This "CO_2 fertilization" effect is used commercially in tomato production where the air in greenhouses is enriched in CO_2 to greatly increase yields (Jaggard et al. 2010). CO_2 at high concentrations causes plants to close stomata, which are small openings or pores, mostly on the under-surface of plant leaves, that are used for gas exchange. The decrease in the aperture of stomata (i.e., increased stomatal resistance) reduces transpiration and thus water consumption. Jaggard et al. (2010) concluded that the effect of CO_2 on water consumption can only have a positive impact on yield because in many situations crop yields are water limited.

Based on historical observational data, Kimball (1983) concluded that a doubling of the earth's CO_2 concentration would probably increase agricultural yields by about 33%, with a 95% confidence range from 28 to 39%. These earlier results indicated that the response of C4 plants (e.g., maize, sorghum, millet, sugar cane) to CO_2 is likely about one fourth the response of C3 plants (e.g., small grains, legumes, most trees; Kimball 1983). The response of plants to CO_2 also depends on other factors such as nutrients and solar radiation. Crops grown on infertile soils without added fertilizer will likely benefit less from increased CO_2.

Bazzaz (1990) observed the following pattern from an extensive literature review of the response of photosynthesis to elevated CO_2:

- elevated CO_2 reduces or completely eliminates photorespiration
- C3 plants are more responsive to elevated CO_2 levels than C4 plants, especially to concentrations above ambient levels
- photosynthesis is enhanced by CO_2 but the enhancement may decline with time
- the response to CO_2 is more pronounced under high levels of other resources (e.g., water, nutrients, and light)
- adjustment of photosynthesis during growth occurs in some species, and the adjustment may be influenced by resource availability
- species even of the same community may have different responses to CO_2.

Crop water productivity (CWP) is the ratio of crop yield to evapotranspiration. Increases in atmospheric CO_2 concentration would impact CWP by both increasing plant growth rates and decreasing transpiration. Deryng et al. (2016) evaluated the global impacts of increased CO_2 on CWP for three C3 crops (wheat, rice and soybean) and a C4 crop (maize) under a high-end greenhouse emissions scenario. The investigation utilized an ensemble of six global crops models with climate input data derived from five global climate models. The RCP8.5 emissions scenario was considered, which projects a doubling of CO_2 by 2080. Changes in CWP were evaluated for climate change with elevated CO_2 (CC w/CO_2) and for climate change with CO_2 constant at present-day levels (CC w/o CO_2).

The simulation results indicated that by 2080 under CC w/o CO_2, severe negative impacts would occur on crop yields at the global scale with small reductions in ET, which would result in large reductions in CWP (Deryng et al. 2016). On the contrary, under CC w/CO_2, substantial increases in global CWP would occur with increases of 27(7; 37)% for wheat, 18(−9; 42)% for soybean, 13(3, 22)% for maize, and 10(0; 47)% for rice (the values in parentheses are the interquartile range; Deryng et al. 2016). Crops grown in arid climates under rainfed conditions benefited the most from elevated CO_2, which could lead to substantial increases in crop production and reductions in consumptive water use. Particularly large beneficial impacts were predicted for maize grown in semi-arid regions, including most of southern Africa, the Middle East, parts of central Asia, the western United States, and Iberian Peninsula (Deryng et al. 2016). Deryng et al. (2016, p. 4) concluded that

> Anticipating climate impacts and interaction across the agriculture and water sectors is essential to improve the efficiency and resilience of agricultural systems. Food security, especially in arid and less developed regions, is not only a function of crop productivity and available land, but also of CWP and available water resources. This relationship is strongly affected by elevated CO_2 and demands greater attention in scientific and policy assessment.

Other studies suggest that the beneficial effect of CO_2 on water consumption may be less than previously thought. While it is recognized that CO_2 would contribute to increased plant growth and water use efficiency, other factors also need to be considered in the evaluation of the impacts of climate change on crops. Crop species respond differently to temperature throughout their life cycles, and there are minimum, maximum and optimal temperatures at which plants grow at their maximum rates (Hatfield et al. 2011). Decreases in solar radiation caused by increased cloudiness related to greater atmospheric water vapor could adversely impact crop growth. Evaluation of the potential impacts of climate change on crop ET requires detailed studies of the sensitivity of ET to a combination of weather and plant variables (Hatfield et al. 2011). Several studies have shown that weeds, rather than crops, show the strongest relative response to rising CO_2 (Ziska 2000, 2003a, b, 2004; Hatfield et al. 2011).

Beneficial impacts on crop production and water use efficiency from increasing CO_2 may be offset by the detrimental impacts of rising atmospheric ozone (O_3) concentrations on plant growth and crop yields. (Jaggard et al 2010). Jaggard et al. (2010) noted that by 2050, the impact of rising O_3 is likely to eliminate most of the yield increase due to increasing CO_2 in C3 crops and cause a yield decrease of at least 5% in C4 species. More research is needed to better elucidate the net impacts of changes in climate and atmospheric CO_2 and O_3 concentrations (and other factors) on crop yields and water productivity.

A recurring theme in discussions of the impacts of climate change on water demands is the human response, which is a large uncertainty factor. A farmer faced with increasing water demands may choose to increase groundwater pumping, adopt more water efficient irrigation technologies, or switch to another crop or cultivar that requires less irrigation. Demand responses also depend upon the adaptative capacity of the parties involved. Depending on circumstances, certainly including wealth, some individuals, communities, and countries have more options available to them to respond to climate change.

2.5.2 Climate Change and Domestic and Industrial Water Demands

Municipal water supplies broadly refer to the water supplies for urban and suburban communities that are provided by a private or public utility. Water in rural areas tends to be self-supplied. Municipal water suppliers in some areas of the world face a triple challenge of (1) meeting increasing demands associated with population growth and economic development, (2) addressing threats to their existing water supplies caused by climate change (e.g., decreasing surface water flows), and (3) increasing demands for water associated with hotter and drier conditions. The main industrial water uses sensitive to climate change are for cooling. Increasing temperatures results in decreased cooling efficiencies and thus greater water demands. Higher temperatures also result in increased electrical power demands for indoor cooling.

Climate change may also influence long-term patterns of regional population growth and industrial and agricultural operations, which drive local water demands (Keifer et al. 2013). As an extreme example, flooding from sea level rise may force coastal residents to migrate inland. The increase in water demands caused by the resettlement of coastal population could stress the water resources of inland areas that are not directly impacted by sea level rise (Curtis and Schneider 2011; Hauer et al. 2016; Hauer 2017).

Impacts of climate change on municipal water demands has received much less attention than its impacts on supplies. The Climate Change and Demand for Water (CCDeW) research project was a detailed investigation of the potential impacts of climate change on domestic, industrial, and commercial water use and irrigated agriculture and horticulture in the United Kingdom (Downing et al. 2003). The CCDeW was an update of an earlier study by Herrington (1996). A key observation was that "the extent to which water consumption will be influenced by climate change depends upon the sensitivity of different sectors to specific aspects of climate change as well as potential behavioural and regulatory changes, in part related to different socio-economic and climatic futures" (Downing et al. 2003, p. xv).

Domestic responses to a warmer climate might include increased use of baths and showers, and increased watering of gardens. The CCDeW project results suggest that domestic demand will be sensitive to the interplay of warmer climates, household choices regarding water-using technologies, and the regulatory environment, such as the imposition of (and compliance with) restrictions on outdoors water use (Downing et al. 2003). The effects of climate change on domestic demand was modeled to be modest; an increase of 1.8–3.7% by the 2050s. Industrial and commercial sectors sensitive to climate change include soft drinks, brewing, and leisure (e.g., golf courses). For example, in a warmer climate people may drink more cold soft drinks and beer (Downing et al. 2003).

Agricultural water demand was found to be approximately an order of magnitude more sensitive to climate change than domestic and industrial/commercial demands (Downing et al. 2003). Climate change could affect irrigation water use through changes in plant physiology, soil water balances, crop mixes, cropping patterns that take advantage of longer growing seasons, and changes in demand for different foods (Downing et al. 2003).

Keifer et al. (2013) investigated the potential changes in water use under regional climate change scenarios. Their approach was to estimate the historic weather sensitivity of water demand in six case study utilities (Colorado Springs Utilities, Region of Durham (Ontario, Canada), Massachusetts Water Resources Authority, Southern Nevada Water Authority, San Diego County Water Authority, and Tampa Bay Water) located in different climatic regions and having different water supply portfolios. Downscaled climate projection scenarios from GCMs were translated into implied future values of weather, which were substituted into water demand models to predict potential demand-side impacts.

Methods were evaluated and refined to partition historical water demands into climate and seasonal components and a weather-sensitive component. Variability of weather conditions within a given climate was found to account for most of the interannual variation in demand (e.g., differences in annual demand between consecutive years; Keifer et al. 2013). Fundamental differences in climate across geographical regions were found to explain large differences in average rates of water consumption as well as differences in the pattern of monthly and daily demands within a calendar year (Keifer et al. 2013). Municipal water demands were found to be sensitive to regional differences in climate, with water demands in the hot and dry climates of the western studied utilities being 50–80% higher than those in the humid east. Seasonal estimates of demand impacts generally point to a lengthening of the watering season (Keifer et al. 2013).

Some of the key observations and conclusions of the Keifer et al. (2013, p. 130–131) study are

> Many highly populated areas rely on surface water as the primary source of public water supplies, which are generally more variable than ground water sources, and this reliance is increasing in some places that are already vulnerable with respect to other indicators.
>
> In regions with high and growing urban demands, high rates of agricultural withdrawals and relative reliance on surface water sources, climate changes that result in warmer and drier conditions or changes in seasonal conditions will likely exacerbate existing demand pressures.
>
> Estimated increases in demand among some case studies would be considered equivalent to effects of significant growth in the number of accounts or population under historical normal climate conditions.
>
> Implied weather variability within some climate projection scenarios produces estimates of demand that would be unlikely to be experienced under historical weather. This relates also to hot and dry weather spells and the potential for weather anomalies such as drought. For some cases, the largest absolute average projected change (decrease) in precipitation is paired with relatively large projected changes in seasonal demand.

The major unknown when evaluating the impacts of climate change on water demands is socioeconomic factors. It is possible that the impacts of a drier climate could translate into an increased awareness in society of the importance of water in people's daily lives and thus promote a greater conservation ethic (Keifer et al.

2013). Projected increases in demand relative to available water supplies may force increases in the cost of water to users to finance investments in alternative supplies and system rehabilitation, which may create further incentives for more efficient water use (Keifer et al. 2013).

References

Allen RG, Pereira LS, Raes D, Smith M (1998) Crop evapotranspiration—guidelines for computing crop water requirements, FAO irrigation and drainage paper 56. Food and Agricultural Organization of the United Nations, Rome

Alley WM, Leake SA (2004) A journey from safe yield to sustainability. Ground Water 42(1): 12–16

Alley WM, Reilly TE, Franke OL (1999) Sustainability of ground-water resources. U.S. Geological Survey Circular 1186

Allison GB (1988) A review of some of the physical chemical and isotopic techniques available for estimating groundwater recharge. In: Simmers I (ed) Estimation of natural groundwater recharge. North Atlantic Treaty Organization, Scientific Affairs Division, pp 49–72

Allison GB, Gee GW, Tyler SW (1994) Vadose-zone techniques for estimating groundwater recharge in arid and semiarid regions. Soil Sci Soc Am J 58:6–14

Bazzaz FA (1990) The response of natural ecosystems to the rising global CO_2 levels. Annu Rev Ecol Syst 21(1):167–196

Bourdon DJ (1977) Flow of fossil groundwater. Q J Eng Geol Hydrogeol 10:97–124

Bouwer H (1978) Groundwater hydrology. McGraw-Hill, New York

Constantz J, Thomas CL, Zellweger G (1994) Influence of diurnal variations in stream temperature on streamflow loss and groundwater recharge. Water Resour Res 30:3253–3264

Curtis KJ, Schneider A (2011) Understanding the demographic implications of climate change: estimates of localized population predictions under future scenarios of sea-level rise. Popul Environ 33(1):28–54

Cuthbert MO, Gleeson T, Moosdorf N, Befus KM, Schneider A, Hartmann J, Lehner B (2019) Global patterns and dynamics of climate–groundwater interactions. Nat Clim Change 9(2): 137–141

De Vries JJ, Simmers I (2002) Groundwater recharge: an overview of processes and challenges. Hydrogeol J 10:5–17

Deryng D, Elliott J, Folberth C, Müller C, Pugh TA, Boote KJ, Conway D, Ruane AC, Gerten D, Jones JW, Khabarov N, Olin S, Schaphoff S, Schmid E, Yange H, Rosenzweig C (2016) Regional disparities in the beneficial effects of rising CO_2 concentrations on crop water productivity. Nat Clim Change 6(8):786

Dillon P (2005) Future management of aquifer recharge. Hydrogeol J 13:313–316

Downing TE, Butterfield RE, Edmonds B, Knox JW, Moss S, Piper BS, Weatherhead EK (and the CCDeW project team) (2003) Climate change and the demand for water. Stockholm Environment Institute Oxford Office, Oxford

Eriksson E, Khunakasem V (1969) Chloride concentration in groundwater, recharge rate and rate of deposition of chloride in Israel Coastal Plain. J Hydrol 7:178–197

FAO (2009) Global agriculture towards 2050. In: High level expert forum—how to feed the world in 2050, Rome, 12–23 Oct 2009. http://www.fao.org/fileadmin/templates/wsfs/docs/Issues_papers/HLEF2050_Global_Agriculture.pdf. Accessed 26 May 2020

FAO (n.d.) Water use. In: QUASTAT—FAO's global information system on water and agriculture. Food and Agriculture Organization of the United Nations. http://www.fao.org/aquastat/en/overview/methodology/water-use. Accessed May 26 2020

Flint AL, Flint LE, Kwicklis EM, Fabryka-Martin JT, Bodvarson GS (2002) Estimating recharge at Yucca Mountain, Nevada, USA, comparison of methods. Hydrogeol J 10:180–204
Gee GW, Hillel D (1988) Groundwater recharge of arid regions: review and critique of estimation methods. Hydrol Process 2:255–266
Gee GW, Fayer MJ, Rockhold ML, Campbell MD (1992) Variations in recharge at the Hanford Site. Northwest Sci 66:237–250
Graham RW, Grimm EC (1990) Effects of global climate change on the patterns of terrestrial biological communities. Trends Ecol Evol 5(9):289–292
Gurdak JJ, Hanson RT, McMahon PB, Bruce BW, NcCray JE, Thyne GD, Reedy RC (2007) Climate variability controls on unsaturated water and chemical movement, high plains aquifer, USA. Vadose Zone J 6:533–547
Hatfield JL, Boote KJ, Kimball BA, Ziska LH, Izaurralde RC, Ort D, Thomson AM, Wolfe D (2011) Climate impacts on agriculture: implications for crop production. Agron J 103(2): 351–370
Hauer ME (2017) Migration induced by sea-level rise could reshape the US population landscape. Nat Clim Change 7(5):321–325
Hauer ME, Evans JM, Mishra DR (2016) Millions projected to be at risk from sea-level rise in the continental United States. Nat Clim Change 6(7):691–695
Healy RW, Cook PG (2002) Using ground water levels to estimate recharge. Hydrogeol J 10(1): 91–109
Herczeg AL, Leaney FW (2011) Review: environmental tracers in arid-zone hydrology. Hydrogeol J 19(1):17–30
Herrington P (1996) Climate change and the demand for water. HMSO, London
Holman IP (2006) Climate change impacts on groundwater recharge—uncertainty, shortcomings, and the way forward? Hydrogeol J 14:637–647
Horton RE (1933) The role of infiltration in the hydrologic cycle. Trans Am Geophys Union 14:446–460
Horton RE (1940) An approach towards a physical interpretation of infiltration capacity. Proc Soil Soc Am 5:399–417
Irmak S, Haman DZ (2003) Evapotranspiration: potential or reference?. University of Florida Institute of Food and Agricultural Sciences, Gainesville
Izbicki JA, Johnson RU, Kulongoski J, Predmore S (2007) Ground-water recharge from small intermittent streams in the western Mojave Desert, California. In: Stonestrom DA, Constantz J, Ferré TPA, Leake SA (eds) Ground-water recharge in the arid and semiarid southwestern United States. U.S. Geological Survey Professional Paper 1703, pp 157–184
Jaggard KW, Qi A, Ober ES (2010) Possible changes to arable crop yields by 2050. Philos Trans R Soc London B: Biol Sci 365(1554):2835–2851
Jaynes DB (1990) Temperature variations effects on field-measured infiltration. Soil Sci Soc Am J 54:305–311
Jury WA, Horton R (2004) Soil physics. John Wiley & Sons, Hoboken
Keese KE, Scanlon BR, Reedy RC (2005) Assessing controls on diffuse groundwater recharge using unsaturated flow modeling. Water Resour Res 41:W06010. https://doi.org/10.1029/2004WR003841
Keifer JC, Clayton JM, Dziegielewski B, Henderson J (2013) Changes in water use under regional climate change scenarios. Water Research Foundation, Denver
Kelly AE, Goulden ML (2008) Rapid shifts in plant distribution with recent climate change. Proc Natl Acad Sci 105(33):11823–11826
Kimball BA (1983) Carbon dioxide and agricultural yield: An assemblage and analysis of 430 prior observations. Agron J 75(5):779–788
Lerner DN, Issar AS, Simmers I (1990) Groundwater recharge. A guide to understanding and estimating natural recharge (contributions to Hydrogeology 8). International Associations of Hydrogeologists, Kenilworth

Lerner DN, Issar AS, Simmers I (1997) Groundwater recharge. In: Saether OM, de Caritat P (eds) Geochemical processes, weathering and groundwater recharge in catchments. AA Balkema, Rotterdam, pp 109–150
Lloyd JW, Farag MH (1978) Fossil ground-water gradients in arid sedimentary basins. Ground Water 16(6):388–393
Maliva RG (2019) Anthropogenic aquifer recharge. Springer, Cham
Maliva RG, Missimer TM (2012) Arid lands water evaluation and management. Springer, Berlin
Marshall CH, Pielke RA Sr, Steyaert LT, Willard DA (2004) The impact of anthropogenic land-cover change on the Florida peninsula sea breezes and warm season sensible weather. Mon Weather Rev 132(1):28–52
Monteith JL (1965) Evaporation and environment in the state and movement of water in living organisms In: Fogg GE (ed) Symposium of the society for experimental biology, vol 19, pp 205–234. Academic Press, New York
NRMMC, EPHC, NHMRC (2009) Australian Guidelines for water recycling: managing health and environmental risks (Phase 2), Managed aquifer recharge (July 2009). Natural Resource Management Ministerial Council, Environment Protection and Heritage Council, National Health and Medical Research Council
Penman HL (1948) Natural evaporation from open water, bare soil and grass. Proc R Soc London A193:120–146
Penman HL (1956) Estimating evaporation. Eos, Trans Am Geophys Union 37(1):43–50
Pielke RA Sr, Walko RL, Steyaert LT, Vidale PL, Liston GE, Lyons WA, Chase TN (1999) The influence of anthropogenic landscape changes on weather in south Florida. Mon Weather Rev 127(7):1663–1673
Priestley CHB, Taylor RJ (1972) On the assessment of surface heat flux and evaporation using large-scale parameters. Mon Weather Rev 100(2):81–92
Ronan AD, Prudic DE, Thodal CE, Constantz J (1998) Field study and simulation of diurnal temperature effects on infiltration and variably saturated flow beneath an ephemeral stream. Water Resour Res 34:2137–2153
Russo TA, Lall U (2017) Depletion and response of deep groundwater to climate-induced pumping variability. Nat Geosci 10(2):105–108
Sanford W (2002) Recharge and groundwater models: an overview. Hydrogeol J 10(1):110–120
Sanford W (2011) Calibration of models using groundwater age. Hydrogeol J 19:13–16
Scanlon BR, Tyler SW, Wierenga PJ (1997) Hydrologic issues in arid, unsaturated systems and implications for contaminant transport. Rev Geophys 35:461–490
Scanlon BR, Healy RW, Cook PG (2002) Choosing appropriate techniques for quantifying groundwater recharge. Hydrogeol J 10:18–39
Scanlon BR, Reedy RC, Stonestrom DA, Prudic DE, Dennehy KF (2005) Impact of land use and land cover change on groundwater recharge and quality in the southwestern US. Glob Change Biol 11:1577–1593
Scanlon BR, Keese KE, Flint AL, Flint LE, Gaye CB, Edmunds WM, Simmers I (2006) Global synthesis of groundwater recharge in semiarid and arid regions. Hydrol Process 20:3335–3379
Scanlon BR, Jolly I, Sophocleous M, Zhang L (2007) Global impacts of conversions from natural to agricultural ecosystems on water resources: quantity versus quality. Water Resour Res 43: W03437. https://doi.org/10.1029/2006WR005486
Şen Z (2008) Wadi hydrology. CRC Press, Boca Raton
Simmers I (1990) Aridity, groundwater recharge and water resources management. In: Lerner DN, Issar AS, Simmers I (eds) Groundwater recharge. A guide to understanding and estimating natural recharge (Contributions to Hydrogeology 8, pp 1–20). International Associations of Hydrogeologists, Kennilworth
Simmers I (1998) Groundwater recharge: an overview of estimation "problems" and recent developments. In Robins NS (ed) Groundwater pollution, aquifer recharge and vulnerability (Special Publication 130, pp 107–115). Geological Society, London
Snyder RL (1992) Equation for evaporation pan to evapotranspiration conversions. J Irrig Drainage Eng 118:977–980

Snyder RL, Orang M, Matyac S, Grisner ME (2005) Simplified estimation of reference evapotranspiration from pan evaporation data in California. J Irrig Drainage Eng 131:249–253

Sophocleous M (2004) Groundwater recharge. In: Silveira L, Wohnlich S, Usunoff EJ (eds) Encyclopedia of life support systems (EOLSS). Eolss Publishers, Oxford. http://www.eolss.net

Stephens DB (1996) Vadose zone hydrology. CRC Press, Boca Raton

Stonestrom DA, Harrill JR (2007) Ground-water recharge in the arid and semiarid southwestern United States—climate and geologic framework. In: Stonestrom DA, Constantz J, Ferré TPA, Leake SA (eds) Ground-water recharge in the arid and semiarid southwestern United States, pp 1–27. U.S. Geological Survey Professional Paper 1703

Stonestrom DA, Prudic DE, Walvoord MA, Abraham JD, Stewart-Deaker AE, Glancy PA, Constantz J, Lacniak RJ, Andrasji BJ (2007) Focused ground-water recharge in the Amargosa Desert Basin. In: Stonestrom DA, Constantz J, Ferré TPA, Leake SA (eds) Ground-water recharge in the arid and semiarid southwestern United States, pp 107–136. U.S. Geological Survey Professional Paper 1703

Subyani A, Şen Z (2006) Refined chloride mass-balance method and its application in Saudi Arabia. Hydrol Process 20:4373–4380

Task Committee on Standardization of Reference Evapotranspiration (2005) The ASCE standardized reference evapotranspiration equation. Environmental and Water Resources Institute, American Society of Civil Engineers, Reston

USDA (2019) Irrigation & water use. U.S. Department of Agriculture Economic Research Service. https://www.ers.usda.gov/topics/farm-practices-management/irrigation-water-use/. Accessed 26 May 2020

Wheater HS (2002) Hydrological processes in arid and semi arid area. In: Wheater H, Al-Weshah RA (eds) Hydrology of wadi systems (IHP-V, Technical documents in hydrology 55, pp 5–22). UNESCO, Paris

Wilson JL, Guan H (2004) Mountain-block hydrology and mountain-front recharge. In: Phillips FM, Hogan J, Scanlon B (eds) Groundwater recharge in a desert environment: the Southwestern United States. American Geophysical Union, Washington, D.C., pp 113–127

Wood WW, Rainwater KA, Thompson DB (1997) Quantifying macropore recharge: examples from a semi-arid area. Ground Water 35:1097–1106

Ziska LH (2000) The impact of elevated CO_2 on yield loss from a C3 and C4 weed in field-grown soybean. Glob Change Biol 6:899–905

Ziska LH (2003a) Evaluation of yield loss in field sorghum from a C3 and C4 weed with increasing CO_2. Weed Sci 51:914–918

Ziska LH (2003b) Evaluation of the growth response of six invasive species to past, present and future carbon dioxide concentrations. J Exp Bot 54:395–404

Ziska LH (2004) Rising carbon dioxide and weed ecology. In: Inderjit (ed) Weed biology and management, pp 159–176. Kluwer Academic, Netherlands

Chapter 3
Historical Evidence for Anthropogenic Climate Change and Climate Modeling Basics

3.1 Introduction

The Earth's climate and sea level have always been changing. Since the beginning of the Pleistocene Epoch approximately 2.58 million years ago, the planet has experienced repeated cooling periods, during which glaciers and continental ice sheets expanded, and warmer interglacial periods, which saw a retreat of the ice and associated rise of sea level. The enormous volume of water stored in continental ice sheets during the peak of the last ice age caused sea level to stand approximately 134 m (440 ft) lower than today (Gornitz 2007; Lambeck et al. 2014). The current Holocene Epoch, which began approximately 11,650 yr before present, is a warming period after the last glacial episode (ice age). The cause of the glacial-interglacial cyclicity is now attributed ultimately to cyclicity in external astronomical forcing (insolation—the amount of solar radiation reaching the Earth's surface) combined with complex feedbacks involving atmospheric CO_2 concentration and the Earth's albedo (Past Interglacials Working Group of PAGES 2016). It is uncertain if and when the Holocene will be followed by another glacial period and thus whether it represents just another temporary warming period in the climatic cyclicity that started at the being of the Pleistocene Epoch.

Pleistocene and Holocene climate variability has had profound impacts on terrestrial ecosystems and human dispersion and cultural development. Current anthropogenic acceleration of climate change due to the greenhouse effect is expected to impact natural ecosystems and humans in a myriad of ways. Historical observational data can provide insights on the recent direction, magnitude, and rate of temperature, sea level, and precipitation changes. However, numerical climate modeling is now relied upon for the prediction of future climate change. Historical trends provide some insights on near future changes in climate conditions but projected future temperatures and sea level rise rates diverge greatly (accelerate) from historical trends in scenarios with larger increases in greenhouse gas (GHG) concentrations. It is also important to appreciate that climate describes

R. Maliva, *Climate Change and Groundwater: Planning and Adaptations for a Changing and Uncertain Future*, Springer Hydrogeology,
https://doi.org/10.1007/978-3-030-66813-6_3

average conditions over long periods of time and short-term weather conditions may reflect the natural variability of local climates. For example, several years of drier than normal conditions may be just part of the natural variability of a stable climate regime rather than being evidence for a transition of long-term conditions (i.e., the climate) to a drier regime.

Climate modeling is an extremely complex undertaking and the development of general circulation models (aka global climate models; GCMs) is in the realm of a small number of expert groups. To skeptical members of the public, it can be unsatisfying to be asked to blindly accept modeling results that one does not understand and have no means of personally verifying. Within the climate science (i.e., expert) community, there is virtual unanimity that anthropogenic global warming is true (Powell 2015). It is also recognized that there are inherent uncertainties in the GCM results. Indeed, the Intergovernmental Panel on Climate Change (IPCC) in its reports takes pains to acknowledge the degree of uncertainty in outcomes using qualitative levels of confidence (from very low to very high) and, when possible, provides a quantitative probability ranging from virtually certain (99–100% probability) and very likely (90–100% probability) to extremely unlikely (0–5%; Mastrandrea et al. 2010).

This chapter provides overviews of global scale historical climate change data and the climate modeling process, including GCMs, model ensembles, and downscaling procedures, emissions scenarios, and accessing climate modeling results.

3.2 Historical Climate Trends

Climate and sea level data have historically been obtained primarily from weather stations and tide gauges, and by less formal observations for data from further back in time. Weather stations and tide gauges are still used for direct measurements of climate parameters and sea level but have the limitation of being point measurements. Weather stations tend to be concentrated in developed countries. To reconstruct climate further back in time when direct measurement data are not available, proxies are used, which are physical, biological, and chemical data that can be correlated to climate. Proxies used by paleoclimatologists include tree rings, fossil pollen, and isotopic data from ice cores and marine fossils.

3.2.1 Temperature

Satellite data have been increasingly used for climate monitoring since the late 1970s. Satellites cannot directly measure sea surface and atmospheric temperatures. They instead measure radiances in various wavelength bands, which are mathematically processed to obtain indirect inferences of temperature. For example, microwave sounding units (MSUs) are used on satellites to measure atmospheric temperature from the thermal emissions of molecular oxygen in the atmosphere at four frequencies near 60 GHz. Radiance at microwave frequencies is directly proportional to the temperature of the emitting body (Spencer and Christy 1990; Spencer et al. 1990; Hooker et al. 2018; Zuo et al. 2018). Infrared detectors are used in a similar manner to measure sea surface temperatures.

Some variation occurs between different plots of historical climate parameters versus time, which reflects differences in the data sets and processing methods used. Nevertheless, the data, as a whole, provide a coherent picture of climate change. National Aeronautics and Space Administration (NASA 2020a) data on the change in mean global surface temperature relative to the 1951–1980 average indicate an increase of 0.98 °C (1.76 °F) by 2019 with a clear upward trend since the late 1970s (Fig. 3.1). Nineteen of the twenty warmest years have occurred since 2001 with the exception being 1998 (NASA 2020a).

The IPCC concluded that "it is certain that Global Mean Surface Temperature has increased since the late 19th century. Each of the past three decades has been successively warmer at the Earth's surface than all the previous decades in the instrumental record, and the first decade of the 21st century has been the warmest"

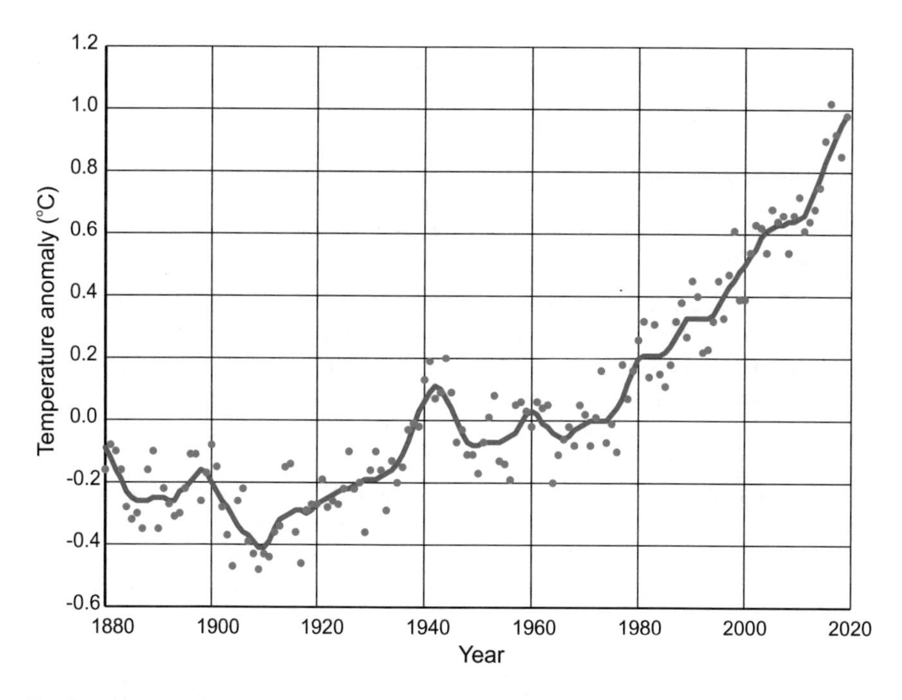

Fig. 3.1 Plot of differences in mean annual global surface temperature relative to 1951–1980 average temperatures. A pronounced warming trend is evident starting in the late 1970s. Nineteen of the 20 warmest years all have occurred since 2001. Red curve is a Lowess smoothing. *Data source* NASA (2020a)

(Hartmann et al. 2013, p. 161). The IPCC also concluded that it is very likely that the numbers of cold days and nights have decreased, and the number of warm days and nights have increased globally since about 1950 (Hartmann et al. 2013).

The data are less clear on changes in the length and frequency of warm spells including heat waves. There is only medium confidence that heat wave frequency has globally increased since the middle of the 20th century, which is mostly due to a lack of data or studies in Africa and South America (Hartmann et al. 2013). It is likely that heatwave frequency has increased during this period in large parts of Europe, Asia and Australia (Hartmann et al. 2013).

Climate data from 1980 through 2011 indicate that high temperature anomalies are becoming more extreme (Hansen et al. 2012). Seasonal mean temperature anomalies have shifted toward higher temperatures and the range of anomalies has increased with the emergence of summertime extremely hot outliers, defined as being more than three standard deviations (3σ) warmer than the climatology of the 1951–1980 base period (Hansen et al. 2012).

3.2.2 Precipitation

The IPCC concluded that there is low confidence that average precipitation over global land areas had changed between 1901 and 1951 and a medium confidence that it had changed afterwards (Hartmann et al. 2013). It was also concluded that (Hartmann et al. 2013, p. 162).

> Averaged over the mid-latitude land areas of the Northern Hemisphere, precipitation has likely increased since 1901 (*medium confidence* before and *high confidence* after 1951). For other latitudinal zones area-averaged long-term positive or negative trends have low confidence due to data quality, data completeness or disagreement amongst available estimates

The historical climate data indicate a general trend of precipitation falling more as heavy events. The IPCC concluded that it likely that since about 1950, the number of heavy precipitation events over land has increased in more regions than it has decreased, with highest confidence for increased frequency or intensity of heavy precipitation, with some seasonal and/or regional variation, for North America and Europe (Hartmann et al. 2013).

A key expectation of the climate change research is that increasing temperatures would result in increased evaporation, which would correspondingly result in increased precipitation as any net changes in atmospheric water content would be small. However, pan evaporation rates from the second half of the 1900s were reported to have actually decreased (Peterson et al. 1995), which was attributed to increasing cloud cover, aerosol concentrations, and/or increasing humidity in the pan vicinity (Peterson et al. 1995; Roderick and Farquhar 2002). Increasing cloud cover is evidenced by decreasing diurnal temperature ranges (i.e., difference between daytime and nighttime temperatures). Roderick et al. (2007) concluded that any changes in temperature and humidity regimes were generally too small to impact pan evaporation rates and that the observed decreases in pan evaporation rates were mostly due to decreasing wind speed with some regional contributions from decreasing solar irradiance.

3.2.3 Drought

Droughts are commonly defined as periods of reduced rainfall that result in a shortage of water. The technical definition of a drought is more nuanced with four types of droughts defined by Wilhite and Glantz (1985): meteorological, hydrological, agricultural, and socioeconomic. Meteorological droughts coincide with the colloquial definition of a drought and are defined in terms of the magnitude and duration of precipitation shortfalls. Hydrological droughts are periods of reduced surface water or groundwater availability. Agricultural droughts are defined based on adverse agricultural impacts, focusing on precipitation shortages, differences between actual and potential evapotranspiration (ET), and soil moisture deficits (Wilhite and Glantz 1985). Decreasing soil moisture can be due to decreases in precipitation or increases in ET driven by rising temperatures, or a combination of both. Socioeconomic droughts are defined based on social and economic impacts of water shortages.

There is controversy over whether there has been an increase in global droughts since the middle of the twentieth century. Evaluation of the occurrence of droughts depends upon how droughts are quantified, and, in particular, on how potential evaporation (PET) is quantified for use in drought indices, such as the Palmer Drought Severity Index (PDSI; Sheffield et al. 2012). Sheffield et al. (2012) concluded that PET—the evaporative demand of the atmosphere—is overestimated when PET is calculated from temperature alone, such as is done in the standard PDSI, rather than also considering radiative and aerodynamic controls.

Dai (2013) reported that using the PDSI method with PET calculated using the Penman-Monteith equation revealed broad patterns characterized by drying over most of Africa, southeast Asia, eastern Australia and southern Europe, and increased wetness over the central United States, Argentina and northern high-latitude areas. Stream flow data show a similar pattern, which suggests that these drying trends are real (Dai 2013). Dai (2013, p. 52) concluded that "the observed global aridity changes up to 2010 are consistent with model predictions, which suggest severe and widespread droughts in the next 30–90 yr over many land areas resulting from either decreased precipitation and/or increased evaporation."

Greve et al. (2014) evaluated whether historical data confirm the "dry gets drier, wet gets wetter" paradigm, which has become a standard catchphrase frequently used in studies and assessments of historical and future climate change. They found that over about three-quarters of the global land area, robust dryness changes cannot be detected and that only 10.8% of the global land area shows a robust "dry gets drier, wet gets wetter" pattern, compared to 9.5% of global land area with the opposite pattern. Greve et al. (2014) concluded that aridity changes over land areas that have the highest potential for direct socioeconomic impacts have not followed a simple intensification of existing patterns. Milly and Dunne (2016) similarly concluded that historical and future tendencies toward continental drying may be considerably weaker and less extensive than previously thought. Nevertheless, it must be stressed that historical drying trends are not a reliable indicator of future changes under more extreme radiative forcings (i.e., high atmospheric GHG concentrations).

3.3 Historic Sea Level Rise

Sea level rise (SLR) is due to mainly to thermal expansion of the oceans, melting of glaciers and the Greenland and Antarctic ice sheets, and, to a lesser degree, decreased terrestrial water storage. Sea level rise has been historically measured using tide gauges. NOAA began continuously monitoring sea level with the TOPEX/Poseidon satellite in 1992, which was followed by the Jason-1, Jason-2, and now the Jason-3 missions, with the latter launched in 2016 (NESDIS 2018). The Jason-3 satellite measures the height of 95 percent of world's ice-free ocean every 10 days to within less than a centimeter.

Global mean sea level (GMSL) has risen by about 0.23 m (9 in) between 1880 and 2019 (Fig. 3.2). The rate of rise has increased since the middle 1990s and is reported to have been approximately 3.3 mm/yr since 1993 (USGCRP 2019). Recent studies indicate that the rate of sea level rise is accelerating. Chen et al. (2017) reported that the GMSL rise has increased from 2.2 ± 0.3 mm/yr in 1993 to 3.3 ± 0.3 mm/yr in 2014. The IPCC (2019) reported a rate of GMSL rise of 3.6 mm/yr over the period 2006–2015. The mass contributions to GMSL (as opposed to the thermal expansion contribution) has increased from about 50% in 1993 to 70% in 2014 with the largest part of the increase being the contribution of the Greenland ice sheet (Chen et al. 2017).

Nerem et al. (2018) estimated the climate change-driven acceleration of global mean sea rise over the last 25 yr to be 0.084 ± 0.025 mm/yr^2 from a 25 yr time series of precision satellite altimeter data from TOPEX/Poseidon, Jason-1, Jason-2, and Jason-3. Nerem et al. (2018) cautioned that their projection of future sea-level rise is based only on the satellite observed changes over the last 25 yr, and that if SLR begins changing more rapidly (e.g., due to rapid changes in ice sheet dynamics), then their simple extrapolation will likely represent a conservative lower bound on future sea level change. Nerem et al. (2018) also cautioned that few potential processes exist to suggest that the estimated rate of acceleration of SLR is too high.

Hybrid sea level reconstructions generated using both tide gauge and satellite data show that the initiation of GMSL rise acceleration was already underway in the 1960s, rather than starting in the early 1990s (Dangendorf et al. 2019). The acceleration of GMSL rise starting in the 1960s was linked to an intensification and a basin-scale equatorward shift of the southern hemispheric westerlies, leading to

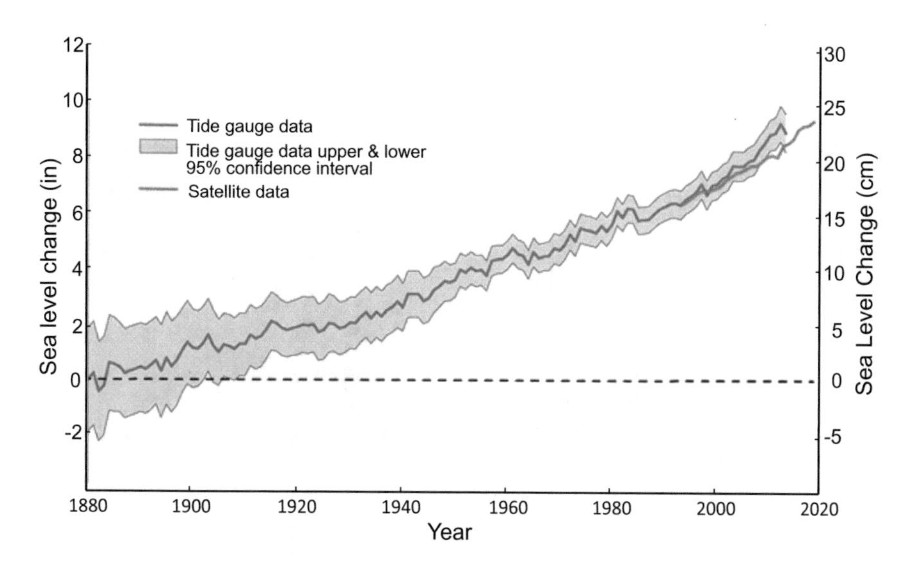

Fig. 3.2 Graph showing the increase in global average sea level relative to 1880. The blue line shows sea level as measured by tide gauges (1880–2013) and the surrounding light blue-shaded area shows upper and lower 95% confidence intervals. The red line shows sea level as measured by satellites. *Source* USGCRP (2019)

increased ocean heat uptake and thermal (i.e., steric) expansion. Dangendorf et al. (2019) calculated an acceleration of 0.09 ± 0.02 mm/yr^2, which they noted was in agreement with the value obtained by Nerem et al. (2018). Dangendorf et al. (2019) reported that only since the 1990s has ice melting (particularly in Greenland and Antarctica) again dominated the GMSL acceleration. Dangendorf et al. (2019, p. 709) also cautioned that "it is likely that the detected steric acceleration over the twentieth century as a whole will therefore emerge again, leading (together with accelerated mass loss) to a further steepening of the rates of GMSL rise."

The IPCC (2019) concluded that global mean sea level (GMSL) is rising (*virtually certain*) and accelerating (*high confidence*) and that the sum of the glacial and ice sheet contributions is now the dominant source of GMSL rise (*very high confidence*). The dominant cause of GMSL rise since 1970 is attributed to anthropogenic forcing (*high confidence*).

Local impacts from sea level rise depend on the rate of relative (local) sea level rise rather than that of GMSL rise alone. For example, the long-term tidal gauge data from Key West, Florida (which has the longest record in the region) has experienced a relative sea level rise trend of 2.42 mm/yr (with a 95% confidence interval of ± 0.14 mm/yr) based on monthly mean sea level data from 1913 to 2018 (Fig. 3.3; NOAA n.d.). The annual rate of rise is equivalent to a change of 0.24 m (0.79 ft) in 100 yr (NOAA n.d.). Local sea level rise depends on other variables in addition to GMSL rise, particularly land subsidence, tectonic or isostatic uplift, and changes in oceanic circulation.

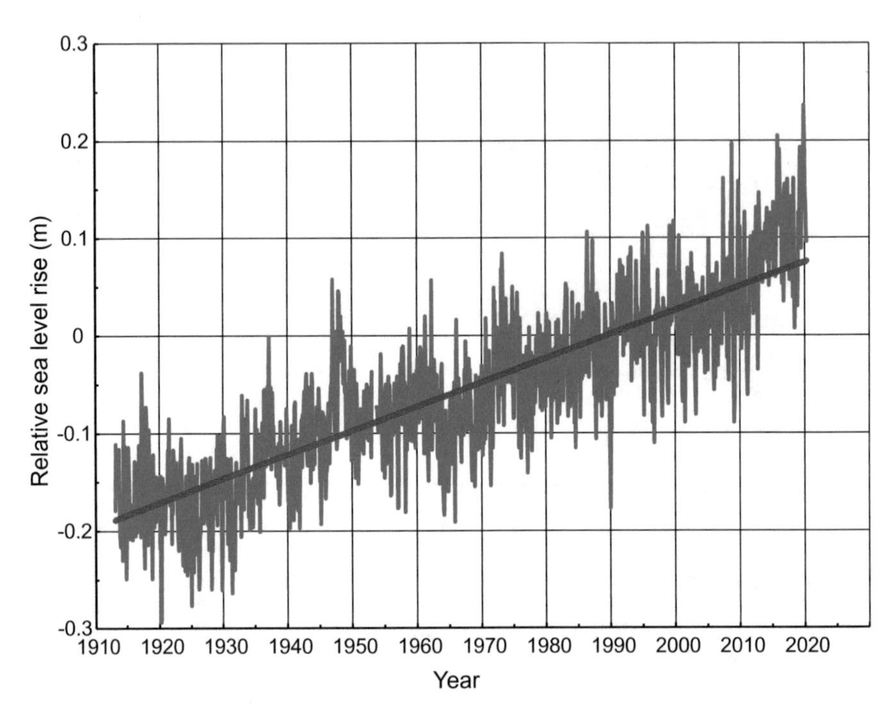

Fig. 3.3 Relative sea level plot for Key West, Florida. Red line shows linear trend. *Source* (NOAA n.d.)

The Atlantic meridional overturning circulation (AMOC) is a system of thermohaline circulation in the Atlantic Ocean, including the Gulf Stream, which involves the northward flow of warm, salty water in the upper layers of the ocean and a southward flow of colder, deep waters. The GS has a crucial impact on the climate of the northern Atlantic Ocean and adjoining land areas as it transports large amounts of heat from the warm tropical regions to the high latitudes. Variations in Gulf Stream flow impact coastal sea levels along the eastern United States. It has long been observed that lower water levels along the Atlantic coast of the United States occur during periods of greater Gulf Stream transport (Blaha 1984).

The Northern American Atlantic coast, especially north of Cape Hatteras, has been identified as a SLR hotspot, which was referred to by Sallenger et al. (2012) as the northeast hot spot (NEH). Between 1950–1979 and 1980–2009, SLR in the NEH was 3 to 4 times the global average (Sallenger et al. 2012). The NEH is consistent with local SLR associated with a slowdown of the GS and AMOC. Ezer et al. (2013) concluded that it appears that the GS has shifted from a 6 to 8 yr oscillation cycle to a continuous weakening trend since about 2004 and that this trend may be responsible for the recent acceleration in local SLR. More recently, Dong et al. (2019) presented data that indicate that changes in GS properties are spatially dependent (a southward shift and weakening of the GS is found east of

65 °W but not west of 70 °W) and suggest that SLR acceleration, as observed by tide gauges along the U.S. East Coast, is not directly linked to a slowdown of the GS.

3.4 Tropical Storm Frequency and Intensity

Vulnerability to tropical cyclones (hurricanes and typhoons) is increasing because of greater populations and property values along coasts. Tropical cyclones can cause erosion of beaches and other coastal environments, and cause damage and loss of life through high winds and, usually to a greater degree, by inland flooding and storm surges. The inland extent of storm surges and flooding will increase as sea level rises and the surges and flooding occur atop a higher baseline. Greater precipitation rates and amounts, especially in mountainous areas, would lead to increased inland flooding with associated damage to property and loss of life (Anthes et al. 2006). Tropical cyclones obtain their energy from warm surface seawater, and it is intuitive that increasing seawater surface temperatures (SSTs) and humidity should be favorable for tropical cyclone development and intensification.

Emanuel (1987) in an early investigation of the impacts of climate change on hurricanes used a simple Carnot energy model to estimate the maximum intensity of tropical cyclones under somewhat warmer conditions. The modeling results indicated that relatively small changes in sea surface temperature can result in large intensity changes, with an increase of 3 °C leading to a 30–40% increase in maximum pressure drop (Emanuel 1987). Maximum sustainable central pressure drop is a measure of **maximum** tropical cyclone intensity. Emanuel (1987, p. 485) cautioned that "This analysis pertains only to the maximum sustainable pressure drop in tropical cyclones and has no direct implications for either the average intensity of cyclones or their frequency of occurrence."

Modeling by Knutson and Tuleya (2004) suggests that atmospheric temperatures, moisture profiles, and SSTs from nine different climate models indicate a CO_2 induced increase in storm intensity and near-storm precipitation rates. The aggregate results project a 14% increase in central pressure fall, a 6% increase in maximum surface water wind speed, and an 18% increase in average precipitation rate within 15 km of the storm center (Knutson and Tuleya 2004). Knutson and Tuleya (2004) concluded that if the frequency of tropical cyclones remains the same over the century, GHG-induced warming may lead to a gradually increasing occurrence of highly destructive category-5 storms.

Hurricanes generally occur in regions of oceans where SSTs exceed 26 °C (Trenberth 2005). Global warming has been increasing SSTs and atmospheric water vapor, so it is reasonable to presume that hurricane activity is also increasing. However, SSTs and hurricane activity vary widely on interannual to multidecadal times scales (Trenberth 2005). Interannual variability in Atlantic hurricanes is related to the El Niño Southern Oscillation (ENSO), which is an irregularly periodic variation in winds and SSTs over the tropical eastern Pacific Ocean. Atlantic hurricanes are suppressed during El Niño periods (Trenberth 2005). Hurricane activity is also suppressed by wind shear, which may also increase in response to global warming, perhaps offsetting to some degree the effects of higher SSTs. Similarly, Atlantic tropical storm activity is suppressed by Saharan dust blown off the northwest coast of Africa (Evan et al. 2006). Climate changes that impact Saharan dust outbreaks could impact future Atlantic tropical storm frequency and intensification.

Although there has been an increase in hurricane activity in recent decades, trends in their number and intensity have not been statistically significant due to the natural variability in hurricane activity (Trenberth 2005). The climate signal is not clearly statistically distinguishable above the noise in the data. There is some difference of opinion on the impacts of climate change on Atlantic hurricane activity. Trenberth (2005) concluded that trends due to anthropogenic environmental changes are now evident in hurricane regions and that climate changes are expected to affect hurricane intensity and rainfall, but the effects on hurricane numbers and tracks remain unclear.

Evaluation of trends in the frequency of extreme tropical cyclones is hampered by subjectivity and variability in the procedures used (Landsea et al. 2006). Tropical cyclone databases in regions that are primarily dependent on satellite imagery for monitoring are inhomogeneous and likely have a bias toward increased intensity over time due to an underestimation of cyclone intensities in the 1970s and 1980s (Landsea et al. 2006). The infrared Dvorak Technique used at that time, in the era of few satellites with low spatial resolutions, tended to underestimate intensities (Landsea et al. 2006).

Emanuel (2005) defined the potential destructiveness index (PDI) of tropical cyclones, which is based on the total dissipation of power integrated over the life of a cyclone. The data show a pronounced increase in the index since the mid-1970s, which is due to both greater storm intensities and longer storm life times. The PDI has more than doubled in the North Atlantic and western Pacific over the 30 yr study period.

Webster et al. (2005) conducted an analysis of global tropical cyclone statistics for the satellite era (1970–2004). The data do not show a statistically significant trend in the number of cyclones and cyclone days for the period 1970–2004. Therefore, against the background of increasing SSTs, no global trend has emerged in the number of tropical storms and hurricanes (Webster et al. 2005). However, a substantial change in global hurricane intensity distribution occurred over the study period. Hurricanes in the strongest categories (4 and 5) were reported to have almost doubled over the study period and have increased as a proportion of all storms and hurricanes (Webster et al. 2005). Webster et al. (2005, p. 1846) concluded that the historical hurricane intensity trend "is not inconsistent with recent climate model simulations that a doubling of CO_2 may increase the frequency of the most intense storms."

Tropical cyclone development depends on other factors in addition to local SSTs. Vecchi and Soden (2007) investigated whether there has been a temporal trend in the tropical cyclone "potential intensity" (PI) parameter that represents a theoretical upper limit on the intensity of tropical cyclones based on SSTs and the local vertical thermodynamic structure of the atmosphere. Calculation of the PI requires vertical temperature and humidity data from storms and varies depending on local SSTs and more remote SSTs. Local SSTs alone were found to be inadequate for characterizing even the sign of changes of the PI (Vecchi and Soden 2007). Vecchi and Soden (2007) reported than even though SSTs have increased everywhere, changes in PI are mixed. For example, although tropical Atlantic SSTs are currently at a historic high, recent PI values are near the historic average. A key conclusion of the Vecchi and Soden (2007, p. 1066) study is that "the response of tropical cyclone activity to natural climate variations, which tend to involve localized changes in sea surface temperature, may be larger than the responses to the more uniform patterns of greenhouse-gas-induced warming."

Several papers and replies were published in the mid-2000s concerning whether there was evidence for climate change affecting the human impacts of tropical cyclones. Pielke et al. (2005, p. 1573) in a peer review of the literature noted, as others have observed, that it is necessarily very difficult to detect trends in the frequency of storms or major hurricanes in the context of their great year-to-year and decade-to-decade variation. They emphasized that the most significant factor underlying trends and projections associated with hurricane impacts on society is societal vulnerability to those impacts, rather than trends or variations in the storms themselves. The growing population and wealth in coastal areas susceptible to tropical cyclones guarantees increased economic damage from storms in the future, which dwarfs the potential impacts from climate-induced changes in storm frequency and intensity, if these will indeed occur (Pielke et al. 2005).

Mendelsohn et al. (2012) investigated the impacts of climate change on the economic damage done by tropical cyclones. A tropical cyclone model was used in conjunction with climate models to predict the frequency, intensity, and location of tropical cyclones in each ocean basin of the world. The path of each cyclone was then tracked until it reached land and damage was estimated using a function related to its intensity and what is in harm's way. An important relationship concerning storm damage is that the probability density function of damage is highly skewed. Under current conditions, 93% of the cyclone damage is caused by 10% of the storms (Mendelsohn et al. 2012). The nonlinear damage function means that climate change-induced increases in the intensity of storms would result in a disproportionate increase in damage. The return period of highly damaging storms becomes shorter (Mendelsohn et al. 2012).

The results of the Mendelsohn et al. (2012) study indicate that the greatest total damage caused by climate change will occur in North America (US $26 billion/yr), East Asia (US $15 billion/yr), and the Central America-Caribbean region (US $10 billion/yr) because in these affluent areas more assets are in harm's way. The Central America-Caribbean region has the highest damage per unit of gross domestic product (GDP) of 0.37%. Countries with the highest damage per unit GDP were found to be tropical islands. Other areas of the world will experience small tropical storm effects from climate change because they are either rarely struck by tropical cyclones (e.g., North Africa, Middle East, and South America), the storms that do strike them tend to be of low intensity (e.g., Europe), or there is less in harm's way (sub-Saharan Africa; Mendelsohn et al. 2012).

3.5 Atmospheric Carbon Dioxide Concentration

Increasing atmospheric CO_2 concentrations are believed to be the primary driver of global climate change. At plot of atmospheric CO_2 concentration over the past 400,000 yr shows that pre-industrial CO_2 levels were below 300 ppm and that a rapid increase in concentrations has occurred over the past 100 yr (Fig. 3.4). More recent data from the Mauna Loa Observatory in Hawaii show a steady progressive rise in atmospheric CO_2 concentrations from 315.7 ppm in March 1958 to 414 ppm in May 2020 (Fig. 3.5, NASA 2020b).

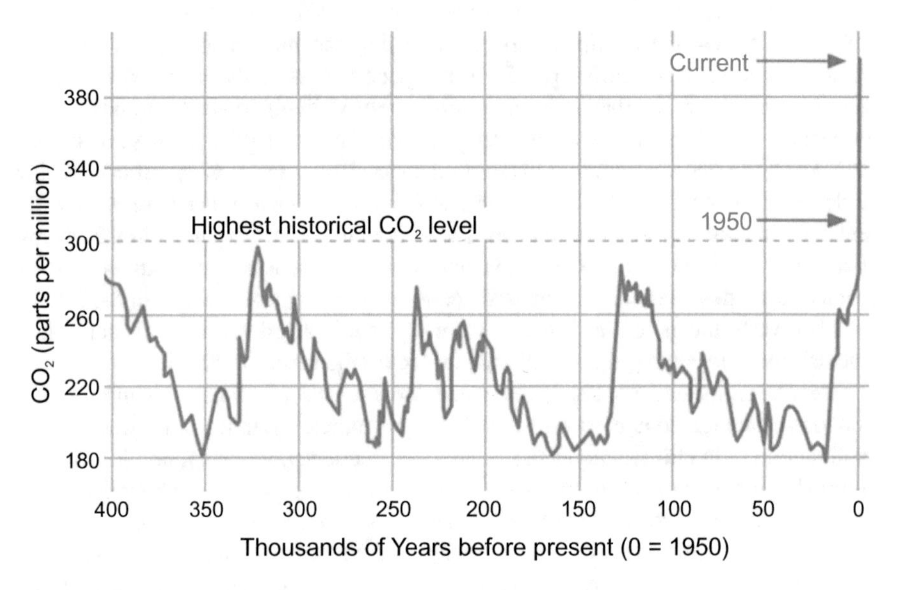

Fig. 3.4 Plot of atmospheric CO_2 concentrations over time derived from ice cores. *Source* NASA (NASA 2020b, credit NOAA)

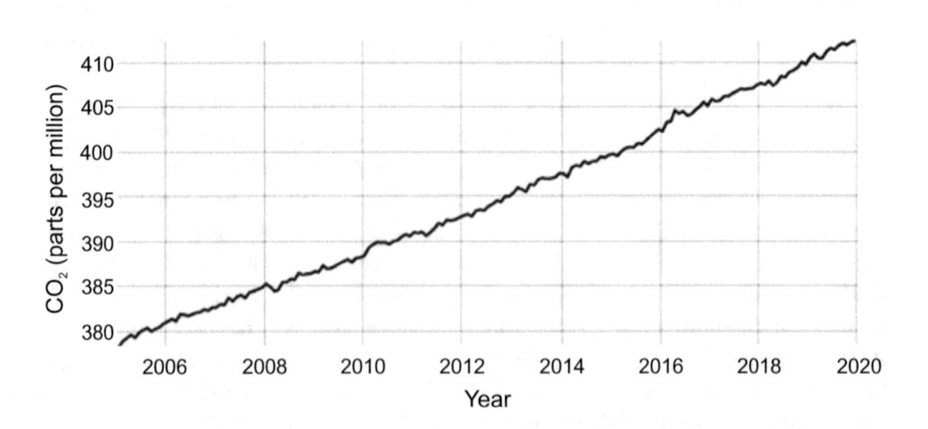

Fig. 3.5 Direct monthly measurements of atmospheric CO_2 concentrations (average seasonal cycle removed) from the Mauna Loa Observatory, Hawaii. *Source* NASA (NASA 2020b, credit NOAA)

3.6 General Circulation Models (GCMs)

3.6.1 GCM History

The foundation for climate change projections is general circulation models (GCMs), which are now also often referred to as global climate models. Grotch and MacCracken (1991, p. 3) described GCMs as

> numerical models that attempt to simulate the global climate by calculating the hour-by-hour evolution of the atmosphere in all three spatial dimensions based on the conservation laws for atmospheric mass, momentum, total energy, and water vapor. CGMs also typically include representations of surface hydrology, sea ice, cloudiness, convection, atmospheric radiation and other pertinent processes.

Weart (2010) reviewed the history of development of GCMs. The first, and by modern standards very primitive, GCM was developed by Dr. Normal Phillips of Princeton in the 1950s. Early climate modeling was severely limited by computation power. With reference to the models in the 1950s and early 1960s, Weart (2010, p.210) observed that

> Even if the computers had been vastly faster, the simulations would still have been unreliable. For they were running up against the famous limitation of computers, 'garbage in, garbage out.' To diagnose the failings that kept GCMs from being more realistic, scientists needed an intensified effort to collect and analyze data showing the actual profiles of wind, heat, moisture, and so forth, at every level of the atmosphere and all around the globe.

From the mid-1950s to the mid-1970s, the computer power available to modelers increased exponentially allowing for the development of more complex models (Weart 2010). Increasing amounts of data on the real world became available during this time period that could be used to constraint and calibrate models. However, through the 1970s and early 1980s, some basic problems remained, particularly predicting cloudiness and constructing a realistic model of the joint ocean-atmosphere system (Weart 2010). Early on it was recognized that "Climate was not changed by any single cause" but rather "It was the outcome of a staggeringly intricate complex of interactions" (Weart 2010, p. 211).

Dr. James Hansen of NASA and his coworkers published a paper in 1988 on what is considered to be the first modern coupled ocean-atmosphere climate model (Hansen et al. 1988). The Goddard Institute for Space Studies (GISS) model II was crude by modern standards with a coarse grid of nine atmospheric layers and a horizontal resolution of 8° latitude by 10° longitude. A 100 yr control simulation with daily calculations was first run with an atmospheric composition fixed at 1958 values (CO_2 = 315 ppm). The global mean surface air temperature had not changed much over the 100-year run but there was substantial unforced variability in the modeled temperature up to the decadal time scale (Hansen et al. 1988). The unforced variability in the model was only slightly less than the variability in global surface air temperatures that had been observed over the previous 100 years.

The next simulations by Hansen et al. (1988) were of three scenarios that included increasing concentrations of atmospheric CO_2, CH_4, N_2O, chlorofluorocarbons, and aerosols. The key conclusion of the Hansen et al. (1988, p. 9359) study was that "global greenhouse warming will soon rise above the level of natural climate variability" and that there is "an urgent need for global measurements in order to improve knowledge of climate forcing mechanisms and climate feedback processes."

Climate modeling continues to be questioned by skeptics with varied motives. Numerical modeling inherently involves simplifications and there are always uncertainties in the results. Weart (2010, p. 213) observed that

> Debates over climate models also helped stimulate philosophers of science, who explained that a computer model, like any other embodiment of a set of scientific hypotheses, could never be "proved" in the absolute sense one could prove a mathematical theorem. What models could do was help people sort through countless ideas and possibilities, offering evidence of which were most plausible.

Climate change skeptics then and still criticize climate models as being inaccurate, and thus unreliable, but none has ever produced a model of their own that shows that historical and anticipated future concentrations of GHGs will not cause a rise in global temperatures (Weart 2010). The predictions of Hansen et al. (1988) and others since the 1980s that an increase of mean global temperatures would emerge from the "noise" of random, background climate shifts around the start of the 21st century have indeed been "fulfilled by a rise ominously faster and higher than anything in the historical record" (Weart 2010, p. 216).

Atmospheric parameters considered in GCMs include surface pressure, two horizontal wind components, temperature, moisture, geospatial height, radiative transfer, boundary layers (i.e., regions where surface friction has a large effect on flow), cloud prediction, convection processes, precipitation, and gravity wave drag (drag of mountains on the atmosphere; McGuffie and Henderson-Sellers 2005). Modeling of oceans involves consideration of the effects of density and salinity and surface exchanges of heat and moisture (McGuffie and Henderson-Sellers 2005). A key issue is accurately simulating the advective flow of energy from the tropics to the poles and the sequestration of heat in the deep ocean (McGuffie and Henderson-Sellers 2005).

New GCMs are being developed and existing models refined by research groups across the world. With progressively increasing computer power, and available atmospheric and ocean data, and scientific progress, the spatial resolution of models has been improved as well as their ability to simulate the various complex interactions within and between the atmosphere and oceans. Stevens and Bony (2013) observed that inadequate representation of clouds and moist convection, or more generally the coupling between atmospheric water and circulation, is the main limitation in current representations of the climate system. The current start-of-the art is Earth system models (ESMs), which expand upon atmosphere–ocean general circulation models (AOGCMs) to include representation of various biogeochemical processes, such as those involving the carbon cycle, the sulfur cycle, and ozone (Flato et al. 2013).

3.6.2 *Coupled Model Intercomparison Project*

Only a very small minority of scientists involved in climate change issues will ever run a GCM or ESM. Instead workers involved in climate change rely upon the results of modeling performed by expert climate modeling groups. Nevertheless, it is important for users of modeling data to have a basic understanding of the modeling process and the limitations of GCMs and ESMs.

Numerical modeling in hydrogeology and the climate sciences usually involve three main steps. An initial model is constructed using the best understanding of the physical system (i.e., conceptual model) and likely values of key parameters. The model is then calibrated to a given period of historical observational data, which involves adjusting model parameters until model output values match corresponding observed values as closely as can be practically achieved. Model calibration is a form of inverse modeling, which inherently does not lead to unique solutions. Validation involves running a model on another historical time period (not used for the model calibration) for which observational data are available to test the predictive ability of the model. Once a model has been calibrated and validation results demonstrate that it provides a good match to historical data, then predictive simulations are performed.

Climate models involves simplifications as to the representation of various physical processes and in model spatial and temporal resolution (McGuffie and Henderson-Sellers 2005). "There is no one 'right' climate model or even one 'best' climate model type. All have the potential to add value if they are honestly evaluated and appropriately applied" (McGuffie and Henderson-Sellers 2005, p. 244). The greatest confidence may be placed on models that best duplicate historical conditions, i.e., have greater skill at matching values of parameters of interest in the area of interest. However, reliance on model skill to evaluate predictive capability "may produce excessive confidence because of the rather narrow climate experience during the observable record" (McGuffie and Henderson-Sellers 2005, p. 72). A model that has a high accuracy at simulating historical climate conditions may have a lesser accuracy at predicting temperature and precipitation under changed future environmental conditions.

The preferred approach now employed for using climate modeling results is to consider the output from multiple models in order to be able to evaluate the uncertainty in the projections. A caveat is that agreement between models may be due to the models having been derived from a common source or using similar modeling approaches. The results of GCM simulations depend upon the emissions scenario used and the climate model itself. The ocean-land-atmosphere system is extraordinarily complex and GCMs vary in their structure and how various process, interactions, and feedbacks are simulated. A result of this variation in GCMs is that there is a spread in projected future climate conditions for each given emissions scenario. It was recognized that the "nature and causes of these disagreements must be accounted for in a systematic fashion in order to confidently use GCMs for simulation of putative global climate change" (PCMDI n.d.).

The Coupled Model Intercomparison Project (CMIP) was initiated in 1995 by the World Climate Research Programme's (WCRP) Working Group on Coupled Modelling (WGCM) as an international effort to improve climate models by comparing multiple model simulations to observational data and to each other. According to the WCRP (n.d.), the purpose of the CMIP is

> to better understand past, present and future climate changes arising from natural, unforced variability or in response to changes in radiative forcing in a multi-model context. This understanding includes assessments of model performance during the historical period and quantifications of the causes of the spread in future projections.

An important part of the CMIP since its inception has been to make the model intercomparison data available to the general scientific community besides just those who run the models (WCRP n.d.). The latest phase of the CMIP, CMIP5, was completed in 2015. An overview of the experimental design and organization of the subsequent CMIP6 was published in 2016 (Eyring et al. 2016).

CMIP5 was managed by the Program for Climate Model Diagnosis and Intercomparison (PCMDI) at the Lawrence Livermore National Laboratory located in the San Francisco Bay area of California. The goals of CMIP5 were "1) assessing the mechanisms responsible for model differences in poorly understood feedbacks associated with the carbon cycle and with clouds, 2) examining climate 'predictability' and exploring the ability of models to predict climate on decadal time scales, and, more generally, 3) determining why similarly forced models produce a range of responses." (Taylor et al. 2012; PCDMI, n.d.). The CMIP5 data can be accessed via the PCMDI website (https://pcmdi.llnl.gov/mips/cmip5/).

The CMIP GCM database is widely used in climate change studies and, of particular importance, in the IPCC assessments. The CMIP3 results were relied upon for the 4th Assessment Report (AR4) of the IPCC and the 5th Assessment Report (AR5) of the IPCC relied heavily on the CMIP5 results. The CMIP5 data considered in the AR5 consisted of 952 simulations from 58 models from 24 institutions (Emori et al. 2016). CMIP6 will inform the upcoming 6th IPCC Assessment Report. Technical studies outside of the IPCC very commonly use outputs from the CMIP because they are available to the scientific community and the impracticality for most scientists to independently develop and run their own GCMs.

3.6.3 CMIP and IPCC Emissions Scenarios

A fundamental aspect of the CMIP program is running simulations with different GCMs using the same set of future emissions scenarios. CMIP3 and the IPCC AR4 used a suite of SRES (Special Report on Emissions Scenarios) scenarios. CMIP5 and IPCC AR5 used four Representative Concentration Pathways (RCPs). Although the SRES scenarios have been superseded by the RCPs, older, but still relevant, papers on climate change used the former, so both need to be understood.

The SRES scenarios are a series of storylines that explore alternative development pathways, covering a wide range of demographic, economic and technological driving forces, and resulting GHG emissions, that are intended to serve as inputs to climate change vulnerability and impact assessments (IPCC 2007). The SRES scenarios are grouped into four scenario families (A1, A2, B1 and B2), of which B1 is the least fossil fuel-intensive future and A2 is the most intensive (Fig. 3.6).

As described by the IPCC (2000, 2007), the A1 storyline assumes a world of very rapid economic growth, a mid-century peak in the global population, and rapid introduction of new and more efficient technologies. The A1 scenarios are divided into three groups that describe alternative directions of technological change: fossil-fuel intensive (A1FI), adoption of non-fossil-fuel energy resources (A1T), and a balance across all sources (A1B). The B1 scenarios describe a convergent world, with the same global population as A1, but with more rapid changes in economic structures toward a service and information economy. The B2 scenarios describe a world with intermediate population and economic growth, emphasizing local solutions to economic, social, and environmental sustainability. The A2 scenarios describe a very heterogeneous world with high population growth, slow economic development, slow technological change, and continued high levels of

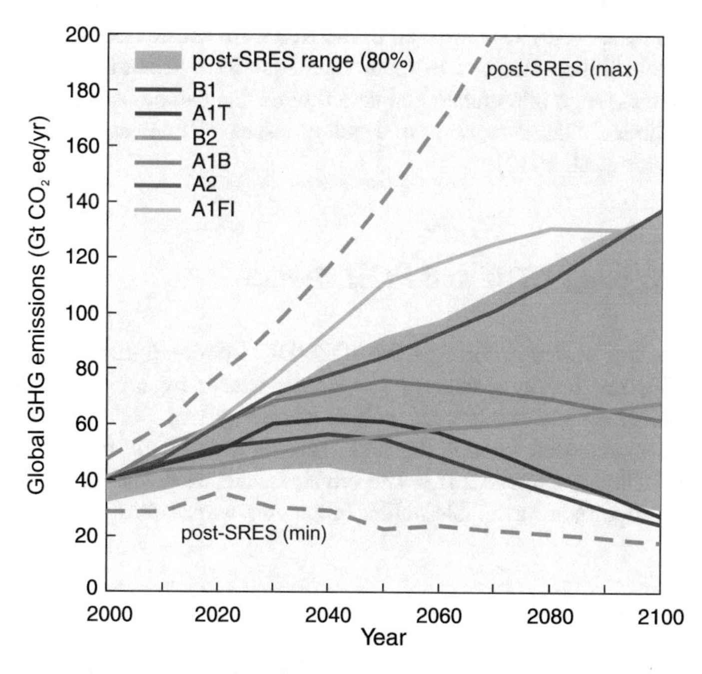

Fig. 3.6 Global GHG emissions (in $GtCO_2$-eq per year) in the absence of additional climate policies. Six illustrative SRES marker scenarios (colored lines) and 80th percentile range of recent scenarios published since SRES (post-SRES) (gray shaded area) are shown. Dashed lines show the full range of post-SRES scenarios. The emissions considered include CO_2, CH_4, N_2O and F-gases. *Source* IPCC (2007, Fig. 3.1)

fossil fuel use. No likelihood was attached to any of the SRES scenarios (IPCC 2000, 2007), which were intended to envelope the range of possible future GHG emissions.

The RCPs (RCP2.6, RCP4.5, RCP6, and RCP8.5) used for CMIP5 are defined based on radiative forcings in the year 2100 of 2.6, 4.5, 6.0, and 8.5 W/m^2, respectively. The RCPs represent a range of future GHG emissions in the wider literature from a stringent mitigation scenario (RCP2.6), to two intermediate scenarios (RCP4.5 and RCP6.0) and a scenario with very high GHG emissions (RCP8.5; IPCC 2014). The RCPs cover a wider range of GHC concentrations, and thus radiative forcings, then the previously used SRES scenarios. In terms of overall forcings, RCP8.5 is broadly comparable to the SRES A2 and A1FI scenarios, RCP6.0 and RCP4.5 to the B2 and B1 scenarios, respectively. There is no equivalent in the SRES scenarios to RCP2.6 (IPCC 2014). RCP2.6 is a high mitigation scenario that has much lower emissions than any SRES scenario because it includes the use of policies to achieve net negative carbon dioxide emissions before the end of the century, while the SRES scenarios do not. Graphs of atmospheric GHG (CO_2, CH_4, and NO_2) concentrations versus time for the RCPs, SRES scenarios, and an earlier IS92A scenario are provided in Fig. 3.7.

The wider range of future emissions represented by the RCPs results in a greater variation in the magnitude of the AR5 climate change projections (IPCC 2014). As is the case for the SRES scenarios, all of the RCPs are considered plausible and do not have probabilities attached to them. The four RCP scenarios run in CMIP5 provide a broad range of simulated climate futures that can be used as the basis for exploring climate change impacts and policy issues of interest and relevance to society (Taylor et al. 2012).

3.6.4 Accessing GCM and RGM Results

CMIP5 data are available through the (PCMDI) website (https://pcmdi.llnl.gov/mips/cmip5/), but not in a manner practicably usable by lay (i.e., non-climate modeler) scientists and the general public. Summary maps of the results of the CMIP5 can be accessed through the IPCC reports, particularly the supplementary material to "Climate Change 2013: The Physical Science Basis" (IPCC 2013). All of the IPCC reports are available online (https://www.ipcc.ch/reports/) at no cost.

3.6.4.1 Climate Wizard

Climate Wizard is a web-based tool developed through a collaboration between the Nature Conservancy, the University of Washington, and the University of Southern Mississippi to provide technical and non-technical audiences access to leading climate change information and visualization of predictions throughout the world (Girvetz et al. 2009). Climate Wizard was launched on-line in 2009 and taken

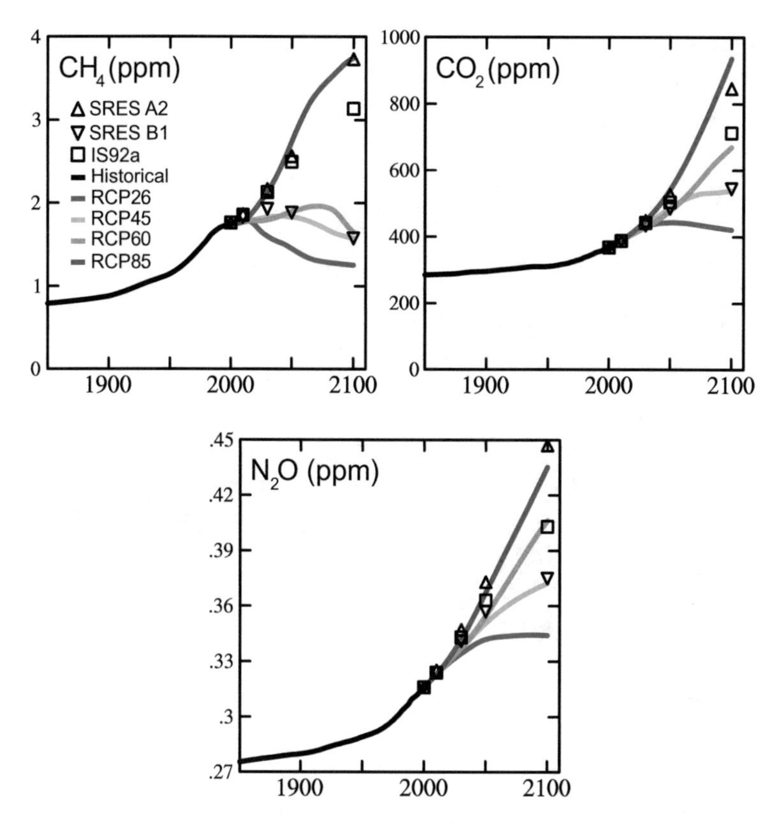

Fig. 3.7 Time evolution of global-averaged atmospheric concentrations of long-lived species from 1850 to 2100 following each RCP. Earlier SRES and IS92A emissions scenarios are also plotted. *Source* Myhre et al. (2013, Fig. 8.5)

offline by the developers in 2015. However, Climate Wizard is still accessible through the CGIAR website (https://climatewizard.ciat.cgiar.org/). Climate Wizard is based on the IPCC AR4 modeling and allows users to choose among a suite of precipitation and temperature parameters, GCMs, and between the SRES A2 (high), A1B, and B1 (low) emissions scenarios. As an example, maps of projected changes in annual average precipitation in Saudi Arabia obtained from Climate Wizard are provided in Fig. 3.8.

3.6.4.2 U.S. Geological Survey Viewers

The U.S. Geological Survey developed a series of viewers that allow for the visualization of projected future temperature and precipitation (and other parameter) changes in the United States and across the world (USGS n.d.) The Global Climate Change Viewer (GCCV) allows for the visualization of future temperature and

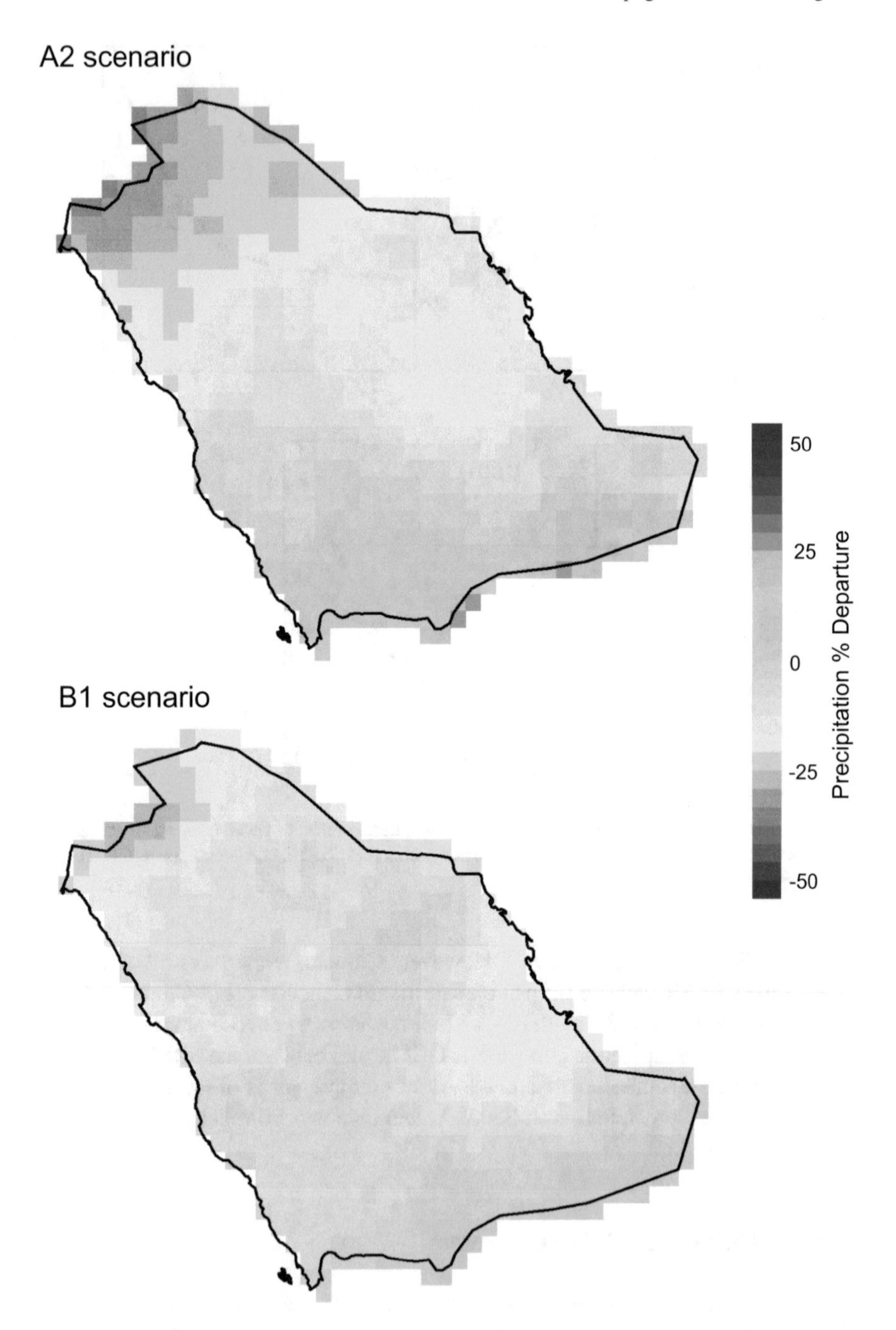

Fig. 3.8 Maps generated using Climate Wizard of projected changes in mean annual precipitation for Saudi Arabia for 2070–2099 compared to 1961–1990

precipitation changes simulated by CMIP5 models (Alder et al. 2013; Alder and Hostetler 2013). The viewer provides a coarse-scale image of changes (anomalies) across the world. The user specifies country, annual or specific month time scale, CMIP5 mean or a specific GCM, RCP, and future time period. The viewer provides average present and projected temperature or precipitation values for the selected country. A limitation of the GCCV is a lack of zoom capability. The National Climate Change Viewer (NCCV) provides state-level imaging and county-scale average change data from the CMIP5 models for RCP4.5 and RCP8.5.

The U.S. Geological Regional Climate Change Viewer (RCCV) allows users to visualize and download data from the Dynamical Downscaling project. The data set has more than 60 variables available, including air temperature, precipitation, soil moisture, snow water equivalent (SWE), growing degree days, and evapotranspiration (USGS n.d.). Users can view averages over states, counties, and hydrologic units. The Dynamic Downscaling project used the regional climate model RegCM3. The simulations were run using output from four GCMs and the IPCC AR4 A2 emissions scenario (Hostetler et al. 2011). The A2 scenario was used because it provides an upper bound on future emissions and because it is similar to the RCP8.5 scenario developed for the IPCC AR5. Simulations were run over 50 and 15 km model grids in an attempt to capture more of the climatic detail associated with processes such as topographic forcing than can be captured by GCMs (Hostetler et al. 2011). The RCCV is very user friendly and provides climate projections on the county level in the United States. The USGS (n.d.) cautioned that "these data should be considered to be approximate and should not be used in research, planning, or policy documents." Nevertheless, all climate modeling results are approximations and users need to carefully consider their inherent uncertainties.

References

Alder JR and Hostetler SW (2013) CMIP5 global climate change viewer. US Geological Survey http://regclim.coas.oregonstate.edu/gccv/index.html, https://doi.org/10.5066/f72j68w0. Accessed 27 May 2020

Alder JR, Hostetler SW, Williams D (2013) An interactive web application for visualizing climate data. Eos Trans Am Geophys Union 94:197–198

Anthes RA, Corell RW, Holland G, Hurrell JW, Mac Cracken MC, Trenberth KE (2006) Hurricanes and global warming—potential linkages and consequences. Bull Am Meteor Soc 87(5):623–628

Blaha JP (1984) Fluctuations of monthly sea level as related to the intensity of the Gulf Stream from Key West to Norfolk. J Geophys Res Oceans 89(C5):8033–8042

Chen X, Zhang X, Church JA, Watson CS, King MA, Monselesan D, Legresy B, Harig C (2017) The increasing rate of global mean sea-level rise during 1993–2014. Nat Clim Change 7(7): 492–495

Dai A (2013) Increasing drought under global warming in observations and models. Nat Clim Change 3(1):52–58

Dangendorf S, Hay C, Calafat FM, Marcos M, Piecuch CG, Berk K, Jensen J (2019) Persistent acceleration in global sea-level rise since the 1960s. Nat Clim Change 9(9):705–710

Dong S, Baringer MO, Goni GJ (2019) Slow down of the Gulf Stream during 1993–2016. Sci Rep 9(1):6672

Emanuel KA (1987) The dependence of hurricane intensity on climate. Nature 326:482–485

Emanuel K (2005) Increasing destructiveness of tropical cyclones over the past 30 years. Nature 436(7051):686–686

Emori S, Taylor K, Hewitson B, Zermoglio F, Juckes M, Lautenschlager M, Stockhause M (2016) CMIP5 data provided at the IPCC data distribution centre. Fact Sheet, Task Group on Data and Scenario Support for Impact and Climate Analysis (TGICA) of the Intergovernmental Panel on Climate Change (IPCC)

Eyring V, Bony S, Meehl GA, Senior CA, Stevens B, Stouffer RJ, Taylor KE (2016) Overview of the coupled model intercomparison project phase 6 (CMIP6) experimental design and organization. Geosci Model Dev, 9 (LLNL-JRNL-736881)

Evan AT, Dunion J, Foley JA, Heidinger AK, Velden CS (2006) New evidence for a relationship between Atlantic tropical cyclone activity and African dust outbreaks. Geophys Res Lett 33(19)

Ezer T, Atkinson LP, Corlett WB, Blanco JL (2013) Gulf Stream's induced sea level rise and variability along the US mid-Atlantic coast. J Geophys Res Oceans 118(2):685–697

Flato G, Marotzke J, Abiodun B, Braconnot P, Chou SC, Collins W, Cox P, Driouech F, Emori S, Eyring V, Forest C, Gleckler P, Guilyardi E, Jakob C, Kattsov V, Reason C, Rummukainen M (2013) Evaluation of climate models. In: Stocker TF, Qin D, Plattner G-K, Tignor M, Allen SK, Boschung J, Nauels A, Xia Y, Bex V, Midgley PM (eds), Climate change 2013: the physical science basis. Contribution of working group I to the fifth assessment report of the Intergovernmental Panel on Climate Change. Cambridge University Press, Cambridge, UK, pp 741–866

Girvetz EH, Zganjar C, Raber GT, Maurer EP, Kareiva P, Lawler JJ (2009) Applied climate-change analysis: the climate wizard tool. PLoS One 4(12)

Gornitz V (2007) Sea level rise, after the ice melted and today. National Aeronautics and Space Administration Goddard Institute for Space Studies, Jan 2007. https://www.giss.nasa.gov/research/briefs/gornitz_09/. Accessed 27 May 2020

Greve P, Orlowsky B, Mueller B, Sheffield J, Reichstein M, Seneviratne SI (2014) Global assessment of trends in wetting and drying over land. Nat Geosci 7(10):716–721

Grotch SL, MacCracken MC (1991) The use of general circulation models to predict regional climate change. J Clim 4:286–303

Hansen J, Fung I, Lacis A, Rind D, Lebedeff S, Ruedy R, Russell G, Stone P (1988) Global climate changes as forecast by Goddard Institute for Space Studies three-dimensional model. J Geophys Res Atmos 93(D8):9341–9364

Hansen J, Sato M, Ruedy R (2012) Perception of climate change. Proc Natl Acad Sci 109(37): E2415–E2423

Hartmann DL, Klein Tank AMG, Rusticucci M, Alexander LV, Brönnimann S, Charabi Y, Dentener FJ, Dlugokencky EJ, Easterling DR, Kaplan A, Soden BJ, Thorne PW, Wild M, Zhai PM (2013) Observations: atmosphere and surface. In: Stocker TF, Qin D, Plattner G-K, Tignor M, Allen SK, Boschung J, Nauels A, Xia Y, Bex V, Midgley PM (eds) Climate change 2013: the physical science basis. Contribution of working group I to the fifth assessment report of the intergovernmental panel on climate change. Cambridge University Press, Cambridge, UK

Hooker J, Duveiller G, Cescatti A (2018) A global dataset of air temperature derived from satellite remote sensing and weather stations. Sci Data 5(1):1–11

Hostetler SW, Alder JR, Allan AM (2011) Dynamically downscaled climate simulations over North America: methods, evaluation and supporting documentation for users. U.S. Geological Survey Open-File Report 2011–1238

IPCC (2000) In: Nakicenovic N, Swart R (eds) Emissions scenarios. Cambridge University Press, Cambridge, UK

IPCC (2007) Climate change 2007: synthesis report. In: Pachauri RK, Reisinger A (eds) Contribution of working groups I, II and III to the fourth assessment report of the Intergovernmental Panel on Climate Change. IPCC, Geneva

IPCC (2013) Climate change 2013: the physical science basis. In: Stocker TF, Qin D, Plattner G-K, Tignor M, Allen SK, Boschung J, Nauels A, Xia Y, Bex V, Midgley PM (eds) Contribution of working group I to the fifth assessment report of the Intergovernmental Panel on Climate Change. Cambridge UK: Cambridge University Press

IPCC (2014) Climate change 2014: synthesis report. In: Pachauri RK, Meyer LA (eds) Contribution of working groups I, II and III to the fifth assessment report of the Intergovernmental Panel on Climate Change. IPCC, Geneva

IPCC (2019) Technical summary. In: Pörtner H-O, Roberts DC, Masson-Delmotte V, Zhai P, Tignor M, Poloczanska E, Mintenbeck K, Alegría A, Nicolai M, Okem A, Petzold J, Rama B, Weyer NM (eds) IPCC special report on the ocean and cryosphere in a changing climate (In press)

Knutson TR, Tuleya RE (2004) Impact of CO_2-induced warming on simulated hurricane intensity and precipitation: sensitivity to the choice of climate model and convective parameterization. J Clim 17(18):3477–3495

Lambeck K, Rouby H, Purcell A, Sun Y, Sambridge M (2014) Sea level and global ice volumes from the last glacial maximum to the holocene. Proc Natl Acad Sci 111(43):15296–15303

Landsea CW, Harper BA, Hoarau K, Knaff JA (2006) Can we detect trends in extreme tropical cyclones? Science 313(5786):452–454

Mastrandrea MD, Field CB, Stocker TF, Edenhofer O, Ebi KL, Frame DJ, Held H, Kriegler E, Mach KJ, Matschoss PR, Plattner G-K, Yohe GW, Zwiers FW (2010). Guidance note for lead authors of the IPCC fifth assessment report on consistent treatment of uncertainties. Intergovernmental Panel on Climate Change (IPCC)

McGuffie K, Henderson-Sellers A (2005) A climate modelling primer, 3rd edn. John Wiley, Chichester

Mendelsohn R, Emanuel K, Chonabayashi S, Bakkensen L (2012) The impact of climate change on global tropical cyclone damage. Nat Clim Change 2(3):205

Milly PC, Dunne KA (2016) Potential evapotranspiration and continental drying. Nat Clim Change 6(10):946–949

Myhre G, Shindell D, Bréon F-M, Collins W, Fuglestvedt J, Huang J, Koch D, Lamarque J-F, Lee D, Mendoza B, Nakajima T, Robock A, Stephens G, Takemura T, & Zhang H (2013) Anthropogenic and natural radiative forcing. In: Stocker TF, Qin D, Plattner G-K, Tignor M, Allen SK, Boschung J, Nauels A, Xia Y, Bex V, Midgley PM (eds) Climate change 2013: the physical science basis. In: Contribution of working group I to the fifth assessment report of the Intergovernmental Panel on Climate Change. Cambridge University Press, Cambridge, UK, pp 659–740

NASA (2020a) Global climate change. https://climate.nasa.gov/ Accessed 26 May 2020

NASA (2020b) Carbon dioxide. https://climate.nasa.gov/vital-signs/carbon-dioxide/. Accessed 27 May 2020

Nerem RS, Beckley BD, Fasullo JT, Hamlington BD, Masters D, Mitchum GT (2018) Climate-change–driven accelerated sea-level rise detected in the altimeter era. Proc Natl Acad Sci 115(9):2022–2025

NESDIS (2018) A decade of global sea level measurements: Jason-2 marks tenth year in Orbit. NOAA. National Environmental Satellite, Data, and Information Service (NESDIS). https://www.nesdis.noaa.gov/content/decade-global-sea-level-measurements-jason-2-marks-tenth-year-orbit. Accessed 6 Feb 2020

NOAA (n.d.) Tides and currents. Relative sea level trend 8724580 Key West, Florida. https://tidesandcurrents.noaa.gov/sltrends/sltrends_station.shtml?id=8724580. Accessed 2 Feb 2020

Past Interglacials Working Group of Pages (2016) Interglacials of the last 800,000 years. Rev Geophys 54(1):162–219

PCDMI (n.d.) CMIP5—coupled model intercomparison project phase 5—overview. https://pcmdi.llnl.gov/mips/cmip5/. Accessed 27 May 2020

Peterson TC, Golubev VS, Groisman PY (1995) Evaporation losing its strength. Nature 377 (6551):687–688

Pielke RA Jr, Landsea C, Mayfield M, Layer J, Pasch R (2005) Hurricanes and global warming. Bull Am Meteor Soc 86(11):1571–1576
Powell JL (2015) Climate scientists virtually unanimous: anthropogenic global warming is true. Bull Sci Technol Soc 35(5–6):121–124
Roderick ML, Farquhar GD (2002) The cause of decreased pan evaporation over the past 50 years. Science 298(5597):1410–1411
Roderick ML, Rotstayn LD, Farquhar GD, Hobbins MT (2007) On the attribution of changing pan evaporation. Geophys Res Lett 34(17):L17403
Sallenger AH, Doran KS, Howd PA (2012) Hotspot of accelerated sea-level rise on the Atlantic coast of North America. Nat Clim Change 2(12):884
Sheffield J, Wood EF, Roderick ML (2012) Little change in global drought over the past 60 years. Nature 491(7424):435–438
Spencer RW, Christy JR (1990) Precise monitoring of global temperature trends from satellites. Science 247(4950):1558–1562
Spencer RW, Christy JR, Grody NC (1990) Global atmospheric temperature monitoring with satellite microwave measurements: method and results 1979–84. J Clim 3(10):1111–1128
Stevens B, Bony S (2013) What are climate models missing? Science 340:1053–1054
Taylor KE, Stouffer RJ, Meehl GA (2012) An overview of CMIP5 and the experiment design. Bull Am Meteor Soc 93(4):485–498
Trenberth K (2005) Uncertainty in hurricanes and global warming. Science 308:1753–1754
USGCRP (2019) Sea level rise. U.S. Global Change Research Program (USGCRP). https://www.globalchange.gov/browse/indicators/global-sea-level-rise. Accessed 27 May 2020
USGS (n.d.) Regional and climate change. Visualization. http://regclim.coas.oregonstate.edu/visualization/gccv/. Accessed 27 May 2020
Vecchi GA, Soden BJ (2007) Effect of remote sea surface temperature change on tropical cyclone potential intensity. Nature 450(7172):1066–1070
Weart S (2010) The development of general circulation models of climate. Stud Hist Philos Mod Phys 41:208–217
Webster PJ, Holland GJ, Curry JA, Chang HR (2005) Changes in tropical cyclone number, duration, and intensity in a warming environment. Science 309(5742):1844–1846
Wilhite DA, Glantz MH (1985) Understanding the drought phenomenon: the role of definitions. Water Int 10(3):111–120
WCRP (n.d.) WCRP Coupled Model Intercomparison Project (CMIP). https://www.wcrp-climate.org/wgcm-cmip. Accessed 10 Feb 2020
Zou CZ, Goldberg MD, Hao X (2018) New generation of US satellite microwave sounder achieves high radiometric stability performance for reliable climate change detection. Sci Adv 4(10):eaau0049

Chapter 4
Intergovernmental Panel on Climate Change and Global Climate Change Projections

4.1 Intergovernmental Panel on Climate Change

The Intergovernmental Panel on Climate Change (IPCC) was established in 1988 by the United Nations Environment Programme (UNEP) and the World Meteorological Organization (WMO). According to the Principles Governing IPCC Work (IPCC 2013a):

> The role of the IPCC is to assess on a comprehensive, objective, open and transparent basis the scientific, technical and socio-economic information relevant to understanding the scientific basis of risk of human-induced climate change, its potential impacts and options for adaptation and mitigation. IPCC reports should be neutral with respect to policy, although they may need to deal objectively with scientific, technical and socio-economic factors relevant to the application of particular policies.

and

> Review is an essential part of the IPCC process. Since the IPCC is an intergovernmental body, review of IPCC documents should involve both peer review by experts and review by governments.

Freudenburg and Muselli (2010, p. 484) further observed that

> the IPCC is not so much a scientific organization as a distinctive way of dealing with the interface between science and policy. In the interest of ensuring that the Panel's work is serving the needs of government and policy, it includes governmental as well as scientific representatives, and it is charged with producing assessments that are relevant to policy, without being policy prescriptive.

The peer-review process for IPCC assessments is extensive and the reports are generally taken in the scientific community as objective and authoritative. Climate change projections in the latest Fifth Assessment (AR5) reports are derived largely from the Coupled Model Intercomparison Project 5 (CMIP5) ensemble of general circulation model (GCM) results (Sect. 3.6.2). The IPCC peer-review process has been largely effective although there have been some isolated examples of errors

R. Maliva, *Climate Change and Groundwater: Planning and Adaptations for a Changing and Uncertain Future*, Springer Hydrogeology,
https://doi.org/10.1007/978-3-030-66813-6_4

subsequently discovered in reports. In one instance, a 2007 report contained unfounded claims about the rate of melting of Himalayan glaciers (Berini 2010). In another instance, the area of the Netherlands below sea level was incorrectly calculated as also including land prone to flooding (Reuters 2010). The very small number of errors in many thousands of pages of heavily scrutinized reports is testimony to the effectiveness of the IPCC peer-review process rather than being evidence for the unreliability of the reports as some climate change skeptics claim.

The existence of biases in the IPCC process has been raised across the political spectrum and within scientific communities. Within the climate change skeptic or contrarian community, the charge is frequently made that the IPCC reports are exaggerated and overly pessimistic (Freudenburg and Muselli 2010). Some scientists have claimed that the IPCC reports are biased against nuclear power as a mitigation option due to the "the ideology of the environmental movement," a claimed denied by the IPCC (Shellenberger 2018). It has also been claimed that the IPCC reports neglect the benefits of global warming (Drawall 2014). Inasmuch as the current climate of the Earth is not optimal (at least for humans) everywhere, some regions would be expected to experience net benefits from rising temperatures and increasing rainfall.

On the other side of the spectrum, some scientists claim that the consensus building goal of the IPCC results in a tendency to underplay climate change impacts to avoid the risk of appearing biased and overly negative (Freudenburg and Muselli 2010; Scherer 2012; Brysse et al. 2013). Indeed, comparison of the predictions of the initial IPCC assessments with subsequently collected GHG emissions and climate data found no evidence of exaggeration, but instead revealed that in many instances IPCC predictions tended to have been underestimations (Freudenburg and Muselli 2010; Brysse et al. 2013). It has been suggested that the "outspoken arguments of the conservative think tanks" have tended to cause scientists to underestimate their results and the actual degree of climate disruption that is taking place to avoid the charge of being alarmists (Freudenburg and Muselli 2010).

Sound science, in general, tends to be quite conservative in that scientists are trained to not make conclusions beyond what they can rigorously defend with the data. Brysse et al (2013) argued that

> the scientific values of rationality, dispassion, and self-restraint tend to lead scientists to demand greater levels of evidence in support of surprising, dramatic, or alarming conclusions than in support of conclusions that are less surprising, less alarming, or more consistent with the scientific status quo.

Brysse et al. (2013) described the conservativism in the climate change community as "erring on the side of least drama."

Criticism of the IPCC tends to focus on the interpretation of the data rather than on the raw data itself. There is no technical basis for questioning the CMIP5 projections beyond consideration of the acknowledged limitations of climate modeling in general. Based on past experiences, the IPCC AR5 projections are more likely to turn out to be underestimations than exaggerations.

Climate change skeptics have not produced their own alternative model calibrated to historical GHG concentration and climate data that shows that progressively increasing GHG concentrations will not cause climate warming. Weart (2010, p. 214) observed that

> Those who still denied any serious risk of climate change could not reasonable dismiss computer modeling in general. That would throw away much of the past few decades's work in many fields of science and engineering, and even key business practices. The challenge to them was to produce a simulation that did not show global warming. Now that personal computers were far more powerful than the most expensive computers of earlier decades, it was possible to explore thousands of combinations of parameters. But no matter how people fiddled with climate models, the answer was the same. If your model could reproduce something resembling the present climate, and you added some greenhouse gases, the model showed serious global warming.

The journalistic doctrine of "balance" requires both sides in a debate get equal coverage (Edwards and Schneider 2001), which can give the public and uniformed politicians the impression that there is great uncertainty. Edwards and Schneider (2001) put forth that "news stories are grossly misleading and irresponsible if they present the unrefereed opinions of skeptics as if they were comparable in credibility to the hundred-scientists, thousand-reviewer documents released by the IPCC. The general public—or lay politicians—cannot be expected to determine for themselves how to weigh these conflicting opinions."

The CMIP5 and IPCC climate change projections provide a sound foundation for evaluating the impacts of potential climate changes and for adaptation planning. It is important to always keep in mind that there is a wide range in projected future climate conditions and that users of projections should consider their inherent uncertainties. As more data become available and climate models are further developed, climate projections in future IPCC assessments will be further refined but there will always be considerable uncertainties because some key parameters, particularly future GHG emissions, are poorly constrained or unknowable.

Summary climate change projections from the CMIP5 are included in the IPCC reports, which can be downloaded at no charge from the IPCC website (https://www.ipcc.ch/reports/). The reports include both a "Summary for Policymakers" and a "Technical Summary". Maps summarizing the CMIP5 temperature and precipitation projections are available in the Supplementary Material to the report "Climate Change 2013: The Physical Science Basis" (IPCC 2013b). For each Representation Concentration Pathway (RCP), an "Atlas of Global and Regional Climate Projections" is provided. Maps are provided for each RCP scenario of annual and seasonal precipitation changes for the periods 2016–2035, 2046–2065 and 2081–2100 with respect to the 1986–2005 period. Individual maps are provided for each region and time period for the 25th, 50th and 75th percentiles of the distribution of the CMIP5 ensemble results (e.g., Fig. 4.1). More up to date modeling results will be available in the upcoming Sixth Assessment (AR6) report "Climate Change 2021: The Physical Science Basis."

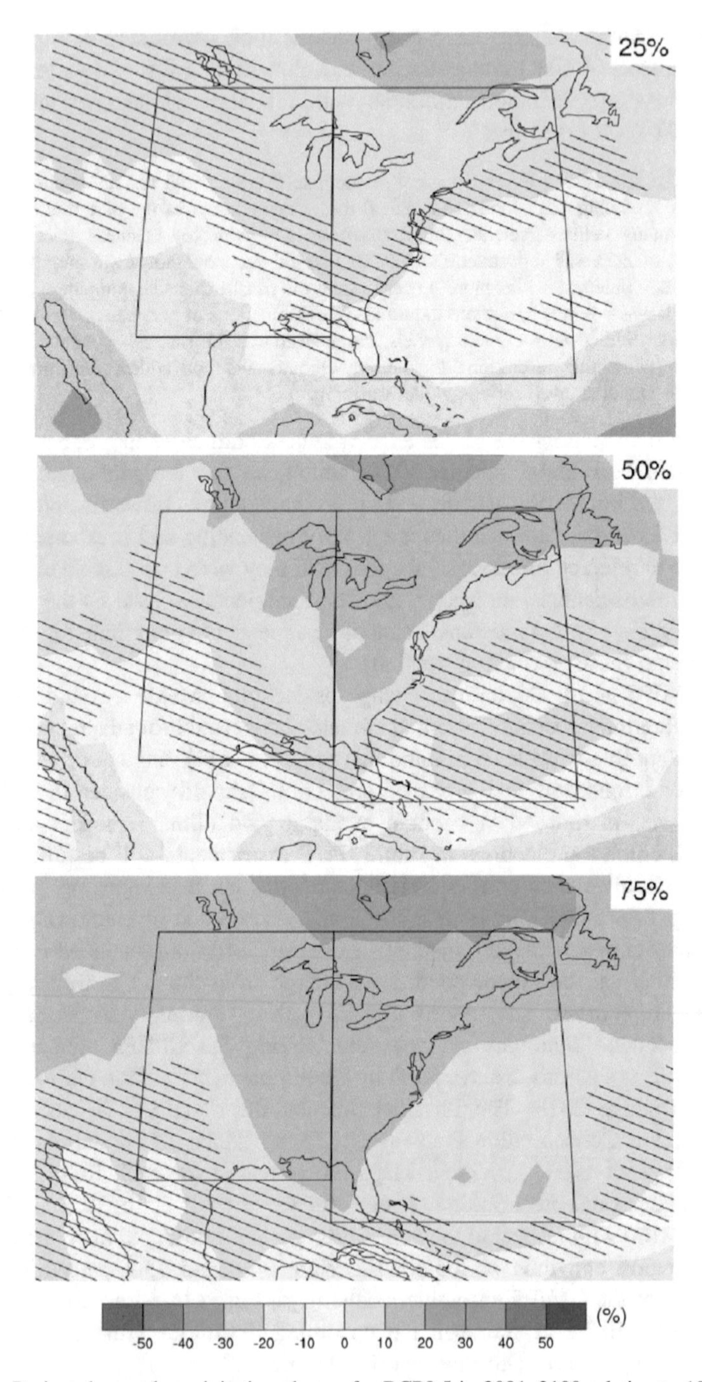

Fig. 4.1 Projected annual precipitation change for RCP8.5 in 2081–2100 relative to 1985–2005 in eastern North America. The 25, 50, and 75 percentiles of the CMIP5 ensemble are shown. Hatching indicates regions where the magnitude of the change of the 20 yr mean is less than 1 standard deviation of the model-estimated present-day natural variability of 20 yr mean differences (i.e., the projected change is relatively small or there is little agreement between models on the sign of the change). *Source* IPCC (2013c, Fig. AI.SM8.5.041)

4.2 Global Climate Change Predictions

4.2.1 Introduction

Historic and on-going anthropogenic increases in atmospheric GHG concentrations are increasing the temperature of the Earth's surface and lower atmosphere and the increases will continue into the foreseeable future. The rise in atmospheric temperatures is causing sea levels to rise primarily through the thermal expansion of the oceans and the melting of continental glaciers and polar ice sheets. Warmer temperatures result in increased evaporation rates and an increase in atmospheric water content as warm air can hold more water at a given relative humidity. Inasmuch as over the long-term (i.e., under steady-state atmospheric conditions) total precipitation must equal total evapotranspiration, based on atmospheric equilibrium and mass balance, greater surface water and soil evaporation and plant evapotranspiration are expected to result in increased precipitation. Changes to atmospheric circulation patterns are expected to result in changes in the temporal and spatial distribution of local precipitation.

Future climate conditions are predicted using GCMs. A fundamental part of the climate change adaptation and resilience planning processes is extracting from the climate modeling results an envelope of future climate conditions that is plausible for the geographic area of interest. It is recognized that the technical and financial resources available to those responsible for planning in the water area are limited. Collaboration between decision makers and the research community is one means to bridge the technical gap. A plethora of coarse geographic-scale information on projected climate change is available in the IPCC reports, governmental reports (e.g., U.S. National Climate Assessment reports), and published technical papers.

The greatest uncertainty associated with climate change projections is future GHG emissions, which depend on the degree of implementation of mitigation measures. GHG emissions is perhaps the ultimate "tragedy of the commons." As described by Hardin (1968) is his influential essay, the tragedy of the commons occurs when individuals who share a limited (common-pool) resource have an incentive to maximize their exploitation of the resource even though such actions will be to the detriment of all users. Hardin used cattle grazing on a commons pasture as a metaphor for people sharing a common finite resource. The commons is open to all herdsmen. Too many animals would result in over-grazing and the ruin of the commons. As a rational being, each herdsman seeks to maximize his gain, which would occur by increasing the size of his herd. By adding an additional animal, a herdsman receives the full benefit from the additional animal, while the cost of the additional animal in terms of damage to the resource is borne by all users of the commons. It is, therefore, the rational conclusion of each herdsman that the sensible action is to increase his herd. The tragedy, as explained by Hardin, is that each herdsman becomes locked into a system that compels him to increase his herd without limits, which ultimately leads to the ruin of the shared resource upon which all depend. In the case of GHG emissions, everyone is ultimately harmed by too much GHG emissions, but the economic benefits for some countries to continue expanded fossil fuel use (or to maintain high current use rates) with associated GHG emissions is greater than their immediate harm from climate change.

Garrett Hardin's solution to the tragedy of the commons is "coercion" that is mutually agreed upon by the majority of the people involved. The Paris Climate Agreement, which was adopted by nearly every nation in 2015, includes commitments from all major GHG-emitting countries to cut their emissions, but there is no guarantee that such commitments will be met. No sound technical basis exists for predicting whether future emissions will fall closer to the RCP2.6 (low) or RCP8.5 (high) emissions pathways, since this will involve decisions to be made by governments, businesses, and individuals. For each RCP, the ensemble of GCMs predict a range of future conditions adding another layer of uncertainty. From a practical perspective, decision makers need to consider the plausible range of future local climate conditions rather than trying to predict and plan based on a single most-likely scenario or an average or median modeling result.

Climate modeling results indicate that increasing global atmospheric temperatures will result in spatial changes in average annual precipitation with the general pattern that wetter regions will tend to getter water and arid and semiarid regions will tend to become drier. Mean annual precipitation will very likely increase in high and some of the mid latitude areas and will more likely than not decrease in the subtropics (IPCC 2013b). At more regional and local scales, precipitation changes may be influenced by anthropogenic aerosol emissions and will be strongly influenced by natural internal variability (Kirtman et al. 2013). Projected changes in the water cycle include (Trenberth et al. 2007; Bates et al. 2008; Karl et al. 2009):

- changes in precipitation patterns and intensity
- changes in the incidence, duration, and geographic extent of droughts
- widespread melting of "permanent" snow and ice that reduce the volume of glaciers
- more precipitation falling as rain rather the snow
- reductions in annual winter snow accumulation
- changes in the timing and duration of spring snowmelt flows
- increases in atmospheric water vapor and cloudiness
- increases in evaporation and plant evapotranspiration
- increases in water and soil temperatures
- reductions in lake and river ice cover
- changes in soil moisture and runoff.

Climate models predict that, in general, precipitation in coming decades will tend to be concentrated into more frequent intense events, with longer periods of little precipitation in between (Bates et al. 2008; Seneviratne et al. 2012; IPCC 2014). Areas expected to experience an intensification of rainfall include the high latitudes and tropical regions, and northern mid-latitude regions in the winter (Seneviratne et al. 2012).

The IPCC (2014, p. 13) concluded that.

> Climate change over the 21st century is projected to reduce renewable surface water and groundwater resources in most dry subtropical regions (*robust evidence, high agreement*), intensifying competition for water among sectors (*limited evidence, medium agreement*).

This chapter provides an overview of projected global climate (temperature and precipitation) and sea level changes derived largely from the CMIP5 results summarized in the IPCC AR5 documents. Such projections can provide general insights into both the direction and potential magnitude of local climate changes, with the caveat that local precipitation changes can be strongly influenced by local topography, land cover, and atmospheric conditions. The results of some more recent climate change prediction studies are also presented. In particular, recent studies suggest more rapid sea level rise (SLR) due to accelerated melting of the continental ice sheets.

4.2.2 Global Temperature Change

The global climate will experience continued warming caused by past anthropogenic emissions as well as from additional future anthropogenic emissions. The results of the IPCC AR5 are summarized below (IPCC 2014). Future changes are relative to a 1986–2005 baseline and the IPCC qualitative level of confidence is given in italics. In the near term, the changes in global mean surface temperature for all four RCPs are projected to likely be in the range of 0.3–0.7 °C for the period 2016–2035 (*medium confidence*), assuming there will be no major volcanic eruptions, changes in some natural sources (e.g., CH_4 and N_2O), or unexpected changes in total solar irradiance (Fig. 4.2). By mid-century, the temperature projections start to diverge depending on which emissions pathway unfolds. By the end of the 21st century (2081–2100), the global mean surface temperature increase is projected to likely be in the range of 0.3–1.7 °C under RCP2.6, 1.1–2.6 °C under RCP4.5, 1.4–3.1 °C under RCP6.0, and 2.6– 4.8 °C under RCP8.5. Warming is expected to likely exceed 2 °C for RCP6.0 and RCP8.5 (*high confidence*), more likely than not to exceed 2 °C for RCP4.5 (*medium confidence*), and unlikely to exceed 2 °C for RCP2.6 (*medium confidence*).

The IPCC (2014, p. 60) concluded that it

> is virtually certain that there will be more frequent hot and fewer cold temperature extremes over most land areas on daily and seasonal timescales, as global mean surface temperature increases. It is very likely that heat waves will occur with a higher frequency and longer duration. Occasional cold winter extremes will continue to occur.

Global warming will not be even across the world. The temperature increase over land will be greater than that over the oceans (Fig. 4.2). The Arctic and much of the boreal regions will warm more rapidly than the global mean. The far northern, subpolar Atlantic Ocean is simulated to cool or warm less. This so-called

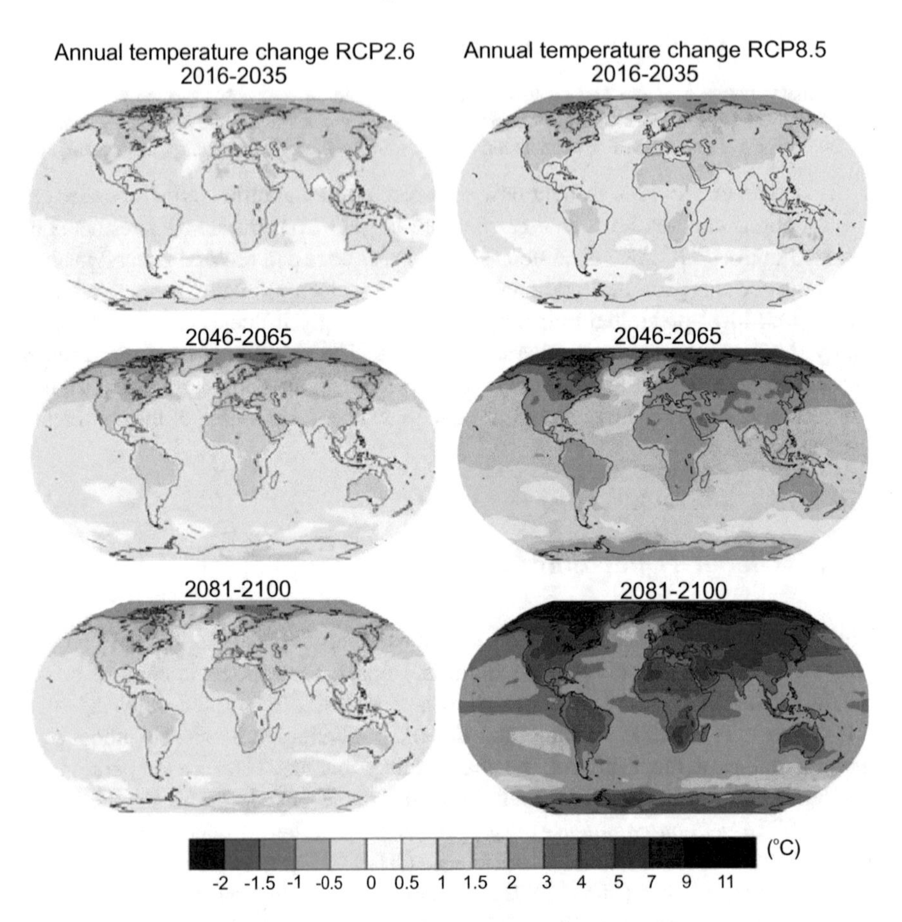

Fig. 4.2 Projected annual global temperature change for RCP2.6 and PCP8.5 relative to 1985–2005; median (50%) percentiles of CMIP ensembles are shown. *Source* IPCC (2013c, Fig. AI. SM8.5.4; 2013d, Fig. AI.SM2.6.4)

"cold blob," which currently exists, has been attributed to a weakening of the Atlantic meridional overturning circulation (AMOC) and the influx of cold water from the melting of the Greenland ice sheets (Rahnstorf et al. 2015). The Southern Ocean is also projected to experience a lesser warming. Simulation results project that subtropical continental interior areas will tend to have greater than mean warming rates.

The AMOC transports warm surface water from the tropics to the subpolar Atlantic where heat loss to the atmosphere increases its density, causing it to sink and flow southward (Chen and Tung 2018). When the AMOC is weaker, less water is transported north, less heat is subducted to the deeper ocean, and greater warming of the atmosphere and the upper 200 m of the global oceans occurs (Chen and Tung 2018).

Weakening of the AMOC, which includes the Gulf Stream, can have large climate consequences as it would decrease the northward flow of warm water, which has a profound impact on the climate of the northeastern United States and northwestern Europe. As summarized in Sect. 3.3, change in the rate of flow of the Gulf Stream may also impact sea level along the eastern coast of the United States. The AMOC has decreased since the mid-1970s (Rahnstorf et al. 2015). A key question is whether the decrease in the AMOC is part of its natural cyclicity or is the result of climate change. Chen and Tung (2018) interpreted that the AMOC changes since the 1940s are best explained by natural variability rather than anthropogenic forcing and that the current AMOC decline is ending. Chen and Tung (2018) predicted the current AMOC decline will last about two decades and will be manifested as a period of more rapid global surface warming.

4.2.3 Precipitation

Global warming is anticipated to cause an acceleration of the hydrological cycle as the result of increasing evaporation rates causing corresponding increases in precipitation rates. The IPCC (2014, p. 40) observed that averaged over the mid-latitude land areas of the Northern Hemisphere, precipitation has increased since 1901 (*medium confidence* before and *high confidence* after 1951). Elsewhere, area-averaged long-term positive or negative trends have low confidence.

The IPCC (2014) noted that model projections indicate that changes in precipitation will not be uniform. High latitude areas and the equatorial Pacific are likely to experience an increase in annual mean precipitation under the RCP8.5 scenario. Mean precipitation will likely decrease in many mid-latitude and subtropical dry regions, whereas mean precipitation will likely increase under the RCP8.5 scenario in many mid-latitude wet regions. A key conclusion is that "Extreme precipitation events over most of the mid-latitude land masses and over wet tropical regions will very likely become more intense and more frequent" (IPCC 2014, p. 11).

The CMIP5-based precipitation maps in the IPCC Fifth Assessment "Climate Change 2013: The Physical Science Basis" report (IPCC 2013b) show that the Mediterranean Basin and North Africa, southern Africa, the Caribbean and Central America, Mexico and the southwestern United States, and much of Australia will likely become drier (Fig. 4.3). Areas that are simulated to experience greater annual average precipitation include the northern and southern polar and boreal regions, most of the eastern and central United States and Canada, and South and East Asia. Most ominous is that some areas already facing water scarcity are projected to become even drier.

The IPCC AR5 climate modeling results show that for all RCPs, the area encompassed by monsoon systems is likely to increase and monsoon precipitation is likely to intensify. El Niño-Southern Oscillation (ENSO) related precipitation variability on regional scales will likely also intensify (IPCC 2014). "Extreme El Niño and La Niña events are projected to likely increase in frequency in the 21st

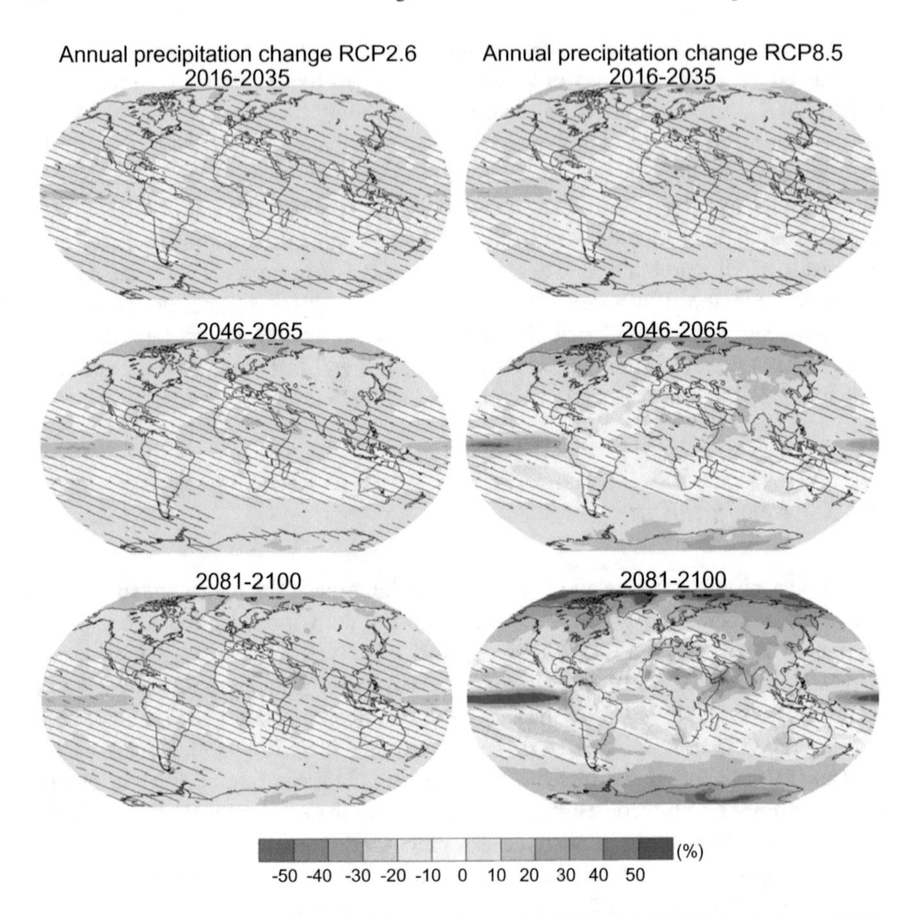

Fig. 4.3 Projected annual global precipitation change for RCP2.6 and PCP8.5 relative to 1985–2005; median (50%) percentiles of CMIP5 ensembles are shown. *Source* IPCC (2013c, Fig. AI. SM8.5.9, 2013d, Fig. AI.SM2.6.9)

century and to likely intensify existing hazards, with drier or wetter responses in several regions across the globe. Extreme El Niño events are projected to occur about as twice as often under both RCP2.6 and RCP8.5 in the 21st century when compared to the 20th century" (IPCC 2019a, p. 68).

4.2.4 Droughts and Aridity

The relationship between climate change and drought is unclear. A drought can be broadly defined as a temporary, recurring reduction in the precipitation in an area. Palmer (1965, p. 3) defined a drought as an "interval of time, generally on the order of months or years in duration, during which the actual moisture supply at a given place rather consistently falls short of climatically expected or climatically appropriate moisture supply."

Aridity is a measure of long-term average climate conditions. A permanent shift in local climate conditions to a drier state is not a drought. Both humid and arid regions experience droughts, but the interannual variation in precipitation is greater in arid regions and there is a greater probability of below average precipitation in any particular year (Smakhtin and Schipper 2008). Arid regions are thus more prone to droughts and may experience more severe impacts from droughts.

The IPCC Fourth Assessment Report (AR4) projected an increase in drought frequency with the increase in the number of consecutive dry days projected to be most significant in North and Central America, the Caribbean, northeastern and southwestern South America, southern Europe and the Mediterranean, southern Africa, and Western Australia (Bates et al. 2008). It was concluded in a subsequent IPCC report that there is medium confidence that droughts will intensify in the 21st century in some seasons due to reduced precipitation and/or increased evapotranspiration, and that vulnerable areas include the Mediterranean region, central Europe, central North America, Central America and Mexico, northeast Brazil, and southern Africa (Seneviratne et al. 2012). However, "Definitional issues, lack of observational data, and the inability of models to include all the factors that influence droughts preclude stronger confidence than medium in the projections" (Seneviratne et al. 2012, p. 114). Elsewhere in the world there are inconsistent projections of changes in drought frequency and severity (Seneviratne et al. 2012).

The IPCC AR5 concluded that "There is low confidence in a global-scale observed trend in drought or dryness (lack of rainfall), owing to lack of direct observations, dependencies of inferred trends on the index choice and geographical inconsistencies in the trends" (Stocker et al. 2013, p. 50). Nevertheless, the frequency and intensity of drought have likely increased in the Mediterranean and West Africa and likely decreased in central North America and northwest Australia since 1950 (IPCC 2014). The IPCC (2014, p. 69) concluded that

> Climate change over the 21st century is projected to reduce renewable surface water and groundwater resources in most dry subtropical regions (*robust evidence, high agreement*) intensifying competition for water among sectors (*limited evidence, medium agreement*). In presently dry regions, the frequency of droughts will likely increase by the end of the 21st century under RCP8.5 (*medium confidence*).

Projected water availability changes were mapped as precipitation minus evaporation (P − E) from the CMIP5 ensemble (Fig. 4.4) with the results consistent with the wet-get-wetter and dry-get-drier pattern (IPCC 2013b). Water availability is projected to increase in the northern and southern polar and boreal regions, most of Asia, and the eastern United States. The Mediterranean region, most of Europe, the southwestern North America, Central America, and southern and interior South America are projected to experience lesser water availability. Decreases in water availability would be expected to favor decreases in runoff and aquifer recharge.

The pattern of projected change in soil moisture is similar to that of water availability. The southwestern United States, northern Mexico, southern Africa,

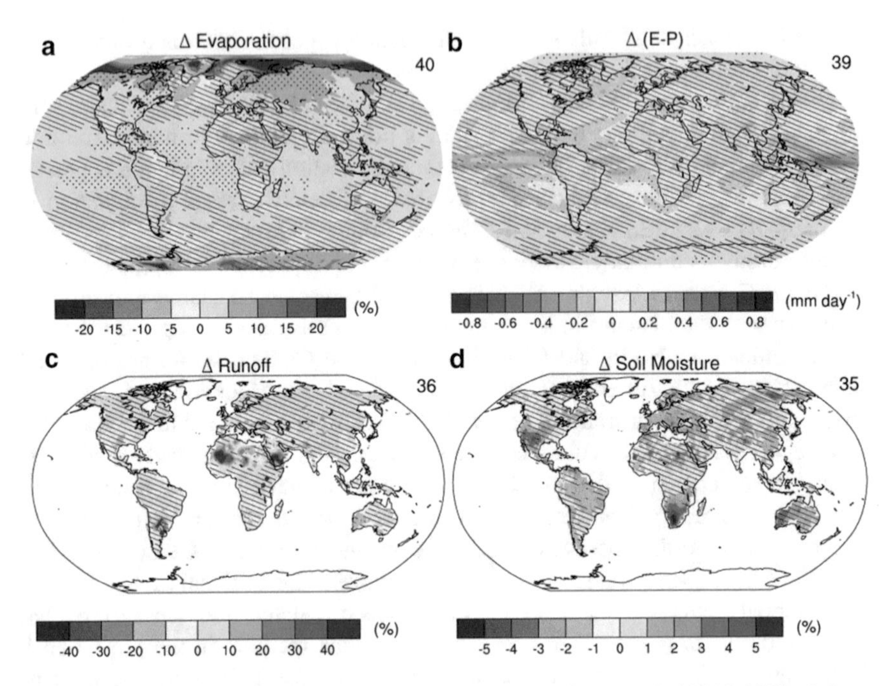

Fig. 4.4 CMIP5 multi-model annual mean projected changes for the period 2016–2035 relative to 1986–2005 under RCP4.5 for: **a** evaporation (%), **b** evaporation minus precipitation (E − P, mm day^{-1}), **c** total runoff (%), and **d** soil moisture in the top 10 cm (%). The number of CMIP5 models used is indicated in the upper right corner of each panel. *Source* Kirtman et al. (2013, Fig. 11.10)

most of Australia, and Europe are projected to experience declining soil moisture (Fig. 4.4). The IPCC (2013b) cautioned that "Owing to the simplified hydrological models in many CMIP5 climate models, the projections of soil moisture and runoff have large model uncertainties" (Kirtman et al. 2013, p. 988).

Cook et al. (2014) investigated projected 21st century drying and wetting trends using the output of the CMIP5 simulations, and the Palmer Drought Severity Index (PDSI) and Standardized Precipitation Evapotranspiration Index (SPEI). PDSI and SPEI projections using precipitation and Penman Monteith-based potential evapotranspiration (PET) changes from the GCMs generally agree in showing robust cross-model drying in western North America, Central America, the Mediterranean, southern Africa, and the Amazon, and robust wetting occurring in the northern hemisphere high latitudes and east Africa (PDSI only). Increased PET was found to intensify drying in areas where precipitation is already reduced and also drives areas into drought that would otherwise experience little drying or even wetting from precipitation trends alone. The PET amplification effect was found to be largest in the northern hemisphere mid-latitudes, and is especially pronounced in western North America, Europe, and southeast China (Cook et al. 2014).

4.2.5 *Snow and Glacier Dominated Water Systems*

Glaciers and seasonal snowpacks are major components of the water supply in some mountainous areas. Annual snowfall that seasonally melts is a renewable resource, whereas much of the ice in glaciers is essentially non-renewable. Increasing global temperatures will continue to impact regions were supplies are dominated by either snowfall or glaciers. The impacts of a warming climate on water availability in snow- and glacier-dominated regions were reviewed by Barnett et al. (2005). Winter snowpacks in snow-dominated hydrologic regimes, such as the western United States, Andes, Alps, and Himalayas provide seasonal storage of water. Precipitation that falls as snow is stored at high altitudes and is released in the spring and summer as the snow melts. In a warmer world, less winter precipitation will occur as snow and the accumulated snow will melt earlier in the spring. The peak river run-off will shift to the early spring, away from the summer and fall when it is often most needed. Floods may become more frequent and intense during the spring.

The IPCC (2019a, p. 47) reported that

> Observations show general decline in low-elevation snow cover (*high confidence*), glaciers (*very high confidence*) and permafrost (*high confidence*) due to climate change in recent decades. Snow cover duration has declined in nearly all regions, especially at lower elevations, on average by 5 days per decade, with a likely range from 0–10 days per decade. Low elevation snow depth and extent have declined, although year-to-year variation is high.

Winter runoff was reported to have increased in recent decades due to more precipitation falling as rain (*high confidence*) (IPCC 2019a).

Glacier-dominated hydrologic regimes, such as western South America and the Himalaya-Hindu Kush region, face even more serious water supply challenges (Barnett et al. 2005). As essentially fossil water, there is no replacement for the glacial water supply once the glaciers melt. In the short term (next several decades), glacier-dominated regions may see an increase in water supply because of the accelerated melting of the glaciers and associated run-off. Any increases in water supply are expected to be followed by abrupt decreases in water supply as the glaciers disappear (Barnett et al. 2005). Observations support that some glacier-fed rivers now have increased summer and annual runoff due to intensified glacier melting, but other rivers have decreased flows where glacier melt water has lessened as glacier area shrinks (IPCC 2019a). Decreases in river flows were especially observed in regions dominated by small glaciers, such as the European Alps (*medium confidence*; IPCC 2019a).

Shifts in seasonal river and stream flows without adequate storage (i.e., reservoirs) will lead to regional water shortages that may impact agricultural and municipal water supplies, hydroelectric production, and environmental flows. Some regions dependent on glacial water are expected, based on current projections, to experience heavily depleted dry season water resources once the glaciers have disappeared (Barnett et al., 2005; Bates et al. 2008).

4.2.6 Global Sea Level Rise

The total global mean sea level (GMSL) rise for 1902–2015 was reported to be 0.16 m (*likely range* 0.12–0.21 m; IPCC 2019a). The rate of GMSL rise has been accelerating and for 2006–2015 the rate of GMSL rise was 3.6 mm/yr (3.1–4.1 mm/yr, *very likely range*), which is unprecedented over the last century (*high confidence*; IPCC 2019a). The IPCC (2019a) concluded that the sum of ice sheet and glacier contributions over the 2006–2015 period was 1.8 mm/yr (*very likely range* 1.7–1.9 mm/y), which exceeded the contribution from the thermal expansion of ocean water (1.4 mm/yr, *very likely range* 1.1–1.7 mm/yr; *very high confidence*).

The IPCC projected rate of sea level rise was raised for the higher emission RCPs between the AR5 and the 2019 "Special Report on the Ocean and Cryosphere in a Changing Climate" (Fig. 4.5; IPCC 2019a). Modeling using the high emissions RCP8.5 pathway project that the rate of GMSL rise may reach 15 mm/yr (10–20 mm/yr, *likely range*) in 2100, and to exceed several cm/yr in the 22nd century (IPCC 2019b). The IPCC (2019b, p. 20) cautioned that

> Processes controlling the timing of future ice-shelf loss and the extent of ice sheet instabilities could increase Antarctica's contribution to sea level rise to values substantially higher than the likely range on century and longer time-scales (*low confidence*). Considering the consequences of sea level rise that a collapse of parts of the Antarctic Ice Sheet entails, this high impact risk merits attention.

Simulations using the RCP8.5 pathway indicate several meters of GMSL in 2200 to 2300, which would have obvious profound impacts on coastal regions.

GMSL rise is projected to also cause an increase in the frequency of extreme sea level events at most locations. Local high sea levels that historically occurred once per century are projected to occur at least annually at most locations by 2100 under all RCP scenarios (*high confidence*; IPCC 2019b). The IPCC (2019a, b, b) also concluded that "Significant wave heights (the average height from trough to crest of the highest one-third of waves) are projected to increase across the Southern Ocean

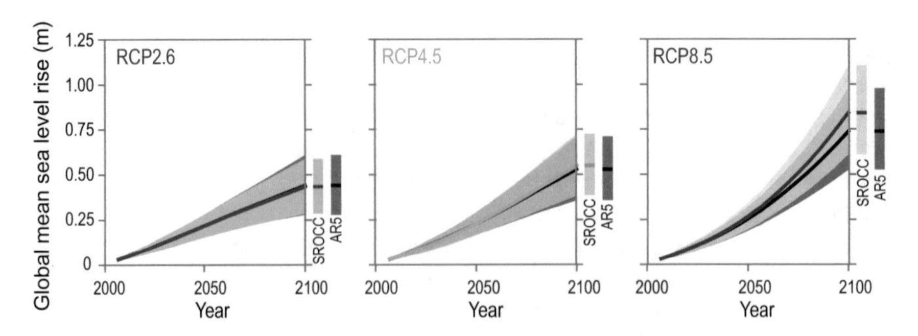

Fig. 4.5 Time series of projected GMSL rise for RCP2.6, RCP4.5 and RCP8.5 from the *Special Report on the Ocean and Cryosphere in a Changing Climate* (SROCC) and earlier IPCC AR5. The shaded region is the likely range. *Source* Oppenheimer et al. (2019, Fig. 4.9)

and tropical eastern Pacific (*high confidence*) and Baltic Sea (*medium confidence*) and decrease over the North Atlantic and Mediterranean Sea under RCP8.5 (*high confidence*)." Low-lying areas that are not permanently inundated with sea water will experience more frequent and greater temporary inundations (IPCC 2019a, b).

4.2.7 Extreme Storms

The IPCC AR4 concluded that it is likely that future tropical cyclones will become more intense with greater peak wind speeds and heavier precipitation associated with ongoing increases in tropical sea surface temperatures (Kundzewicz et al. 2007; Committee on Environment and Natural Resources 2008; Bates et al. 2008). Huntington (2006) reviewed the data on the frequency of extreme weather events, such as hurricanes, typhoons, floods and droughts. No evidence was found for increases in the frequency of flooding, tropical storm frequency and intensity, and the duration of the storm season. However, Huntington (2006) cautioned that the lack of detected increase in frequency and intensity of tropical storms during the 20th century should not be taken as evidence that further warming will not lead to such changes in the future.

The IPCC concluded in the AR5 that there is now low confidence in basin-scale projections of changes in the intensity and frequency of tropical cyclones in all basins to the mid-21st century. The low confidence was said to reflect the small number of studies exploring near-term tropical cyclone activity, the differences across published projections of tropical cyclone activity, and the large role of natural variability and non-GHG forcing on tropical cyclone activity up to the mid-21st century (IPCC 2013b). Confidence also remains low for long-term (centennial) changes in tropical cyclone activity after accounting for past changes in observational capabilities (IPCC 2013b). However, the IPCC (2013b) noted that since the 1970s, it is virtually certain that the frequency and intensity of storms in the North Atlantic have increased although the reasons for the increase are debated.

The IPCC (2013b) also concluded that there is low confidence of large-scale trends in storminess over the last century and that there is still insufficient evidence to determine whether robust trends exist in small-scale severe weather events, such as hailstorms or thunderstorms. In a subsequent report, the IPCC (2019a, p. 21) projected that

> the average intensity of tropical cyclones, the proportion of Category 4 and 5 tropical cyclones and the associated average precipitation rates are projected to increase for a 2 °C global temperature rise above any baseline period (*medium confidence*). Rising mean sea levels will contribute to higher extreme sea levels associated with tropical cyclones (*very high confidence*). Coastal hazards will be exacerbated by an increase in the average intensity, magnitude of storm surge and precipitation rates of tropical cyclones.

References

Barnett TP, Adam JC, Lettenmaier DP (2005) Potential impacts of a warming climate on water availability in snow-dominated regions. Nature 438:303–309

Bates B, Kundzewicz ZW, Wu S, Palutikof J (eds) (2008) Climate change and water, Intergovernmental Panel on Climate Change Technical Paper VI. IPCC, Geneva

Berini N (2010) Himalayan glaciers: how the IPCC erred and what the science says. https://skepticalscience.com/IPCC-Himalayan-glacier-2035-prediction.htm. Accessed 28 May 2020

Brysse K, Oreskes N, O'Reilly J, Oppenheimer M (2013) Climate change prediction: erring on the side of least drama? Glob Environ Change 23(1):327–337

Chen X, Tung K-K (2018) Global surface warming enhanced by weak Atlantic overturning circulation. Nature 559(7714):387–391

Committee on Environment and Natural Resources (2008) Scientific assessment of the effects of global climate change on the United States: a report on the Committee on Environment and Natural Resources. National Science and Technology Council, Washington, DC

Cook BI, Smerdon JE, Seager R, Coats S (2014) Global warming and 21st century drying. Clim Dyn 43(9–10):2607–2627

Drawall R (2014) Why the IPCC report neglects the benefits of global warming. Nat Rev (April 1, 2014). https://www.nationalreview.com/2014/04/why-ipcc-report-neglects-benefits-global-warming-rupert-darwall/. Accessed 28 May 2020

Edwards PN, Schneider SH (2001) Self-governance and peer review in science-for policy: the case of the IPCC Second Assessment Report. In: Miller C, Edwards PN (eds) Changing the atmosphere: expert knowledge and environmental governance. MIT Press, Cambridge, MA, pp 219–246

Freudenburg WR, Muselli V (2010) Global warming estimates, media expectations, and the asymmetry of scientific challenge. Glob Environ Change 20(3):483–491

Hardin G (1968) The tragedy of the commons. Science 162:1243–1248

Huntington TG (2006) Evidence for intensification of the global water cycle: review and synthesis. J Hydrol 319:83–95

IPCC (2013a) Principles governing IPCC work. https://archive.ipcc.ch/pdf/ipcc-principles/ipcc-principles.pdf. Accessed 28 May 2020

IPCC (2013b) Climate change 2013: the physical science basis. In: Stocker TF, Qin D, Plattner G-K, Tignor M, Allen SK, Boschung J, Nauels A, Xia Y, Bex V, Midgley PM (eds) Contribution of working group I to the Fifth Assessment Report of the Intergovernmental Panel on Climate Change. Cambridge University Press, Cambridge, UK

IPCC (2013c) Annex I: atlas of global and regional climate projections supplementary material RCP8.5 (van Oldenborgh GJ, Collins M, Arblaster J, Christensen JH, Marotzke J, Power SB, Rummukainen M, Zhou T (eds)). In: Stocker TF, Qin TD, Plattner G-K, Tignor M, Allen SK, Boschung J, Nauels A, Xia Y, Bex V, Midgley PM (eds) Climate change 2013: the physical science basis. Contribution of working group I to the fifth assessment report of the Intergovernmental Panel on Climate Change. Cambridge University Press, Cambridge, UK

IPCC (2013d) Annex I: atlas of global and regional climate projections supplementary material RCP2.6 (van Oldenborgh GJ, Collins M, Arblaster J, Christensen JH, Marotzke J, Power SB, Rummukainen M, Zhou T (eds)). In: Stocker TF, Qin TD, Plattner G-K, Tignor M, Allen SK, Boschung J, Nauels A, Xia Y, Bex V, Midgley PM (eds) Climate change 2013: The physical science basis. Contribution of working group I to the fifth assessment report of the Intergovernmental Panel on Climate Change. Cambridge University Press, Cambridge, UK

IPCC (2014) In: Pachauri RK & Meyer LA (eds) Climate change 2014: synthesis report. Contribution of working groups I, II and III to the fifth assessment report of the Intergovernmental Panel on Climate Change. IPCC, Geneva

IPCC (2019a) Technical summary. In: Pörtner H-O, Roberts DC, Masson-Delmotte V, Zhai P, Tignor M, Poloczanska E, Mintenbeck K, Alegría A, Nicolai M, Okem A, Petzold J, Rama B, Weyer NM (eds) IPCC special report on the ocean and cryosphere in a changing climate (In press)

IPCC (2019b) Summary for policymakers. In: Pörtner H-O, Roberts DC, Masson-Delmotte V, Zhai P, Tignor M, Poloczanska E, Mintenbeck K, Alegría A, Nicolai M, Okem A, Petzold J, Rama B, Weyer NM (eds) IPCC special report on the ocean and cryosphere in a changing climate (In press)

Karl TR, Melillo JM, Peterson TC (eds) (2009) Global climate change impacts in the United States. Cambridge University Press, Cambridge, UK

Kirtman B, Power SB, Adedoyin JA, Boer GJ, Bojariu R, Camilloni I, Doblas-Reyes FJ, Fiore AM, Kimoto M, Meehl GA, Prather M, Sarr A, Schär C, Sutton R, van Oldenborgh GJ, Vecchi G, Wang HJ (2013) Near-term climate change: projections and predictability. In: Stocker TF, Qin D, Plattner G-K, Tignor M, Allen SK, Boschung J, Nauels A, Xia Y, Bex V, Midgley PM (eds) Climate change 2013: the physical science basis. Contribution of working group I to the fifth assessment report of the Intergovernmental Panel on Climate Change. Cambridge University Press, Cambridge, UK

Kundzewicz ZW, Mata LJ, Arnell NW, Döll P, Jimenez B, Miller K, Oki T, Sen Z, Shiklomanov I (2007) Freshwater resources and their management. In: Parry ML, Canziani OF, Palutikof JP, van der Linden PJ, Hansen CE (eds) Climate change 2007: impacts, adaptation and vulnerability. Contribution of working group II to the fourth assessment report of the Intergovernmental Panel on Climate Change. Cambridge University Press, Cambridge, UK, pp. 173–310

Oppenheimer M, Glavovic BC, Hinkel J, van de Wal R, Magnan AK, Abd-Elgawad A, Cai R, Cifuentes-Jara M, DeConto RM, Ghosh T, Hay J, Isla F, Marzeion B, Meyssignac B, Sebesvari Z (2019) Sea level rise and implications for low-lying islands, coasts and communities. In: Pörtner H-O, Roberts DC, Masson-Delmotte V, Zhai P, Tignor M, Poloczanska E, Mintenbeck K, Alegría A, Nicolai M, Okem A, Petzold J, Rama B, Weye NM (eds) IPCC special report on the ocean and cryosphere in a changing climate (In press, pp 332–445)

Palmer WC (1965) Meteorological drought. Weather Bureau, Research Paper No. 45. United States Department of Commerce

Rahmstorf S, Box JE, Feulner G, Mann ME, Robinson A, Rutherford S, Schaffernicht EJ (2015) Exceptional twentieth-century slowdown in Atlantic Ocean overturning circulation. Nat Clim Change 5(5):475–480

Reuters (2010) U.N. climate panel admits Dutch sea level flaw (February 13, 2010). Reuters. https://www.reuters.com/article/us-climate-seas/u-n-climate-panel-admits-dutch-sea-level-flaw-idUSTRE61C1V420100213. Accessed 28 May 2020

Scherer G (2012) Climate science predictions prove too conservative. Scientific American. https://www.scientificamerican.com/article/climate-science-predictions-prove-too-conservative/. Accessed 28 May 2020

Seneviratne SI, Nicholls N, Easterling D, Goodess CM, Kanae S, Kossin J, Luo Y, Marengo J, McInnes K, Rahimi M, Reichstein M, Sorteberg A, Vera C, Zhang X (2012) Changes in climate extremes and their impacts on the natural physical environment. In: Field CB, Barros V, Stocker TF, Qin D, Dokken DJ, Ebi KL, Mastrandrea MD, Mach KJ, Plattner G-K, Allen SK, Tignor M, Midgley PM (eds) Managing the risks of extreme events and disasters to advance climate change adaptation (A special report of working groups I and II of the Intergovernmental Panel on Climate Change). Cambridge University Press, Cambridge UK, pp 109–230

Shellenberger M (2018) Top climate scientists warn governments of 'blatant anti-nuclear bias' in latest IPCC climate report. https://www.forbes.com/sites/michaelshellenberger/2018/10/29/top-climate-scientists-warn-governments-of-blatant-anti-nuclear-bias-in-latest-ipcc-climate-report/#31e302523973. Accessed 28 May 2020

Smakhtin V, Schipper ELF (2008) Droughts: the impact of semantics and perceptions. W Policy 10(2):131–143

Stocker TF, Qin D, Plattner G-K, Alexander LV, Allen SK, Bindoff NL, Bréon F-M, Church JA, Cubasch U, Emori S, Forster P, Friedlingstein P, Gillett N, Gregory JM, Hartmann DL, Jansen E, Kirtman B, Knutti R, Krishna Kumar K, Lemke P, Marotzke J, Masson-Delmotte V, Meehl GA, Mokhov II, Piao S, Ramaswamy V, Randall D, Rhein M, Rojas M, Sabine C, Shindell D, Talley LD, Vaughan DG, Xie S-P (2013) Technical summary. In: Stocker TF, Qin D, Plattner G-K, Tignor M, Allen SK, Boschung J, Nauels A, Xia Y, Bex V, Midgley PM (eds) Climate change 2013: the physical science basis. Contribution of working group I to the fifth assessment report of the Intergovernmental Panel on Climate Change. Cambridge University Press, Cambridge, UK

Trenberth KE, Jones PD, Ambenje P, Bojariu R, Easterling D, Klein Tank A, Parker D, Rahimzadeh F, Renwick JA, Rusticucci M, Soden B, Zhai P (2007) Observations: surface and atmospheric climate change. In: Solomon SD, Qin D, Manning M, Chen Z, Marquis M, Avery KB, Tignor M, Miller HL (eds) Climate change 2007: The physical science basis. Contribution of working group I to the fourth assessment report of the Intergovernmental Panel on Climate Change. Cambridge University Press, Cambridge, UK, pp 235–336

Weart S (2010) The development of general circulation models of climate. Stud Hist Philos Mod Phys 41:208–217

Chapter 5
Modeling of Climate Change and Aquifer Recharge and Water Levels

5.1 Introduction

A fundamental question water users and managers have with respect to climate change is how future climate conditions will impact their supplies and demands. Climate change will affect aquifer water budgets, and thus the amount of stored water and aquifer water levels, primarily through changes in recharge and groundwater pumping rates. Groundwater pumping rates are a human decision that is ultimately based on water needs (e.g., crop irrigation requirements), the availability and costs of alternative water sources, user preferences and financial capabilities, and groundwater governance. For example, a transition to warmer and drier conditions would be expected to increase plant irrigation requirements. Farmers would likely elect to pump more groundwater so long as their wells did not go dry or salty, they can afford the additional pumping costs, and there are no enforced regulatory restrictions against the additional pumping.

Groundwater recharge is impacted by climate change mainly through available water (precipitation—evapotranspiration; P-ET). Recharge rates tend to be broadly corelated to available water with greater rates occurring with increasing annual precipitation rates. However, groundwater recharge rates in some situations may be controlled to a greater degree by the seasonality of precipitation, and the intensity and duration of individual rainfall events. Recharge rates also depend on soil and surficial rock properties (e.g., vertical hydraulic conductivity and the degree of development of secondary porosity), which control the rates of infiltration and percolation through the vadose zone. Land use and land cover (LULC) changes, either natural or anthropogenic, can also impact recharge rates. In the case of confined and semi-confined aquifers, climate changes in distant upland recharge areas may be of much greater importance than climate changes in basinal extraction areas.

With respect to surface water systems, important issues are future average and variability in stream flows, such as the probabilities of flows in the future falling below critical threshold values and disrupting water supplies. Similarly, key issues

R. Maliva, *Climate Change and Groundwater: Planning and Adaptations for a Changing and Uncertain Future*, Springer Hydrogeology,
https://doi.org/10.1007/978-3-030-66813-6_5

for groundwater users are future recharge rates and the sustainability of supplies. Sustainable use (or safe yield) is broadly defined as the amount of water that can be extracted without causing unacceptable impacts whether they be declines in aquifer water levels and water quality, impacts to surface water users, or environmental impacts (e.g., dehydration of wetlands and declines in spring flows).

Predicting the potential impacts of climate change on aquifer recharge and groundwater levels requires projections of local changes in temperature and precipitation, which are ultimately derived from large-scale general circulation (or global climate) models (GCMs). Downscaling is required to improve the spatial resolution of GCM outputs, particularly changes in the mean annual and temporal distribution of precipitation. A main modeling challenge is accurately simulating the precipitation-recharge relationship, which requires consideration of how and where recharge occurs across a landscape (e.g., direct versus indirect, matrix or diffuse versus focused).

5.2 Modeling Approaches

Modeling of the impacts of climate change on groundwater (and water resources in general) has often been performed using what is referred to as the "top-down" approach starting with greenhouse gas (GHG) emissions scenarios and GCMs (Fig. 5.1). The GCM simulation results are then downscaled to provide greater geographic spatial resolution. Precipitation and temperature data are next inputted into a hydrologic model that partitions precipitation between evapotranspiration (ET), runoff, and aquifer recharge. Finally, recharge data are inputted into a groundwater flow model, which is used to evaluate changes in aquifer water levels and discharge and, in the case of solute-transport modeling, aquifer salinity changes. Integrated surface water/groundwater model codes are available that combine the hydrologic and groundwater modeling steps.

The top-down approach has several major limitations. First is that it is very complex and time consuming, and thus expensive to perform on the applied level (as opposed to as an academic research study). Typically, existing GCM simulation results are utilized (now commonly from the CMIP5 database), but modelers need expertise in working with GCM outputs and performing downscaling and hydrologic modeling. An inherent limitation of the top-down modeling approach is that it is subject to a "cascade of uncertainties," which flows from uncertainties in each step of the modeling process (Foley 2010; Falloon et al. 2014). There is great uncertainty in which emissions scenario (and thus future atmospheric GHG concentrations) will come to pass, in the GCM simulations of the climatic responses to given emissions scenarios, and then in the downscaling of the GCM data to local scales and in local hydrological and groundwater flow modeling.

Several studies examined the relative contributions of each element of the cascade of uncertainty to the total uncertainty in projected recharge rates. Crosbie et al. (2011) in an investigation of the impacts of climate change on groundwater in

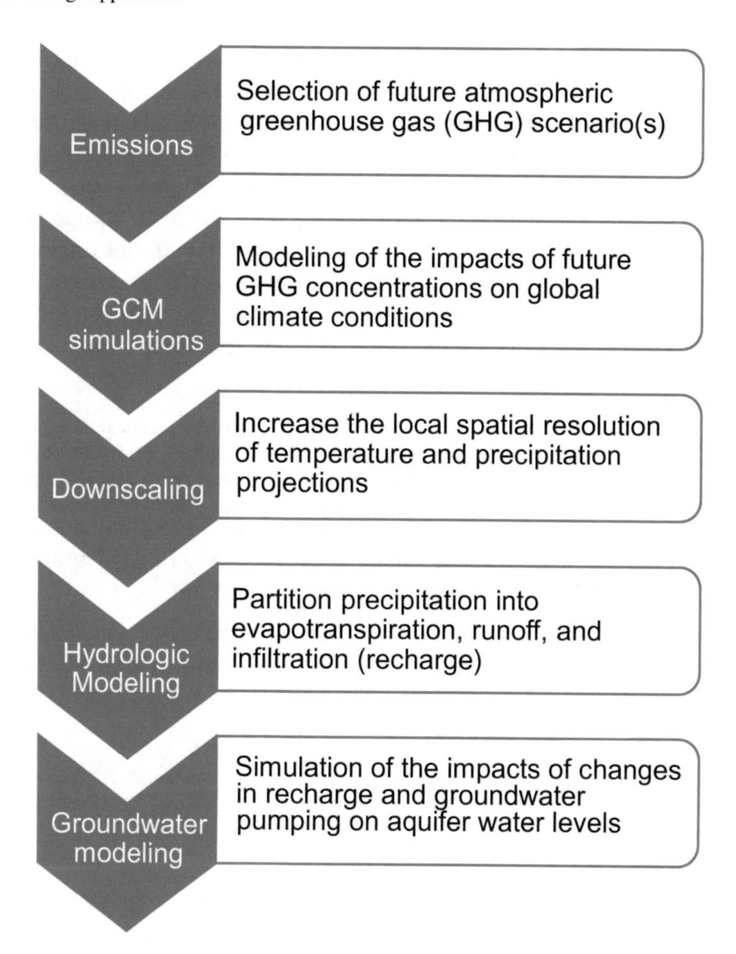

Fig. 5.1 Flow diagram of top-down modeling approach

Australia showed that GCMs account for the largest uncertainty in recharge projections (a median range between the highest and lowest GCM of 53% of historical recharge), followed by downscaling methods (a median range of 44% of historical recharge), and then hydrologic (recharge) modeling (a median range of 24% of historical recharge). The impacts of the choice of emissions scenario were not evaluated.

Allen et al. (2010) investigated the effects of the choice of GCM on estimated recharge in the Abbotsford-Sumas aquifer located in British Columbia (Canada) and Washington State (U.S.A.). Four GCMs were considered, the results from which did not agree in the predicted magnitude or direction of change of precipitation (and consequently recharge). The modeled difference in recharge in the 2080s relative to the base case using the outputs from the four GCMs ranged from −1.5 to +23.2%.

Bosshard et al. (2013) quantitatively evaluated the impacts of uncertainties in projections of climate change on river streamflow in the Alpine Rhine (Eastern Switzerland). Climate models were found to generally be the dominant uncertainty source in the summer and autumn. Uncertainties in the hydrological modeling and the statistical post processing gain importance and even partly dominate in the winter and spring. The individual uncertainties from the three components were found to not be additive. The associated interactions among the climate models, the statistical post-processing scheme, and the hydrologic model accounted for about 5–40% of the total ensemble uncertainty.

Downscaling and hydrologic modeling are data intensive. Wilby and Dessai (2010, p 180) observed that

> It is becoming apparent, however, that downscaling also has serious practical limitations, especially where the meteorological data needed for model calibration may be of dubious quality or patchy, the links between regional and local climate are poorly understood or resolved, and where technical capacity is not in place. Another concern is that high-resolution downscaling can be misconstrued as accurate downscaling (Dessai et al. 2009). In other words, our ability to downscale to finer time and space scales does not imply that our confidence is any greater in the resulting scenarios.

Additional limitations of the top-down approach are that the underlying GCMs do not fully sample the full range of possible climate changes (Brown 2011; Brown and Wilby 2012) nor provide insights into what will likely happen. Brown and Wilby (2012, p. 401) observed that

> it is difficult to communicate exactly what climate projections mean from a decision standpoint—they simulate what might happen under some conditions but do not preclude other outcomes. In fact, climate analysts are often reluctant to say that one future is more or less likely than others.

Despite its limitations, the top-down approach has been invaluable in providing insights on how global climate change could impact surface and groundwater resources. However, it is an open question as to the role of GCM-based, top-down modeling in water management. Specifically, is the top-down modeling approach a worthwhile investment as a decision-making tool for a local water supplier or perhaps should it be used primarily as a part of larger-scale planning and be performed through a government-academia collaboration?

An alternative to the GCM-based, top-down approach is what has been referred to as the "bottom-up," "scenario-neutral" (Prudhomme et al. 2010), "decision-scaling" (Brown 2011; Brown et al. 2012; Brown and Wilby 2012), and sensitivity analysis approach (Fig. 5.2). The basic steps in these assessment strategies are (Brown and Wilby 2012):

- identification of the problem, including defining objectives and performance measures
- performance of a stress test to evaluate the performance of the system under a wide range of nonclimatic and climate variability and change
- evaluation of risks using climate information including model projections.

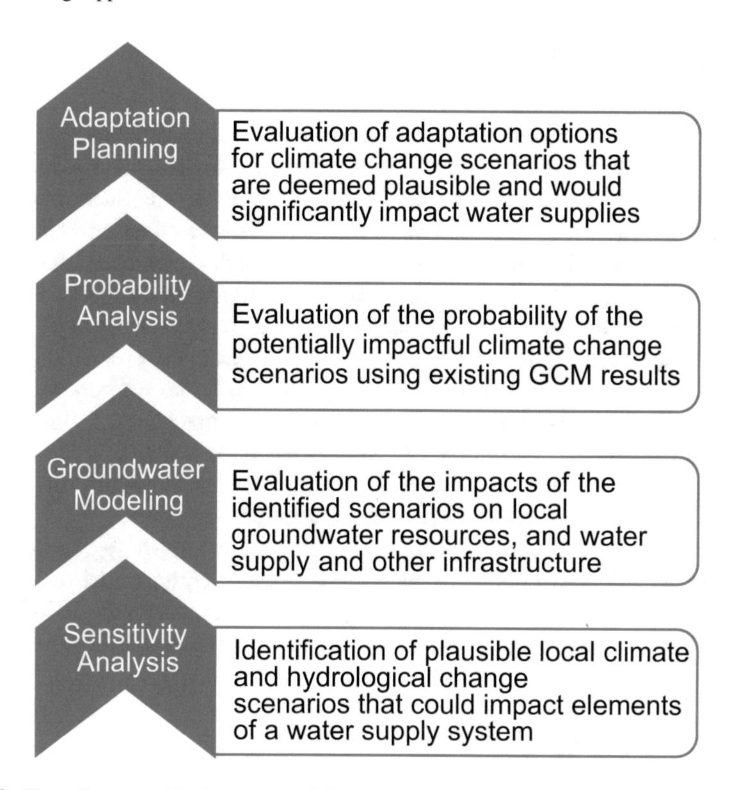

Fig. 5.2 Flow diagram of bottom-up modeling approach

Bottom up approaches begin with an assessment of the vulnerabilities or risks of the system of interest related to climate change and sea level rise (Brown 2011). The vulnerability assessment could involve an evaluation of historical climate extremes and their impacts on the system of interest. The next step is a climate sensitivity analysis, which involves some type of modeling of the response of the system to a range of possible future climate conditions (Brown 2011). If the performance of the system was determined to be sensitive to (i.e., the system would perform unacceptably in) an identified envelope of climate conditions, then GCM projections are considered to evaluate the probability of the identified adverse climate conditions (Brown 2011).

Follows is a review of basic concepts involved in the top-down and bottom-up approaches to assessing the impacts of climate change on groundwater recharge and aquifer water levels. Top-down type modeling is a specialized discipline and requires specific expertise that many hydrogeologists and groundwater modelers may not possess and, therefore, requires a team approach. Examples are given of a range of different modeling approaches that have been employed to assess the impacts of climate change on groundwater.

5.3 Bias Correction

GCMs are tested by performing simulations using historical GHG concentration data and comparing the results to concurrent historical climate data. GCMs often exhibit systematic errors (biases) relative to historical observations. Such errors could compromise the accuracy of predictive simulations of future climate conditions. Bias correction is described simply as correcting the raw model output for a future period using the differences (Δ) between historical observational data for a past reference period and the model output for the reference period (Navarro-Racines et al. 2015). A GCM or regional climate model (RCM) is run for a past reference period and local correction functions are calculated and then applied to future projections. The correction functions can be a simple linear function or more complex functions that capture the distribution of the data (mean, variance, and shape). A key issue in climate modeling, in general, is preserving extreme values. Bias-correction assumes that biases in both the historical and future simulations are equal.

Bias-correction is conceptually simple, but the more complex approaches require expertise in statistics. A number of papers have been recently published addressing bias-correction in climate change modeling (e.g., Mauer and Pierce 2014; Cannon et al. 2015; Navarro-Racines et al. 2015; Heo et al. 2019).

5.4 Downscaling

The typical horizontal resolution (horizontal grid spacing) for current GCMs is approximately 1–2° for the atmospheric component and around 1° for the ocean (Flato et al. 2013). GCMs do not accurately predict local climate, but the internal consistency of these physically based models provides most likely estimates of ratios and differences from historical to predictive scenarios (Loáiciga 2003; Scibek and Allen 2006). A key technical issue is linking the results from large-scale GCMs to much smaller-scale watershed or aquifer models.

Downscaling has been defined as "the general name for a procedure to take information known at large scales to make predictions at local scales" (UCAR n.d.). Downscaling techniques used for hydrological studies were reviewed by Fowler et al. (2007) and others. The simplest method used by hydrologists is to simply interpolate GCM outputs on to a finer grid that is more appropriate for the study (Prudhomme et al. 2002). This method retains the coarse spatial pattern of the GCM but may have a low accuracy on the local level. Three main more-refined approaches are now commonly used for downscaling GCM climate projections to more local scales: change or scaling factors, dynamical, and statistical.

All downscaling methods retain uncertainties associated with the underlying GCM, particularly the emissions scenario used and the accuracy of GCM projections, i.e., the ability of models to accurately simulate the climate response of the oceanic, land, and atmospheric system to changes in GHGs and other climate relevant substances.

5.4.1 Scaling and Change Factors

The scaling (or change) factor method is a relatively simple technique for downscaling GCM and RCM data. The scaling factor method is based on either the ratio of or the difference between GCM or RCM modelled future conditions and current conditions. In the example provided by Loáiciga (2003), the scaling factor for precipitation is the GCM derived ratio of the precipitation rate with twice the atmospheric CO_2 concentration ($2 \times CO_2$) to the rate under current conditions ($1 \times CO_2$) in each GCM grid cell. For locations within each GCM grid cell, future precipitation at a given location and time period (e.g., seasons or months) is the product of the historical precipitation at the location and the GCM grid cell-specific scaling factor (Loáiciga 2003).

Temperature scaling was performed using the difference between GCM/RCM predicted temperatures according to the equation (Loáiciga 2003):

$$T_{2\times CO_2 Scenario} = \{T_{2\times CO_2} - T_{1\times CO_2}\} + T_{Historical}$$

where $T_{1\times CO2}$ and $T_{2\times CO2}$ are GCM modeled temperatures, $T_{Historial}$ is the historical temperature at a given location, and $T_{2\times CO2Scenario}$ is the future temperature at the location. The scaling factors are applied to time series of weather data to obtain equivalent time series of data under future climate conditions for modeling purposes. Historical precipitation and temperature data from weather stations in the United States were extrapolated (kriged) at 0.5° latitude by 0.5° longitude resolution and scaling factors were generated from several GCMs nested within the National Center for Atmospheric Research (NCAR) RCM, a part of the NCAR Vegetation Ecosystem Modeling and Analysis Project (VEMAP; Kittel et al. 1995; Loáiciga 2003).

Loáiciga (2003) estimated indirect recharge (R) in streams using the equation

$$R_{2\times CO_2 Scenario} = \frac{Q_{2\times CO_2}}{Q_{1\times CO_2}} \cdot R_{Historical}$$

where $Q_{1\times CO2}$ and $Q_{2\times CO2}$ are GCM/RCM estimates of stream flow under different atmospheric CO_2 concentrations. A limitation of the scaling factor method is that it assumes that the future spatial and temporal distribution of precipitation (and other climate variable distributions) will be similar to the historic distributions (Serrat-Capdevila et al. 2007).

5.4.2 Dynamical Downscaling

Dynamical downscaling involves creating a finer-resolution regional climate model (RCM) using the coarser output from GCMs as initial and boundary conditions

(Loáiciga 2003). Development of RCMs is within the realm of climate modeling experts. Independent development of RGMs is typically beyond the technical and financial resources of most water decision makers. Instead, RCMs prepared by climate modeling groups may be accessed for local predictions. For example, the US Geological Survey Regional Climate Change Viewer provides access to a dataset comprised of three GCMs that have been dynamically downscaled in the RegCM3 regional climate model (USGS n.d.). The simulations are for the IPCC 4th Assessment A2 scenario. Climate change projections from RCMs will depend on the GCM and emissions scenario used.

The RCM output are then input into hydrologic models. As described by Loáiciga (2003), the key output variables of the RCMs, such as precipitation, surface-air temperature, ground-level radiant-energy fluxes, water-vapor pressure, and wind speed, become the forcing input variables to hydrologic models, which are then used to calculate the dependent hydrologic variables of primary interest (e.g., streamflow and groundwater recharge).

5.4.3 Statistical Downscaling

Statistical downscaling includes a range of methods that are based on the concept that regional climate is related to the large-scale atmospheric state, expressed as a function between the large-scale atmospheric variables (predictors) and local or regional climate variables (predictands; Holman 2006). Statistical downscaling is a two-step process (UCAR n.d.). First is the determination of the statistical relationships between local climate variables (e.g., surface air temperature and precipitation) and large-scale predictors (e.g., pressure fields). Second is the application of such statistical relationships to the output of CGM experiments to simulate local climate characteristics in the future (UCAR n.d.). The objective is to capture changes in both future mean values and variation in conditions, such as extreme events.

A basic requirement of statistical downscaling is a historical dataset from which to derive the correlations between the large-scale climate field and the local climatological data (Yates and Miller 2011). Various types of statistical methods are used to determine the relationship including regression analysis and neural network techniques. Statistical methods assume stable relationships between the large-scale predictor variables (e.g., indices of circulation patterns) and the climate variables being predicted at the local scale (e.g., daily precipitation amounts; Yates and Miller 2011). Care is required in selecting variables and in selecting and developing the statistical model to ensure that the relationship is likely to remain physically sensible in a warmer climate (Yates and Miller 2011).

Statistical downscaling methods are more commonly used for climate change modeling than dynamic downscaling because they are much less technically and computationally demanding. Statistical methods can be difficult to apply because they require large historical data sets and considerable expertise to perform the statistical analyses.

5.4.4 *Stochastic Weather Generators*

Scaling factor and statistical downscaling methods require long continuous time series of data on precipitation, temperature, and other climate variables. Such continuous datasets are often unavailable. Stochastic weather generators are used to infill missing data or produce indefinitely long synthetic times series of data. Daily weather simulators are most commonly used both because of the wide availability of weather data on this timescale and hydrologic models frequently require daily weather inputs (Wilks and Wilby 1999).

The synthetic times series of data are generated to match both the historic mean and variability of the parameters of interest. Precipitation statistics evaluated may include monthly mean precipitation amount, standard deviation of monthly precipitation, mean length of the dry series (consecutive dry days), mean length of the wet series (consecutive wet days), and mean number of wet days (Semenov and Barrow 1997). Extreme weather events can profoundly affect water availability and demands compared to changes in mean values, so it is important that the synthetic time series adequately capture these events.

Semenov and Barrow (1997) clarified that stochastic weather generators are not predictive tools that can be used in weather forecasting, but rather are a means of generating time-series of synthetic weather that are statistically identical to observations. Statistical "identity" depends on the number of statistics used for the comparison (Semenov and Barrow 1997). Time series of weather are generated, and their statistics are compared with the observed historical data. Once synthetic times series are generated that satisfactory match (i.e., are calibrated) to historical climate statistics, then GCM-derived changes (e.g., in mean and variability of parameters) are applied to the parameters obtained using the weather generator for each site to generate projected future time series (Semenov and Barrow 1997). The synthetic time series can be perturbed using either GCM projections directly or after downscaling.

5.5 Hydrologic Modeling

Downscaling processes and stochastic weather generators can provide time series of projected climate data. The next step in groundwater investigations involves hydrologic modeling that partitions precipitation into runoff, infiltration and recharge, and evapotranspiration (ET). A key issue is transferring of recharge rates obtained from a hydrologic model into a groundwater flow model. The impacts of changes in precipitation can be more seamlessly simulated using integrated surface/

water groundwater models (e.g., MIKE SHE; DHI Software 2007). MIKE SHE allows for the simulation of the major processes in the hydrologic cycle and includes process models for evapotranspiration, overland flow, unsaturated flow, and groundwater flow (DHI Software 2007).

Several U.S. governmental agencies independently developed their own distributed-parameter hydrologic modeling programs, such as the U.S. Army Corps of Engineers (2009) HEC-HMS program, the U.S. Department of Agriculture SWAT program (Neitsch et al. 2005), the U.S. Geological Survey Precipitation-Runoff Modeling System (PRMS; Leavesley et al. 1983, 2005), and the U.S. Environmental Protection Agency (USEPA) Hydrologic Evaluation of Landfill Performance (HELP) model (USEPA 1994). The HELP program was developed to design and evaluate landfills but has capabilities to simulate vadose zone processes that have been taken advantage of in climate change and groundwater modeling. Transmission losses or recharge simulated using the hydrologic models is input as recharge into groundwater flow models, such as the MODFLOW family of codes. To adequately simulate storm events, hydrologic models need to have short time steps (1 day or less), whereas groundwater flow models are usually run to examine long-term changes in water levels and often have time steps of a month or longer. Hence, output from hydrologic models often requires processing to match the time steps of numerical groundwater flow models.

It must be emphasized that accurate representation of the rainfall-recharge relationship is data intensive. Barron et al. (2011), in a study of climate change impacts on recharge in Australia, pointed out some key difficulties. With respect to diffuse recharge, prior knowledge of the relationship and elasticity between rainfall and recharge is required to adequately incorporate climate change into a groundwater model. A simplistic model input describing net recharge as a fixed percentage of total rainfall would underestimate changes in recharge (Barron et al. 2011). With respect to focused (localized) recharge, the specifics of stream and aquifer connectivity need to be addressed along with its changes under future climate conditions (Barron et al. 2011).

5.6 Aquifer Heterogeneity and Modeling Results

Landscapes and underlying aquifers have varying degrees of heterogeneity in physical, biological, and hydraulic properties. Numerical models are inherently gross simplifications of natural systems. Modeling recharge is particularly complex because of spatial variability in topography, land cover and land use, and soil properties. Local recharge may be some combination of direct (diffuse) and indirect (focused) recharge rather than being exclusively one or the other.

Heterogeneity occurs at multiple scales in aquifer systems ranging from microscopic (i.e., the scale of individual pores and sand grains) to gigascopic (regional scale; Maliva 2016). A key part of aquifer characterization and groundwater modeling is evaluating and capturing heterogeneity on the spatial scale relevant to the study at hand. Climate change impact studies tend to focus on large-scale impacts, such as overall aquifer water levels and safe yield. Hence, a coarse-scale focus on land surface and aquifer heterogeneity is appropriate. Nevertheless, conceptual models need to correctly capture the primary type and location of recharge. Key distinctions are whether recharge occurs mainly throughout a groundwater basin or predominantly in higher elevations around the basin or in more distal areas (e.g., mountain front recharge) and whether recharge is predominantly direct or indirect. In arid regions where recharge is predominantly indirect, modeling should focus on how climate change impacts runoff into ephemeral channels or dry playas, which are the primary loci of recharge.

5.7 Bottom-Up (Decision-Scaling, Sensitivity Analysis) Approach

The bottom-up (aka decision-scaling and sensitivity analysis) approach is based on sensitivity analyses, which are essentially evaluations of the degree to which changes in model inputs impact model outputs. A model is said to be sensitive to a given parameter if changes in the value of the parameter cause correspondingly large changes in the model output. With respect to climate modeling, Brown (2011) referred to the sensitivity or responsiveness of a system to climate change conditions in terms of a "climate response function." As described by Brown (2011, p. 5)

> To develop the climate response function, the climate conditions are systematically varied to diagnose how such climate changes affect the system. For example, the mean climate (e.g., temperature and precipitation averages) is repeatedly varied over a plausible range of possible climate changes, say between +20 and 20% (IPCC regional reports are a good source for choosing the possible range) and for each "scenario" the climate response function is used to calculate the values of the performance metrics. If a threshold for acceptable performance has been established, the climate conditions under which such a threshold is not met are noted.

For example, a vulnerability assessment of aquifers used for local water supply might evaluate the effects of future decreases in precipitation on aquifer recharge and, in turn, aquifer water levels or saline water intrusion. The assessment could be performed by systematically applying a range of recharge rates to an existing or newly constructed numerical groundwater model of the study area. In many regions of developed countries that are reliant on groundwater, numerical groundwater models have already been developed by water utilities, regional suppliers, environmental or water management regulatory agencies, geological surveys, or academia. If it is determined, for example, that reductions in recharge rates of 20% of greater would cause unacceptable impacts, then projections of future precipitation changes for the study area from GCMs (e.g., from the CMIP5 database) or existing RCMs would be reviewed to assess whether such changes are plausible and, if so, to estimate their probability. Brown (2011, p. 7) cautioned that these "estimates are best considered subjective probabilities." Professional judgement is inherently an important part of the vulnerability assessment process.

Cromwell et al. (2007, p. 12) explained that

> The central idea of this approach is that utilities can work with their own water resources planning models to assess the vulnerability of their 20–50 year supply plans to climate change. Based on the general findings of climate change research, utilities can identify the likely cause-effect pathways that could prove troublesome.

and that

> The "bottom-up" analysis enables a utility to test the robustness of current plans to upsets from changes in key climate-related variables. Once the thresholds or tipping points of a utility's plans have been identified in this manner (using familiar models in which a utility has relatively good confidence), it is then possible to turn to the climate scientists and ask how plausible such breaking point scenarios seem in light of the results of broader research with the various GCMs.

From an applied perspective, the "bottom-up" approach is much more practicable than the "top-down" approach because of its much lesser technical expertise requirements and effort involved (and thus time and cost). A utility's own surface water and groundwater resource modeling tools can be applied to examine the effects of extreme scenarios, involving such conditions as decreased surface water flows, decreased recharge, increased evaporative losses, and seasonal shifts in supplies and demands (Cromwell et al. 2007).

5.8 Published Modeling Studies

Published modeling studies of the impacts of climate change on groundwater recharge provide insights on the range of available approaches, limitations of modeling strategies, and the potential extent of climate impacts. Studies have examined the sources and magnitude of uncertainty in climate change projections by comparing results obtained using different modeling methods. This section provides a summary of a selected suite of published studies, noting the approaches used concerning selection of emissions scenarios, choice of GCM(s), and the downscaling and hydrologic modeling methods employed.

5.8.1 Edwards Aquifer, Texas

Loáiciga et al. (2000) investigated the impacts of climate change and groundwater pumping on the Edwards Balcones Fault Zone (EBFZ) aquifer in Texas. Groundwater use in the aquifer is regulated to maintain minimum flows in the Comal and Sam Marcos springs and related spring runs, lakes, rivers, and caves that are host to a number of endangered species. Recharge of the Edwards Aquifer occurs primarily from stream transmission losses in the "recharge area" where the aquifer is unconfined. Loáiciga et al. (2000) used the Geophysical Fluid Dynamics Laboratory of NASA GFDL R30 GCM and monthly average scaling factors for temperature, precipitation, and stream flow to quantify the impacts of climate change using 1 × CO_2 (1990; 355 ppm) and 2 × CO_2 GCM-simulated temperature and precipitation. Recharge was calculated using the same scaling factor as stream flow.

Modeled changes in recharge were input into the EBFZ aquifer two-dimensional, finite-difference, groundwater simulation program (GWSIM IV), which is a modification of an earlier model developed by the Texas Water Development Board to simulate groundwater flow and springflows in the EBFZ aquifer. The simulation results indicate that 2 × CO_2 climatic conditions could exacerbate negative impacts and water shortages in the EBFZ aquifer even if pumping is maintained at its present average level. The historical evidence and the results of the modeling indicate that without proper consideration of variations in aquifer recharge and sound pumping strategies, the water resources of the EBFZ aquifer could be severely impacted under a warmer climate (Loáiciga et al. 2000).

5.8.2 Rhenish Massif, Germany

Eckhardt and Ulbrich (2003) simulated the effects of changes in temperature and precipitation on groundwater recharge in a central European low mountain range (Rhenish Massif, Germany) catchment. The study considered the effects of elevated ambient CO_2 concentrations on stomatal conductance and leaf area, and thus evapotranspiration. The output from five GCMs for the B1-low and A2-high emissions scenarios were utilized with projected year 2070–2099 conditions compared to a 1961–1990 baseline period. The scaling factor method was used; the observed time-series of daily climate for a baseline period were adjusted by the estimated differences between GCM simulations of current and future climate (Eckhardt and Ulbrich 2003). A revised version of the Soil and Water Assessment Tool (SWAT) was developed to better simulate climate change impacts to recharge and streamflow.

The modeling results suggest that warming would result in a smaller proportion of the winter precipitation falling as snow and, therefore, the spring snowmelt peak would be reduced while the flood risk in winter would probably increase. In the summer, mean monthly groundwater recharge and streamflow are projected to be reduced by up to 50%, which would potentially lead to problems concerning water quality, groundwater withdrawals, and hydropower generation (Eckhardt and Ulbrich 2003).

5.8.3 Southern High Plains, New Mexico and Texas

Ng et al. (2010) employed a probabilistic approach to model groundwater recharge in the Southern High Plains of the western United States, which is a low-relief semiarid region where annual recharge rates are typically small and most recharge occurs episodically. The ensemble forecasting procedure used meteorological time series realizations generated with a stochastic weather generator (LARS-WG) for a range of GCM predictions. Recharge in the study site was simulated using a one-dimensional vertical Richards-based model. As described by Ng et al. (2010, p. 5)

> Ensemble forecasting is a form of Monte Carlo simulation that reveals the range of possible outcomes that could occur in situations where uncertainty is significant. The basic idea is to perform a large number of model simulations, each based on a different sample (or realization) from the physically probable distribution of uncertain inputs. If the input samples are equiprobable, the results of these simulations can be viewed as equally likely alternative futures. The ensemble of simulated realizations can be used to construct probability densities and various statistical measures of variability.

Ng et al. (2010) used LARS-WG to generate daily time series realizations of precipitation, maximum and minimum air temperature (T_{max} and T_{min}), and solar radiation, which were calibrated to historical data. Monthly change factors derived from GCM outputs for future and base case periods were applied to the LARS-WG data to generate time series compatible with a changed climate.

Predicted changes in average recharge for most climate alternatives (spanning −75 to +35%) were found to be larger than the corresponding changes in average precipitation (spanning −25 to +20%), which suggests that amplification of climate change impacts may occur in groundwater systems (Ng et al. 2010). The change in the temporal distribution of precipitation explains most of the variability in predictions of total recharge (Ng et al. 2010).

5.8.4 High Plains Aquifer, Western United States

Crosbie et al. (2013) used 16 global climate models (GCMs) and three global warming scenarios to investigate changes in diffuse groundwater recharge rates in the High Plains region of the western United States. Their methodology consisted of: (1) point-scale numerical modeling of recharge under historical and future climates, (2) upscaling the point results to the entire High Plains Aquifer, and (3) aggregating the results from the 48 future climate variants down to three to facilitate effective communication (Crosbie et al. 2013).

As described by Crosbie et al. (2013), point-scale modeling was conducted at 17 sites (stations) across the High Plains selected to cover the rainfall gradient. The historical climate (1982–2011) was used as a baseline that was assumed to be representative of the 1990 climate. For each of the 17 points, daily rainfall and minimum, maximum, and dew point temperatures were obtained from the National Climate Data Center. As described by Crosbie et al. (2013, p. 3939):

> Archived monthly simulations from the 16 GCMs were analyzed to estimate the change in rainfall, temperature, humidity, and solar radiation per degree of global warming on a seasonal basis. The percent changes in the climate variables per degree of global warming for each of the four seasons from the 16 GCMs were then multiplied by the three levels of global warming to obtain 48 sets of seasonal scaling factors. These seasonal scaling factors were then used to scale the historical daily climate data from 1982 to 2011 to obtain 48 future climate variants, each with 30 years of daily climate data.

The temporal sequencing of rainfall was retained from the historical time series, but changes in the daily rainfall intensity were simulated by scaling different rainfall amounts differently (Crosbie et al. 2013).

Groundwater recharge, assumed to be drainage below the base of a 4 m soil column, was modeled using a slightly modified version of the WAVES model (Zhang and Dawes 1998). WAVES is a soil-vegetation-atmosphere-transfer model that requires three main data sets: climate, soils, and vegetation. The modeling results for the North High Plains was a median recharge scaling factor (RSF) of 1.08 (i.e., an 8% increase in recharge) with a range between the dry and wet scenarios of 0.76 and 1.32, respectively. For the Central High Plains, the median modeled RSF was 0.97 (range 0.63–1.23) and for the South High Plains the median RSP was 0.90 (range 0.50–1.25; Crosbie et al. 2013). The key conclusion of the study is that projections of future recharge for the High Plains encompass both increases and decreases in recharge and that management responses will need to be flexible enough to accommodate the uncertainty in recharge projections (Crosbie et al. 2013).

5.8.5 Serral-Salinas Aquifer, Southeastern Spain

Pulido-Velazquez et al. (2015) modeled the impacts of climate change on groundwater recharge in the Serral-Salinas aquifer, southeastern Spain. RCMs from EU projects were first evaluated for the goodness of their calibrations for the period 1961–1990 in the study area. To generate future time series for precipitation and temperature in the study area, the "delta-change" approach was used to modify the mean and standard deviation of the original historical climate series according to the RCM simulations. RCM results were used only for estimating the relative change expected rather than to provide absolute values. The A2 and A1B emissions scenarios were used. The impacts of climate change (precipitation and temperature) on groundwater recharge was quantified using a hydrological model, developed by the University of La Coruña (Spain), which solves the water balance equations in the soil, the unsaturated zone, and the aquifer.

Key results of the Pulido-Velazquez et al. (2015) study are that there were significant differences in simulated changes in recharge rates depending on the RCM used and whether both means and standard deviations are altered when generating rainfall and temperature series rather than only means. For the period 2071–2100, the ensemble of predictions estimated a reduction in mean annual recharge (with respect to the historical period 1961–1990) of 14.0% (9.9 mm) for scenario A2 and of 57.7% (40.6 mm) for scenario A1B. Lower recharge values were obtained when only the means were changed (decreases of 21.8% for scenario A2 and 65.8% for scenario A1B; Pulido-Velazquez et al. 2015).

5.8.6 *Galicia-Costa, Spain*

Raposo et al. (2013) assessed the potential impacts of climate change on groundwater recharge in fissured aquifers in Galicia-Costa, Spain. The aquifers are expected to be very sensitive to climate change due to their low storage capacity and the short residence-time of groundwater. Climate projections from two GCMs and eight different RCMs were used for the assessment and two climate-change scenarios were evaluated (A2 and B2). The Soil and Water Assessment Tool (SWAT) model was used to simulate recharge within the watersheds. In calculating evapotranspiration (ET) rates, SWAT takes into account variations of radiation-use efficiency, plant growth, and plant transpiration caused by changes in atmospheric CO_2 concentrations (Raposo et al. 2013).

The SWAT models were calibrated to historical stream flows using a daily time-step in four representative catchments in the district. Recharge was calculated as the residual of the water balance. Water-table elevation data were used for model validation. The changes in precipitation and temperature from existing GCM and RCM regional projections were used without further area-specific downscaling. A control scenario for the period 1961–1990 and two warming scenarios for the period 2071–2100 were run. The model results show that the projected annual precipitation decrease will be reflected in a smaller decrease of annual groundwater recharge of 12.68% for the A2 scenario and 6.03% for the B2 scenario. Only six models predicted slight increases in recharge, whereas 50 models predicted decreases (Raposo et al. 2013). The temporal pattern of recharge may change, with recharge mainly concentrated in the winter season and dramatically decreasing in the spring–autumn seasons (Raposo et al. 2013). The SWAT modeling results are sensitive to the CO_2 influence on plant physiology, which if neglected, according to these modeling results, may lead to an overestimation of the recharge decrease by 11.92–9.82%, depending on the scenario (Raposo et al. 2013). Raposo et al. (2013) concluded that projected annual precipitation decreases will be reflected in smaller decreases of annual groundwater recharge, partly due to the greater stomatal resistance of plants in response to increased CO_2 concentration and that the CO_2 influence on plant physiology must always be considered since its neglect was reported to show a greater average decrease in annual recharge (Raposo et al. 2013).

5.8.7 West Bengal, India

Sahoo et al. (2018) investigated the impacts of climate-induced changes in water demand and land use and land cover (LULC) on the hydrological regime of the Gandherswari River basin, West Bengal, India. Data from the HadCM3 GCM for the A2 and B2 emissions scenarios were used for the climate change scenarios. Statistical downscaling was performed using the Statistical Downscaling Model (SDSM) decision support system, which can produce high-resolution monthly climate information using synthetic weather generator methods. SDSM involves a five-step process: (1) selection of predictor variables, (2) calibration of the model (using observed predictors), (3) validation of the model, (4) generation of future scenarios (using climate model predictors), and (5) analysis of outputs (Wilby and Dawson 2013).

Dynamic conversion of land use and its hydrological effects were simulated using the Dyna-CLUE model (Verburg et al. 2002), which considers land use demands, location suitability, neighborhood suitability, spatial restrictions, and conversion parameters. The Soil and Water Assessment Tool (SWAT) was used to estimate future streamflow and spatiotemporally distributed groundwater recharge. The modeling results show increased recharge in the monsoon season and decreased recharge in the non-monsoon season for the years 2030 and 2050. A reverse trend was obtained for the year 2080, and an overall increase in groundwater recharge was simulated for all the years (Sahoo et al. 2018).

5.8.8 Grand Forks, South Central British Columbia, Canada

Scibek and Allen (2006) developed a methodology for linking climate and groundwater models. An unconfined aquifer near Grand Forks, south central British Columbia, Canada, was used to test the methodology. Climate conditions for the modeled present and future conditions were obtained from the CGCM1 model (Canadian Global Coupled Model) for the IPCC IS92a greenhouse gas plus aerosol (GHG + A1) scenario. Two downscaling methods were utilized: the statistical downscaling model (SDSM) and principal component K-nn. A basic problem encountered is the inability of the very coarse-scale CGCM1 model to adequately model precipitation in areas such as Grand Forks where local precipitation occurs as convective events in the summer and there are valley-mountain rain shadow effects. Precipitation was reported to be underestimated by roughly 40% during the summer even after downscaling with an adequately calibrated model (Scibek and Allen 2006). Neither the SDSM or PCA k-mm methods performed very well, but a better calibration was obtained with the SDSM. The approach used to simulate climate change was to compute change factors from the SDSM for precipitation and temperature.

The change factors were applied to daily time series obtained using a stochastic weather generator (LARS-WG), which was calibrated to current conditions. Comparison of the results from SDSM and PCA k-nn downscaling revealed different directions and magnitudes of predicted changes in precipitation under future climate scenarios, which demonstrates the uncertainty associated with the downscaling process (Scibek and Allen 2006). Spatially distributed and temporally varying recharge zones were generated using a GIS linked to the one-dimensional USEPA's HELP model. Recharge rates calculated using HELP were applied to a groundwater flow model using the MODFLOW code.

As Scibek and Allen (2006) discussed in the introduction to their paper, their study was motivated by previous work by Allen et al. (2004) in which the sensitivity of the aquifer to future predicted climate change was assessed using a steady-state flow model, with a more simplistic aquifer and river representation, for two extreme scenarios of climate change. Scibek and Allen (2006) noted that their overall results were consistent with those of Allen et al. (2004) in which a simpler approach was used. Scibek and Allen (2006, p. 17) concluded that the "consistency of the results also suggests that even simple approaches to quantifying the impacts of climate change on groundwater have value where detailed data are lacking."

The modeling results indicate that future climate conditions will result in more recharge to the unconfined aquifer during the spring to summer period. The overall effect of recharge on the aquifer water balance was found to be small because of the dominant river-aquifer interactions and river water recharge (Scibek and Allen 2006).

5.8.9 Mediterranean Coastal Aquifers

Stigter et al. (2014) performed a comparative study of climate change impacts on coastal aquifers in three Mediterranean areas. Output from three regional climate models (CNRM-RM5.1, C4IRCA3, and ICTP-REGCM3) available from the ENSEMBLES project (Van Der Linden and Mitchell 2009) for the CO_2 emission scenario A1B were utilized. Temperature (T) and precipitation (P) data were downloaded for selected historical reference and two future climate periods: 2020–2050 and 2069–2099. Bias correction was applied to the data using two different delta approaches to compare their applicability. The corrections involved either (1) calculation of anomalies where trends in P and T in the modeled data were calculated as monthly delta values between the reference and future climate normal periods, or (2) using monthly linear regressions between observed and modeled values to correct values of P and T for both average and extreme values.

The bias-corrected data were used in calculations of total groundwater recharge and net groundwater recharge (total recharge minus water extractions from the aquifer for irrigation) using the water-budget based Thornthwaite-Mather (1957) and Penman-Grindley (1970) methods. Groundwater flow models were developed for the three sites and used to simulate the impacts of future changes in recharge, crop water demand, and sea level rise.

Stigter et al. (2014) reported that short-term predictions had large ranges of projected change in recharge with only the Spanish site showing a consistent decrease in recharge (mean of 23%), particularly due to a reduction in autumn rainfall. More frequent droughts were predicted at the Portuguese and Moroccan sites, but the results were inconclusive for the Spanish site. Toward the end of the century, the modeling results indicate a significant decrease (mean of 25%) in recharge in all areas. The modeled decrease in recharge was most pronounced at the Portuguese site in absolute terms (mean 134 mm/yr) and at the Moroccan site in relative terms (mean 47%). The models also predicted a steady increase in crop water demand with 15–20% additional ET by 2100. Stigter et al. (2014, p. 41) concluded that "Scenario modeling of groundwater flow shows its response to the predicted decreases in recharge and increases in pumping rates, with strongly reduced outflow into the coastal wetlands, whereas changes due to sea level rise are negligible."

5.8.10 Suwannee River Basin, Northern Florida

Swain and Davis (2016) modeled the effects of climate change on groundwater recharge in the Suwannee River Basin, northern Florida. The study utilized dynamically downscaled GCM-simulated precipitation and temperature data for the southeastern United States from the Florida State University Center for Ocean-Atmospheric Prediction Studies (COAPS). The GCM utilized by COAPS was the Community Climate System Model (CCSM) developed by the University Corporation for Atmospheric Research. The A2 emissions scenario was used.

The existing one-layer, steady-state Suwannee River Basin (SRB) MODFLOW model was modified and refined to create a transient version that is suitable for incorporating downscaled GCM rainfall and ET data. The transient model was calibrated for Jan 1970–Dec 2000 (372 months). Future rainfall for Jan 2039–Dec 2069 was projected. Streams were simulated using the Streamflow Routing (SFR2) module.

The Thornthwaite soil-water balance method was used to estimate net recharge (P-ET) for each station. Net recharge values obtained using the Thornthwaite method do not take into account localized topographic dynamics and, therefore, net-recharge values were further adjusted through model experimentation. Multipliers for net recharge were calculated for each rainfall station zone by trial and error using stream discharge and groundwater level data. Net recharge was calculated for the rainfall zones surrounding each rainfall station by multiplying the initial estimate of recharge for the zone by its respective net-recharge multiplier. For the future climate scenario, rainfall and P-ET values obtained at the dynamically downscaled points were interpolated to the SRB model cells.

Simulated average annual ET increased from 36.2 to 37.9 in. due to higher mean temperatures predicted by the CCSM. Although both actual ET and rainfall are projected to increase in the future period, ET as a percentage of rainfall is simulated to decrease slightly from 65 to 62% with a 4.4-in. (11.2 cm) increase in net annual recharge (Swain and Davis 2016). The simulation results indicate that unsaturated zone depths would be more spatially uniform in the future, with groundwater levels rising in areas where they were low in the recent period and declining in areas where they were higher in the recent period. Vegetation that requires a range of conditions (substantially wetter or drier than average) would be detrimentally affected (Swain and Davis 2016).

5.9 Conclusions

The top-down climate modeling approach has provided valuable insights on the potential impacts of climate change on groundwater resources. However, the top-down approach tends to not be practicable in the applied water management realm because of its very high technical (and thus financial) resources requirements and its chain of uncertainty. Each step in the modeling process from the choice of emissions scenario, to the GCM modeling process, and then to downscaling and hydrologic and groundwater modeling introduces uncertainty and thus contributes to a large potential error in the final projections of future groundwater conditions. The potential benefits of such studies for actual water planning tend to not be commensurate with their costs. Top-down modeling studies, if they consider a range of GCMs and multiple emissions scenarios, yield wide ranges of future recharge projections, in some cases from positive to negative values, limiting their actionability in water supply planning.

The bottom-up, sensitivity analysis approach is more congruent with existing water planning and management practices. Sensitivity analyses are a routine element of groundwater modeling studies. The usual practice is to first construct a baseline model utilizing what are considered the most appropriate conceptual model of the system and the most likely parameter values, as determined from either field testing or within the range of likely values based on the hydrogeology of the system. Simulations are then performed in which parameter values are systematically varied to evaluate the sensitivity of model results to uncertainty in the input values. Sensitivity analyses can be performed stochastically with input values automatically obtained from the variables' probability distributions, or consciously performed with the range of values considered based on expert judgement. For example, rigorous sensitivity analyses performed to project salinity changes occurring during long-term groundwater pumping involve consideration of adverse parameter values and alternative conceptual models as judged to be plausible by hydrogeologists with local area expertise (Maliva et al. 2016). If model results are found to be particularly sensitive to certain parameter values or conditions, then expert opinion is called upon to judge the probability of their occurrence.

The bottom-up modeling approach is intended to identify plausible climate change scenarios that could adversely impact the operation of a water supply system, be it owned or operated by an individual user, a public or private water utility, or a regional water supplier. The next step in the assessment process is evaluation the probability of the adverse climate change conditions. Assigning probabilities is ultimately a matter of professional judgement guided by available climate change modeling results (e.g., CMIP5 ensemble results). There will always be substantial uncertainty in projections of future climate because some key variables, particularly future GHG concentrations, are unknowable. Nevertheless, consideration of the results from an ensemble of GCMs using different emissions scenarios can provide reasonable insights as to at least the likely local directions of climate change, which can guide the design of resilient systems that would adequately perform under a wide range of plausible future conditions.

References

Allen DM, Cannon AJ, Toews MW, Scibek J (2010) Variability in simulated recharge using different GCMs. Water Resour Res 46(10). W00F03

Allen DM, Mackie DC, Wei M (2004) Groundwater and climate change: a sensitivity analysis for the Grand Forks aquifer, southern British Columbia, Canada. Hydrogeol J 12(3):270–290

Barron OV, Crosbie RS, Charles SP, Dawes WR, Ali R, Evans WR, Cresswell R, Pollock D, Hodgson G, Currie D, Mpelasoka F, Pickett T, Aryal S, Donn M, Wurcker B (2011) Climate change impact on groundwater resources in Australia: summary report. CSIRO Water for a Healthy Country Flagship, Australia

Bosshard T, Carambia M, Goergen K, Kotlarski S, Krahe P, Zappa M, Schär C (2013) Quantifying uncertainty sources in an ensemble of hydrological climate-impact projections. Water Resour Res 49(3):1523–1536

Brown C, Ghile Y, Laverty M, Li K (2012) Decision scaling: linking bottom-up vulnerability analysis with climate projections in the water sector. Water Resour Res 48(9)

Brown C (2011) Decision-scaling for robust planning and policy under climate uncertainty. World Resources Report, Washington DC. https://wriorg.s3.amazonaws.com/s3fs-public/uploads/wrr_brown_uncertainty.pdf. Accessed 27 May 2020

Brown C, Wilby RL (2012) An alternate approach to assessing climate risks. Eos, Transactions American Geophysical Union 93(41):401–402

Cannon AJ, Sobie SR, Murdock TQ (2015) Bias correction of GCM precipitation by quantile mapping: how well do methods preserve changes in quantiles and extremes? J Clim 28(17): 6938–6959

Cromwell JE, Smith JB, Raucher RS (2007) Implications of climate change for urban water utilities. Association of Metropolitan Water Agencies, Washington, D.C

Crosbie RS, Dawes WR, Charles SP, Mpelasoka FS, Aryal S, Barron O, Summerell GK (2011) Differences in future recharge estimates due to GCMs, downscaling methods and hydrological models. Geophys Res Lett 38(11):L11406

Crosbie RS, Scanlon BR, Mpelasoka FS, Reedy RC, Gates JB, Zhang L (2013) Potential climate change effects on groundwater recharge in the high plains Aquifer, USA. Water Resour Res 49 (7):3936–3951

Dessai S, Hulme M, Lempert R, Pielke R Jr (2009) Climate prediction: a limit to adaptation? In: Adger WN, Lorenzoni I, O'Brien K (eds) Adapting to climate change: thresholds, values, governance. Cambridge University Press, Cambridge UK, pp 64–78

DHI Software (2007) MIKE SHE user's manual, volume 1: user guide. DHI Software

Eckhardt K, Ulbrich U (2003) Potential impacts of climate change on groundwater recharge and streamflow in a central European low mountain range. J Hydrol 284(1–4):244–252

Falloon P, Challinor A, Dessai S, Hoang L, Johnson J, Koehler AK (2014) Ensembles and uncertainty in climate change impacts. Front Environ Sci 2:33

Flato G, Marotzke J, Abiodun B, Braconnot P, Chou SC, Collins W, Cox P, Driouech F, Emori S, Eyring V, Forest C, Gleckler P, Guilyardi E, Jakob C, Kattsov V, Reason C, Rummukainen M (2013) Evaluation of climate models. In: Stocker TF, Qin D, Plattner G-K, Tignor M, Allen SK, Boschung J, Nauels A, Xia Y, Bex V, Midgley PM (eds) Climate change 2013: the physical science basis. Contribution of working group I to the fifth assessment report of the intergovernmental panel on climate change. Cambridge University Press, Cambridge UK, pp 741–866

Foley AM (2010) Uncertainty in regional climate modelling: a review. Prog Phys Geogr 34(5): 647–670

Fowler HJ, Blenkinsop S, Tebaldi C (2007) Linking climate change modelling to impacts studies: recent advances in downscaling techniques for hydrological modelling. Int J Climatol: J Royal Meteorol Society 27(12):1547–1578

Grindley J (1970) Estimation and mapping of evaporation. In: Symposium on world water-balance, no. 92, IAHS-UNESCO, 1, pp 200–213

Heo JH, Ahn H, Shin JY, Kjeldsen TR, Jeong C (2019) Probability distributions for a quantile mapping technique for a bias correction of precipitation data: a case study to precipitation data under climate change. Water 11(7):1475

Holman IP (2006) Climate change impacts on groundwater recharge-uncertainty, shortcomings, and the way forward? Hydrogeol J 14(5): 637–647

Holman IP, Tascone D, Hess TM (2009) A comparison of stochastic and deterministic downscaling methods for modelling potential groundwater recharge under climate change in East Anglia, UK: implications for groundwater resource management. Hydrogeol J 17(7): 1629–1641

Leavesley GH, Lichty RW, Troutman BM, Saindon LG (1983) Precipitation-runoff modeling system: user's manual. U.S. Geological Survey Water-Resources Investigations Report 83–4238

Leavesley GH, Markstrom SL, Viger RJ, Hay LE (2005) USGS modular modeling system (MMS)—precipitation-runoff modeling system (PRMS) NMS-PRMS. In: Singh V, Frevert D (eds) Watershed models. CRC Press, Boca Raton, pp 159–177

Loáiciga HA (2003) Climate change and ground water. Ann Assoc Am Geogr 93(1):30–41

Loáiciga HA, Maidment DR, Valdes JB (2000) Climate-change impacts in a regional karst aquifer, Texas, USA. J Hydrol 227(1–4):173–194

Kittel TGF, Rosenbloom NA, Painter TH, Schimel DS (1995) The VEMAP integrated database for modelling United States ecosystem/vegetation sensitivity to climate change. J Biogeogr 22(4/5):857–862

Maliva RG (2016) Aquifer characterization techniques. Springer, Berlin

Maliva RG, Barnes D, Coulibaly K, Guo W, Missimer TM (2016) Solute-transport predictive uncertainty in alternative water supply, storage, and treatment systems. Ground Water 54(5): 627–633

Maurer EP, Pierce DW (2014) Bias correction can modify climate model simulated precipitation changes without adverse effect on the ensemble mean. Hydrol Earth Syst Sci 18:915–925

Navarro-Racines CE, Tarapues-Montenegro JE, Ramírez-Villegas JA (2015) Bias-correction in the CCAFS-climate portal: a description of methodologies. Decision and Policy Analysis (DAPA) Research Area. International Center for Tropical Agriculture (CIAT), Cali. http://ccafs-climate.org/bias_correction/

Ng GHC, McLaughlin D, Entekhabi D, Scanlon BR (2010) Probabilistic analysis of the effects of climate change on groundwater recharge. Water Resour Res 46(7):W07502

Neitsch SL, Arnold JG, Kiniry JR, Srinivasan R, Williams JR (2005) Soil and water assessment tool theoretical documentation, version 2005. Grassland, Soil and Water Research Laboratory, Agricultural Research Service, Temple, TX

Prudhomme C, Reynard, Crooks S (2002) Downscaling of global climate models for flood frequency analysis: where are we now? Hydrol Process 16(6):1137–1150

Prudhomme C, Wilby RL, Crooks S, Kay AL, Reynard NS (2010) Scenario-neutral approach to climate change impact studies: application to flood risk. J Hydrol 390(3–4):198–209

Pulido-Velazquez D, García-Aróstegui JL, Molina JL, Pulido-Velazquez M (2015) Assessment of future groundwater recharge in semi-arid regions under climate change scenarios (Serral-Salinas aquifer, SE Spain). Could increased rainfall variability increase the recharge rate? Hydrol Process 29(6), 828–844

Raposo JR, Dafonte J, Molinero J (2013) Assessing the impact of future climate change on groundwater recharge in Galicia-Costa, Spain. Hydrogeol J 21(2):459–479

Sahoo S, Dhar A, Debsarkar A, Kar A (2018) Impact of water demand on hydrological regime under climate and LULC change scenarios. Environ Earth Sci 77(9):341

Scibek J, Allen DM (2006) Modeled impacts of predicted climate change on recharge and groundwater levels. Water Resour Res 42(11):W1105

Serrat-Capdevila A, Valdés JB, Pérez JG, Baird K, Mata LJ, Maddock T III (2007) Modeling climate change impacts–and uncertainty–on the hydrology of a riparian system: the San Pedro Basin (Arizona/Sonora). J Hydrol 347(1–2):48–66

Semenov MA, Barrow EM (1997) Use of a stochastic weather generator in the development of climate change scenarios. Clim Change 35(4):397–414

Stigter et al (2014) Comparative assessment of climate change and its impacts on three coastal aquifers in the Mediterranean. Reg Environ Change 14(1):41–56

Swain E, Davis JH (2016) Applying downscaled global climate model data to a groundwater model of the Suwannee River Basin, Florida, USA. Am J Clim Change 04:526

Thornthwaite CW, Mather JR (1957) Instructions and tables for computing potential evapotranspiration and the water balance. Publications Climatol 10(3):185–311

UCAR (n.d.) What is downscaling. University Corporation for Atmospheric Research. https://gisclimatechange.ucar.edu/question/63. Accessed 29 May 2020

U.S. Army of Corps of Engineers (2009) Hydrologic model system HEC-HMS, user's manual, version 3.4, August 2009. Davis CA, U.S. Army Corps of Engineers, Institute for Water Resources, Hydrologic Engineering Center

USEPA (1994) Hydrologic evaluation of landfill performance (HELP) model B. Publication no. EPA/600/R-94/168b. U.S. Environmental Protection Agency, Washington DC

USGS (n.d.) Regional and climate change. Visualization. http://regclim.coas.oregonstate.edu/visualization/gccv/. Accessed 27 May 2020

Van Der Linden P, Mitchell JFB (eds) (2009) ENSEMBLES: climate change and its impacts: summary of research and results from the ENSEMBLES project. Met Office Hadley Centre, Exeter

Verburg PH, Soepboer W, Veldkamp A, Limpiada R, Espaldon V, Mastura SS (2002) Modeling the spatial dynamics of regional land use: the CLUE-S model. Environ Manage 30(3):391–405

Wilby RL, Dawson CW (2013) The statistical downscaling model: insights from one decade of application. Int J Climatol 33(7):1707–1719

Wilby RL, Dessai S (2010) Robust adaptation to climate change. Weather 65(7):180–185

Wilks DS, Wilby RL (1999) The weather generation game: a review of stochastic weather models. Prog Phys Geogr 23(3):329–357

Yates D, Miller K (2011) Climate change in water utility planning: decision analytic approaches. Water Research Foundation and University Corporation for Atmospheric Research, Denver

Zhang L, Dawes W (1998) An integrated energy and water balance model. CSIRO Land and Water, Australia

Chapter 6
Sea Level Rise and Groundwater

6.1 Introduction

Rising sea levels pose a direct threat to coastal communities through permanent inundation of low-elevation areas and greater and more widespread temporary inundation from possibly more intense storms and king tides atop a higher sea level baseline. Small low-lying islands are particularly vulnerable to the loss of already limited habitable land area, if not to eventual total submergence or erosion. Coastal changes associated with sea level rise (SLR) will be more complex than the inundation of a static land surface. Coastal impacts will also be controlled by other factors including underlying geological character (e.g., erodibility), changes in tidal flow and wave energy and direction, and changes in the sediment influx into coastal systems (Williams 2013).

Sea level has varied considerably over the last 2.6 million years (Quaternary Period) as the result of the expansion and contraction of continental ice sheets. Sea level at the Last Glacial Maximum, 19.0–26.5 thousand years ago (ka; Clark et al. 2009), was about 134 m (440 ft) lower than it is today (Lambeck et al. 2014; Fig. 6.1). During the peak of the last interglacial period (124–119 ka), mean sea level was 4–6 m (13–20 ft) higher than at present (Rohling et al. 2008). The rate of sea level change during the transition from glacial to interglacial periods varied considerably. The average rate of rise from $\sim$16.5 to $\sim$8.2 ka was 12 mm/yr, with rates $\geq$40 mm/yr from 14.5 to 14.0 ka (Lambeck et al. 2014). The rate of SLR decreased between 8.5 ka and $\sim$2.5 Ka, after which time ocean volumes remained nearly constant until the start of the recent anthropogenic rise 100–150 yrs ago (Lambeck et al. 2014). Sea level oscillations from 6.0 to 0.15 ka appear to have been no greater than $\sim$15–20 cm in time intervals $\geq$200 yrs (Lambeck et al. 2014).

Past rates of mean SLR have thus at times been an order of magnitude greater than the recent rate of 3.6 mm/yr for the period 1993–2015 (IPCC 2019). Under the RCP8.5 high emissions scenario, the projected rate of SLR is 15 mm/yr (10–20 mm/yr *likely range*) in 2100 (IPCC 2019). Although the rate and

R. Maliva, *Climate Change and Groundwater: Planning and Adaptations for a Changing and Uncertain Future*, Springer Hydrogeology,
https://doi.org/10.1007/978-3-030-66813-6_6

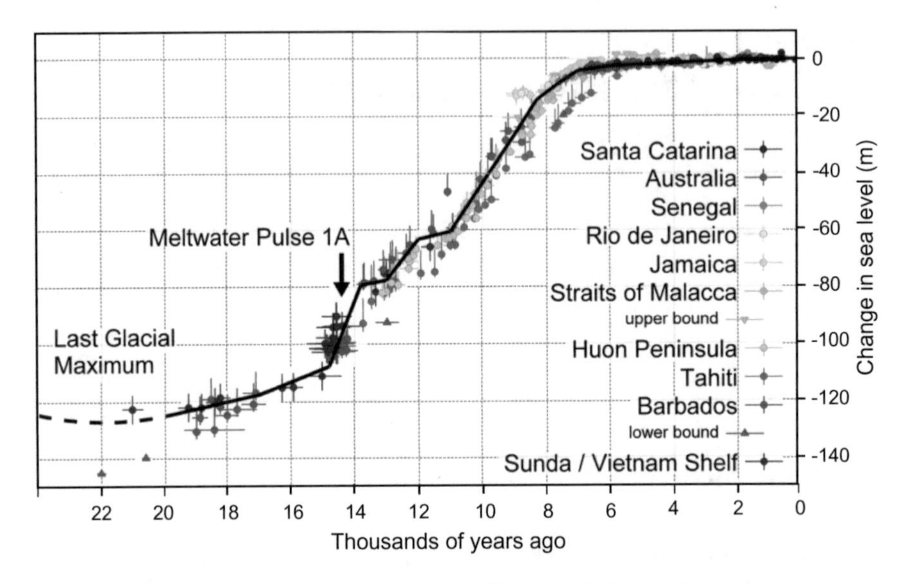

Fig. 6.1 Late Quaternary sea level rise curve. Modified from Rohde (n.d.)

magnitude of anthropogenic SLR over the next century will likely be less than the natural variation at times during the Quaternary Period (Pleistocene and Holocene Epochs), it by no means detracts from its significance. The global human population has expanded enormously since the last ice age and the population is disproportionately located in coastal areas. Many of the largest cities in the world are located near the coast. Estimates of direct human exposure to SLR vary based on criteria for exposure and SLR scenarios. One estimate indicates a conceivable risk of forced displacement of up to 187 million people (up to 2.4% of the global population) within this century (Nicholls et al. 2011; Bamber and Aspinall 2013). The coastal zone also makes a disproportionate contribution to the global economy. Approximately 27% of global population lives in the near-coast zone, but it generates 42% of the world's gross domestic production (Kummu et al. 2016).

A great detail has been written about the potential impacts of SLR on coastal communities and ecosystems, which include (Hay and Mimura 2005; Nicholls et al. 2007; Williams 2013; Wong et al. 2014; Oppenheimer et al. 2019):

- permanent inundation and/or erosion of low-lying coastal areas and low-elevation islands
- storm impacts from surge and waves will potentially be greater and reach further inland
- drowning of coastal wetlands and conversion of near shore land to tidal marsh and salt flats
- dying of coastal vegetation (forests) from salt exposure
- ecosystem changes
- increased saline water intrusion into estuaries and upstream in coastal rivers

- increased saline water intrusion in coastal aquifers
- more frequent flooding of coastal roadways, infrastructure, and urban areas
- displacement of populations and economic infrastructure (e.g., ports and airports)
- migration of sea level rise refugees into inland areas with associated demands on resources
- rising groundwater levels in coastal areas and associated impacts to residents, utilities and other infrastructure, and agricultural operations
- depreciation of the value of coastal properties exposed to increased risk of flooding.

The risks posed by rising sea levels are compounded by the large and growing populations of coastal urban areas. Anthropogenic SLR also threatens coastal ecosystems. During pre-development times, as sea level gradually rose, coastal ecosystems were able to migrate landward. However, extensive coastal development in many areas now essentially forms a wall against ecosystem migration, which is referred to as "coastal squeeze" (IPCC 2019).

Local vulnerability to SLR is highly site-specific. Key variables are the rate of local (relative) SLR (also referred to as effective sea level rise) and the local topography. Relative SLR (RSLR) is the sum of global mean SLR, changes in local land surface elevation, and local changes in sea level due to changes in the strength and distribution of ocean currents, atmospheric pressure distribution, and local seawater salinity. Where land surface is rising, RSLR will be less than the global mean sea level (GMSL) rise. Along parts of the northern North American coast (e.g., Alaska and British Columbia) relative sea level is falling due to glacial (isostatic) rebound (i.e., the rise of land surface from the removal of the ice mass) since the end of thm last ice age (NOAA n.d.a). Relative sea level is similarly stable or falling in much of northern Europe, particularly the northern BalTic Sea region (European Environmental Agency 2019).

Areas most vulnerable to rising relative sea level are those where the effects of GMSL rise are compounded by land subsidence. Land subsidence can be due to tectonism, consolidation (compaction) of sediments, and fluid (water, oil, and/or gas) extraction. The Mississippi Delta and parts of the Texas coast of the United States are experiencing subsidence in the 6–12 mm/yr range (NOAA n.d.a). Subsidence in deltaic regions is driven by the compaction of fine-grained fluvial sediments and decreases in sediment deposition. Other deltaic areas facing high rates of RSLR include the Ganges-Brahmaputra-Meghna River Delta (Bangladesh and India), Mekong Delta (Vietnam), Niger Delta (Nigeria), Indus Delta (Pakistan), Yellow and Yangtze River Deltas (China), and the Nile Delta (Egypt). Under baseline conditions, the estimated relative sea level rise in 40 sampled major deltas range from 0.5 to 12.5 mm/yr (Ericson et al. 2006). Under these levels of RSLR, 4.9% of the area of the studied deltas and 8.7 million people could potentially be affected by coastal inundation by 2050 (Ericson et al. 2006).

The IPCC concluded that "by the end of the twenty-first century, it is *very likely* that over about 95% of the world ocean, regional sea level rise will be positive, and most regions that will experience a sea level fall are located near current and former glaciers and ice sheets. About 70% of the global coastlines are projected to experience a relative sea level change within 20% of the global mean sea level change" (Church et al. 2013, p. 1140).

Adaptations to SLR can be categorized as either protection, accommodation, or planned retreat (Nicholls et al. 2011). Two distinct views have emerged in the literature concerning adaption to SLR, which have been referred to as pessimist and optimist views (Nicholls et al. 2011). The pessimist view assumes that protection (e.g., construction of various coastal barriers) is unaffordable or will largely fail, leading to large-scale population displacement on an unprecedented scale (Nicholls et al. 2011). The optimists assume that protection will be widespread and largely successful, and that the residual impacts will be only a small fraction of the potential impacts (Nicholls et al. 2011). The basis of the optimist view is that the cost of protection will be much less than the cost of the potential damage. The main consequence of SLR will be the diversion of investment into new and upgraded coastal defenses and other forms of adaptation (Nicholls et al. 2011).

The adaptive capacity of countries is highly variable. Adaptation to SLR in developing countries will be more challenging than in coastal areas of wealthier developed countries due to constraints from lesser adaptive capacities (Nicholls et al. 2011; IPCC 2019). Small islands, Africa, and south, southeast and east Asia are high vulnerability regions because of their large and growing exposure to SLR due to large populations living in low-lying, often subsiding coastal lowlands (deltas), and more limited financial resources (Nicholls et al. 2011). As observed by Nicholls et al. (2007, p. 317).

> Adaptive capacity is largely dependent upon development status. Developing nations may have the political or societal will to protect or relocate people who live in low-lying coastal zones, but without the necessary financial and other resources/capacities, their vulnerability is much greater than that of a developed nation in an identical coastal setting.

Adaptation to SLR starts with an identification of local vulnerabilities. SLR may impact coastal groundwater in several key manners:

- temporary inundation from storms (e.g., hurricane storm surges) can contaminate shallow aquifers from above
- inundation can reduce the area of freshwater lenses on islands
- SLR can induce increased lateral intrusion of saline waters into coastal aquifers
- rising sea levels can cause the water table to rise inland of the shoreline, which can cause local flooding of low-lying areas and adversely impact on-site sewage treatment and disposal systems, buried utilities, and other infrastructure
- migration away from coastal areas and declines in agriculture due to the salinization of soils and shallow aquifers may reduce local groundwater demands, but the migration of climate change refugees may further stress inland groundwater resources.

6.2 Direct Inundation

6.2.1 Introduction

Rising relative sea level will cause a progressive inundation of coastal areas. Lower-lying areas will first be increasingly impacted by periodic high-water events (e.g., storm surges and king tides), followed by becoming intertidal, and then permanently inundated. Nearshore topography is a critical factor affecting local susceptibility to inundation from SLR. Where the nearshore land elevation is well above sea level and the slope to the sea is steep, SLR will result in minimal inundation. Conversely, inundation will be more extensive in low-lying coastal areas, particularly those experiencing greater RSLR from land subsidence.

Inundation impacts local groundwater resources in several manners. Shallow aquifers that become permanently submerged by seawater are of obvious very great risk of salinization by downward percolation of seawater. Temporary inundation, for example from hurricane storm surges, can impact aquifers by the infiltration and percolation of salt water and through downward leakage through open or improperly sealed wells. Saline water inundation may extend inland through interconnected low-lying areas, greatly increasing the area of surficial (water table) aquifers vulnerable to salinization. Levees, dikes, berms, walls, and other types of hydraulic barriers constructed across low-lying areas are used break the hydraulic connection to inland areas and provide protection against flooding.

Local communities have long been implementing adaptive measures to reduce vulnerability to extreme high-water events and RSLR. Throughout its history, the Netherlands has successfully fought against rising sea levels. Much of the coast of many European and East Asian countries have defenses against flooding and erosion. Many major coastal cities are heavily dependent upon artificial coastal defenses (e.g., Tokyo, Shanghai, Hamburg, Rotterdam and London; Nicholls et al. 2011). Failure of defenses can have catastrophic consequences as witnessed by the severe flooding caused by the breaching of levees and walls in the New Orleans area during Hurricane Katrina in 2005.

The fight to hold back the seas will be very expensive. The rebuilding of the levee system of New Orleans cost about US$ 14 billion dollars and it is now recognized that they may soon become inadequate due to a combination of SLR and subsidence (Frank 2019). The U.S. Army Corps of Engineers is considering an US $ 8 billion project consisting of 10–13 ft (3.0–4.0 m) high walls and surge barriers to protect parts of Miami-Dade County (Florida) from rising sea levels and storm surges from projected more intense hurricanes (Harris 2020). Developed countries have the financial resources to protect their vulnerable cities but constructing sufficiently robust defenses in developing countries and more rural areas of developed countries may not be economically feasible.

6.2.2 Future Inundation Mapping

Areas vulnerable to direct marine inundation from RSLR are most commonly mapped using what is referred to as the "bathtub" approach, which assumes that sea level rises at the same rate everywhere in a coastal study area. Areas with elevations below that of a projected future sea level are assumed to become inundated. A main limitation of bathtub approach is that it does not consider changes in hydrodynamic processes associated with sea level rise, such as changes in wave height and direction (Norheim et al. 2018). Another limitation of the bathtub method is that it cannot account for changes in coastal topography caused by erosion and sedimentation associated with SLR. The results of an investigation in Hawaii demonstrated that using the bathtub approach alone ignores 35–54% of the total land area that is at risk of flooding from combined tidally forced direct inundation, groundwater inundation, seasonal wave inundation, and coastal erosion (Anderson et al. 2018). More accurate hydrodynamic modeling is complex and data intensive, and thus more difficult and expensive to perform.

The bathtub modeling approach (and modifications thereof) can provide an inexpensive initial screening of vulnerability to future SLR inundation (Gesch 2018; Norheim et al. 2018). Uncertainties in the bathtub approach will likely be less than those associated with projections of GHG emissions and associated SLR.

The inundation mapping process starts with a high-resolution topographic map, commonly a detailed remote sensing-generated digital elevation model (DEM). High accuracy local DEMs required for accurate mapping of future inundation are commonly obtained using LiDAR (Light Detection and Ranging). LiDAR is a remote sensing technique in which land surface elevation is measured from the return time of rapid pulses of laser light fired at land surface from an airborne platform. Global DEMs are available but may not be accurate enough for high-confidence mapping of exposure to fine increments of SLR (<1 m) over shorter planning horizons (<100 yrs; Gesch 2018).

Projected future sea level elevations are taken as the sum of current sea level elevation and the projected relative sea level rise. Future relative sea level elevations are subtracted from land surface elevations with negative values indicating inundation. However, inundation mapping is more nuanced. Practical procedures for mapping sea level rise and inundation were summarized by NOAA (2017), Gesch (2018), and Norheim et al. (2018).

A basic issue is the choice of the datum used for current and future sea levels. The National Oceanic and Atmospheric Administration (NOAA; United States) commonly uses "mean higher high water" (MHHW), which is defined as the average height of the highest tide recorded at a tide station each day during a 19-yr recording period. Alternative datums include "mean lower low water" (MLLW), which is the average height of the lowest tide recorded at a tide station each day during a 19-yr recording period, the 10-yr surge (Norheim et al. 2018), and mean sea level (MSL). The sea level elevation geodetic datum used should the same as that used in the DEM, which in the United States is now the North American

Vertical Datum of 1988 (NAVD 88). Current MSL is not equal to 0 ft elevation in NAVD 88. For example, at Miami Beach, Florida, MSL is at −0.29 m (−0.96 ft) NAVD 88 and MHHW is at 0.10 m (0.33 ft) NAVD 88 (NOAA n.d.b).

Inundation is controlled by relative sea level change rather than global mean sea level (GMSL) change, hence project area specific data on the RSLR rate are required. The contribution of vertical land motion (VLM) to RSLR can be calculated as the current difference between the rates of RSLR and GMSL rise. Future RSL can be estimated as the sum of projected GMSL rise and VLM, the rate of which is assumed to be constant.

DEM data should be additionally processed to remove artifacts over water (i.e., hydro-flattening) and to account for hydraulic connectivity (i.e., hydro-enforcement). Hydro-flattening is a modification of LiDAR-derived DEMs to remove elevation values interpolated though water surfaces and to thus make the surfaces appear as flat surfaces and behave as they would in traditional topographic DEMs created from photogrammetric digital terrain models (Sanborn n.d.; Norheim et al. 2018).

A limitation of the bathtub approach is the failure of DEMs to represent the detailed hydraulic connections and barriers that will control local flooding (Gallien et al. 2011; Gesch 2018). Hydro-enforcement is the modification of topographic DEMs to better simulate the hydrologic functioning of landscapes. For example, in a topographic DEM, roadways with culverts and bridges create artificial impediments (digital "dams") and introduce sinks (undrained areas) into the landscape (Sanborn n.d.). LiDAR data will generally not show the correct elevations of dikes and levees. If these features are mapped incorrectly low, then it would result in a greater apparent hydraulic connectivity in the landscape (Norheim et al. 2018).

Inundation projections for Everglades National Park and surrounding areas of south Florida, provided as an example (Fig. 6.2), show that under the high emissions RCP8.5 large parts of the national park and southern Miami-Dade County would become inundated. The 99% percentile map shows the 1% probability "risk intolerant" scenario (Kopp et al. 2014; Park et al. 2016, 2017). However, the SLR projections could be too low if the rate of melting of Greenland and Antarctic ice sheets is more rapid than simulated in the GCMs. The inundation maps provided in Fig. 6.2 do not considered temporary inundation from storm surges, which would also greatly impact local ecosystems.

6.3 Extreme Sea Level Events (Storm Surges)

6.3.1 Climate Change and ESLs

As sea level rises, low-lying areas will first experience periodic inundation during high tides followed by permanent inundation as the low tide elevation exceeds the local land surface elevation. Coastal areas will also experience increased exposure

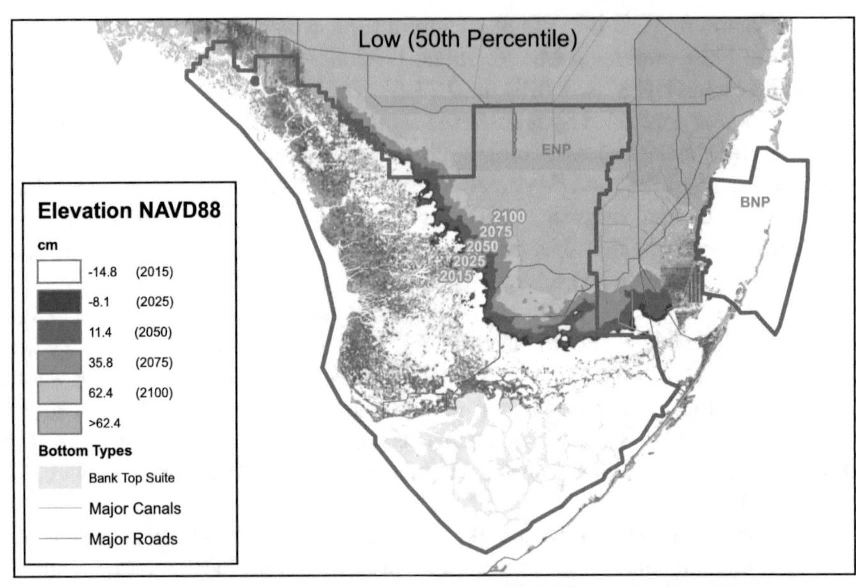

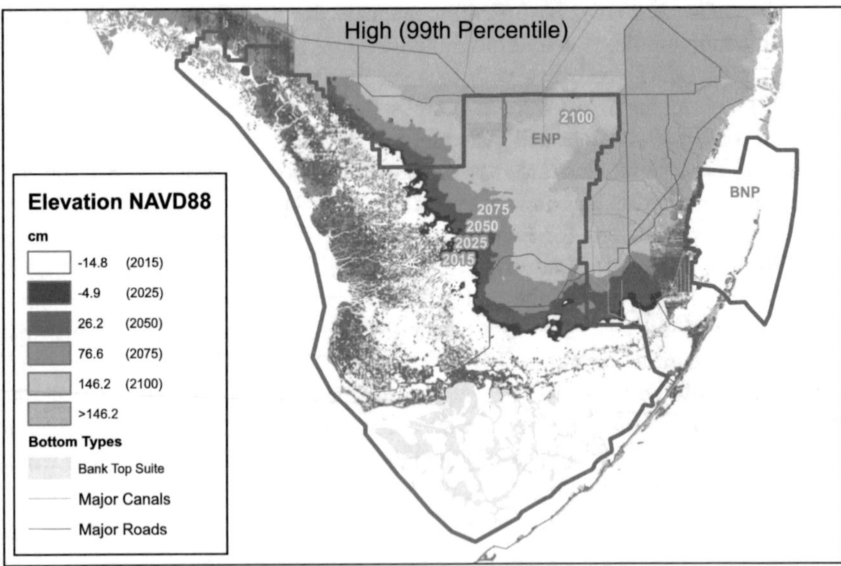

Fig. 6.2 Mean sea-level elevation maps for South Florida including Everglades and Biscayne National Parks for the median (50th, top) and high (99th percentile, bottom) RCP8.5 projections using current topography and NAVD-referenced digital elevation data. Tides and storm surges are not included in the projections. *Source* Park et al. (2016, 2017)

to storm surges with increasing sea level. A storm surge is defined as "an abnormal rise of water generated by a storm, over and above the predicted astronomical tides" (NHC n.d.). A storm tide is defined as "the water level rise due to the combination of storm surge and the astronomical tide" (NHC n.d.). The rise in coastal water levels is amplified when the surge from a storm (e.g., a hurricane or typhoon) coincides with a normal high tide, resulting in storm tides reaching up to 6 m (20 ft) or more in some cases (NHC n.d.). Storm surges and tides are referred to by the Intergovernmental Panel on Climate Change (IPCC) as extreme sea level (ESL) events (Oppenheimer et al. 2019). Storm surges and tides usually pose a greater threat to property and lives during tropical storms than the wind.

Climate change will impact ESLs through both an intensification of storm events and SLR. Small increase in mean sea level can significantly increase the frequency and intensity of flooding because SLR elevates the platform for storm surges, tides, and waves, and there is a log-linear relationship between a flood's height and its occurrence interval (return period; Oppenheimer et al. 2019). The impacts of increased mean sea level can be evaluated from annual exceedance probability curves (Fig. 6.3). Adding the increase in sea level from climate change to the curves provide the projected change in the return period of a given extreme sea level.

The IPCC (Oppenheimer et al. 2019, p. 360) concluded with respect to ESL events that

> as a consequence of SLR, events which are currently rare (e.g., with an average return period of 100 yrs), will occur annually or more frequently at most available locations for RCP8.5 by the end of the century (high confidence). For some locations, this change will occur as soon as mid-century for RCP8.5 and by 2100 for all emission scenarios. The affected locations are particularly located in low-latitude regions, away from the tropical cyclone (TC) tracks. In these locations, historical sea level variability due to tides and storm surges is small compared to projected mean SLR. Therefore, even limited changes in mean sea level will have a noticeable effect on ESLs, and for some locations, even RCP2.6 will lead to the annual occurrence of historically rare events by mid-century.

Storm surges are driven by the surface winds and pressure gradient forces of tropical cyclones. The magnitude of storm surges is determined, in a complex manner, by the characteristics of the storms (e.g., intensity, size, direction and speed) and the bathymetry and topography of the coast (Lin et al. 2012). Mousavi et al. (2011) evaluated the coupled impact of sea level rise and hurricane intensity on storm surges using the Corpus Christi, Texas, area as case study. Flood surge

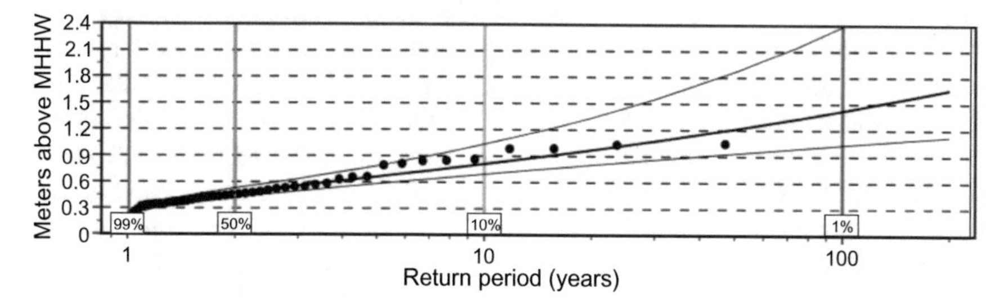

Fig. 6.3 The annual exceedance probability curves for Fort Myers, Florida, with 95% confidence intervals shown, indicate the highest and lowest water levels as a function of return period in years. *Source* NOAA (n.d.c)

elevations were estimated by the 2030s and 2080s for the B1, A1B, and A1F1 future climate scenarios. Local RSLR estimates included the local land subsidence rate (about 0.29 cm/yr) in the Corpus Christi area. Three historical major hurricanes were selected for intensification using a planetary boundary layer (PBL) model, which uses as inputs hurricane central pressure, size, and position in time, among other parameters. Hurricane Beulah (1967), Hurricane Bret (1999) and Hurricane Carla (1961) were selected as being suitable for analysis. The former two were Corpus Christi storms. Hurricane Carla tracked to the north and its position was shifted in the analysis 130 km to the south such that the maximum surge generation occurs at Corpus Christi. Future global warming and eustatic sea level rise for the 2030s and 2080s were predicted for the three IPCC future climate scenarios using the climate model MAGICC/SCENGEN (UCAR n.d.).

Future hurricane intensity scenarios were developed using an equation developed by Knutson and Tuleya (2004) that estimates projected hurricane central pressure from present-day (or historical) central pressure and sea surface temperature (SST) change. The Knutson and Tuleya (2004) equation estimates an average 7% increase in intensity for every 1 °C of SST rise. Flood levels from the projected hurricane intensification scenarios were then estimated using a surge response function (SRF), in which the time history of storm position, size, forward speed, and approach angle were kept constant. The only variables controlling surge level that were changed are hurricane intensity and relative SLR (Mousavi et al. 2011). Hurricane flood levels were simulated using the shallow-water hydrodynamic model ADCIRC (Advanced CIRCulation). The results of the simulations indicate that hurricane flood elevation (storm surge plus SLR) will, on average, rise by 0.3 m by the 2030s and by 0.8 m by the 2080s. For catastrophic-type hurricane surge events, flood elevations are projected to increase by as much as 0.5 and 1.8 m by the 2030s and 2080s, respectively (Mousavi et al. 2011). In the low-lying Corpus Christi area, the projected increase in flood elevations would cause more widespread inundation and damage (Mousavi et al. 2011).

Climate change-induced impacts to the hurricane threat in New York City were investigated by Lin et al. (2012) using a GCM-driven statistical/deterministic hurricane model and two hydrodynamic surge models. Four climate models, CNRM-CM3 (Centre National de Recherches Météorologiques, Météo-France), ECHAM5 (Max Planck Institute), GFDL-CM2.0 (National Oceanic and Atmospheric Administration Geophysical Fluid Dynamics Laboratory) and MIROC3.2 (CCSR/NIES/FRCGC, Japan), were used to generate 5000 New York-region storms under present climate conditions (1981–2000 statistics) and another 5000 New York-region storms under future climate conditions (2081–2100 statistics) for the IPCC A1B emission scenario. The annual storm frequencies from the models were calibrated to the historical frequency of New York region storms. Estimated future storm frequency ranged from a decrease of 15% to an increase of 290%. It was assumed that the annual number of New York-region storms is Poisson-distributed with the annual storm frequency as the mean. Two hydrodynamic models, ADCIRC (Advanced Circulation) and SLOSH (Sea, Lake, and Overland Surges from Hurricanes), were used to predict storm surges.

Lin et al. (2012) reported that some climate models predict an increase in surge level due to the change of storm climatology that is comparable to the projected SLR for New York City. For example, the CNRM and GFDL models predict that by the end of the century the 100 yr and 500 yr storm tide levels will increase by about 0.7–1.2 m. More consequentially, the combined effect of storm climatology change and SLR were found to greatly shorten the surge flooding return periods. The combined effects of storm climatology change and a 1 m SLR was projected to cause the present New York City 100 yr surge flooding to occur every 3–20 yrs and the present 500 yr flooding to occur every 25–240 yrs by the end of the century.

6.3.2 Historical Impacts of ESLs on Fresh Groundwater Resources

Atoll and barrier islands and other low-lying coastal areas are vulnerable to overtopping by storm surges and tsunami waves, which can result in the salinization of freshwater aquifers. Damage to the aquifers, which ofttimes are critical for both sustaining human habitation and ecosystem maintenance is temporary, but the disruption of water supplies can pose a serious hardship to local populations that face the formidable task of repairing or rebuilding their shattered communities (Terry and Falkland 2010). Follows are summaries of case studies that investigated the extent and duration of impacts to groundwater resources resulting from tropical storm surges and tsunami waves.

Terry and Falkland (2010) investigated the effects of storm-surge overwash from Cyclone Perry (February and March 2005) on freshwater lenses on Pukapuka Atoll, Northern Cook Islands. Freshwater occurs in Holocene carbonate sands that overlie more transmissive Pleistocene limestone. The storm surge partially swept over the islets and resulted in seawater penetrating into the porous coralline substrate and contaminating the freshwater aquifers. The freshwater aquifer recovery time was a function of both climate conditions during the overwash event and the subsequent amount of freshwater recharge. Groundwater salinization on Pukapuka Atoll would have been more severe if the overwash had occurred in the absence of heavy rainfall (Terry and Falkland 2010). For all three of the islets on Pukapuka, saline contamination at the surface of the fresh groundwater was observed to have lasted for 11 months or longer. An unexpected finding was that in the thickest aquifer, a well-defined saline plume was present at a 6 m depth that was underlain by a freshwater layer to the base of the aquifer. The remnant of the saline plume was still present 26 months after the overwash event. Saline plume dispersion on Pukapuka may have been relatively slow due to its emplacement toward the start of the dry season and the low tidal range on the atoll (Terry and Falkland 2010). Inasmuch as the freshwater lenses are the sole local source of freshwater on the islets, the overwash event had a severe disruptive impact on potable water and fresh food supplies, necessitating an emergency shipment of freshwater by the Red Cross (Terry and Falkland 2010).

Anderson (2002) investigated the impacts of hurricane storm overwash on the shallow Buxton Woods Aquifer (BWA) on Hatteras Island, a siliciclastic barrier island off the coast of North Carolina. Chloride concentrations in the BWA increased from 40 mg/L to nearly 280 mg/L after flooding caused by Hurricane Emily in August and September 1993. Chloride concentrations did not decrease to below 100 mg/L until March 1994 and had still not reached pre-flooding concentrations by January 1997. Anderson (2002) noted that the transformation of the Cape Hatteras region from a seasonal to year-round tourist destination is placing increasing stress on the aquifer and diminishing the aquifer's ability to dilute saline overwash waters (Anderson 2002).

Saline water on land surface can enter aquifers through unsealed or improperly sealed wells. The north shore of Lake Pontchartrain in southeastern Louisiana was overwashed with storm surges in 2005 from Hurricane Katrina and, to a lesser extent, subsequent Hurricane Rita (Van Biersel et al. 2007). Private and public wells were reported to have been submerged by 0.6–4.5 m of seawater for several hours to days, during which time water entered some wells through broken casings, wellhead openings, and associated plumbing (Van Biersel et al. 2007). The wells were completed in shallow confined aquifers (overlain by clay units). The most vulnerable wells were completed in aquifers with water levels (heads) below land surface into which water could enter under gravity. Saltwater tracer (calcium, magnesium, boron, chloride, silica, Ca/Mg ratio, Cl/Si ratio, and specific conductance) data confirmed that saltwater had entered the wells. Tracer concentrations had subsequently decreased toward pre-Katrina values. The 2006 post-Katrina groundwater consisted of approximately 86% pre-Katrina groundwater and 14% Lake Pontchartrain surface water (Van Biersel et al. 2007). Coliform bacteria (total and/or fecal) were also detected in 67% of the 2005 samples (Van Biersel et al. 2007).

The December 26, 2004, Indian Ocean tsunami was one of the deadliest natural disasters in recorded history, killing over 220,000 people. The tsunami caused widespread contamination of coastal aquifers across southern Asia (Illangasekare et al. 2006). Illangasekare et al. (2006) and Vithanage et al. (2012) documented the impacts of the 2004 tsunami on groundwater in Sri Lanka, which suffered catastrophic losses from the event. Many coastal areas in Sri Lanka rely on individual or community wells for potable water. These wells are typically of large diameter (1–2 m) and 3–8 m deep and are designed to tap the shallow parts of the fresh groundwater lens found in most of the coastal aquifers (Illangasekare et al. 2006). The tsunami was reported to have immediately inundated and contaminated more than 40,000 wells in Sri Lanka (United Nations Environment Programme 2005; Illangasekare et al. 2006). Illangasekare et al. (2006) reported that in most affected areas, the open dug wells were instantly filled with the seawater resulting in large volumes of saltwater having been injected into the freshwater lens. Seawater also inundated low-lying areas resulting in infiltration of saline water through permeable surficial sands and percolation into the underlying aquifers.

Efforts to clean up the wells appears to have exacerbated the saline water contamination problem in some wells. Excessive pumping may have caused more saline water intrusion from below rather than its removal from the upper part of the saturated zone, and the discharge of the purged well water onto the ground allowed the contaminated water to reenter the aquifer and wells after infiltration (Illangasekare et al. 2006).

Illangasekare et al. (2006, p. 4) further noted that the

> data showed that changes in salinity over time were highly variable from well to well, with some wells actually showing an increase in salinity long after the tsunami. Several factors may be responsible for the observed variations including (1) the amount of seawater that initially entered the wells directly through the well opening, (2) the amount of seawater that infiltrated through the vadose zone, (3) the volume and frequency of water that was pumped from each well after the tsunami, (4) well interference and recontamination through the vadose zone due to pumping from closely spaced wells, and (5) variations in monsoon recharge.

Salinity in some wells was reported to have declined rapidly within a few months after the tsunami, but other wells remained unfit for drinking because of residual salinity (Illangasekare et al. 2006). The rate of salinity reduction was posited to depend on the rate of recharge, the accumulation of salinity in the vadose zone, the permeability of the aquifers, and the pumping intensity (Illangasekare et al. 2006).

A monitoring program in a shallow sandy aquifer in a rural area on the east coast of Sri Lanka showed that the upper part of the aquifer (down to 2.5 m) returned to freshwater conditions within 1–1.5 yrs after the 2004 tsunami due to flushing by subsequent recharge of the aquifer from the monsoon (Vithanage et al. 2012). Modeling results support the suggestion of Oostrom et al. (1992) that an unstably stratified plume can undergo significant convective mixing, which can substantially increase the volume of contaminated water (Vithanage et al. 2012). The convective mixing allows saline water to flow downward into the deeper parts of the aquifer, rather than migrating laterally toward the sea, and increases the volume of contaminated water. The simulation results show a rapid initial decrease in salinity just after the tsunami and that it will take about 5 yrs to recover to the ambient water quality in the upper 6 m of the aquifer. The simulation results also indicate that the aquifer should totally recover through flushing out of the saltwater within approximately 15 yrs (Vithanage et al. 2012).

6.4 Saline Water Intrusion

6.4.1 Basics

Anthropogenic saline water intrusion is caused by groundwater pumping (or other activities) altering natural groundwater flow patterns so that saline groundwater migrates inland (and upward) and displaces fresh groundwater. Fresh groundwater is less dense than seawater and thus tends to float atop more saline groundwater in

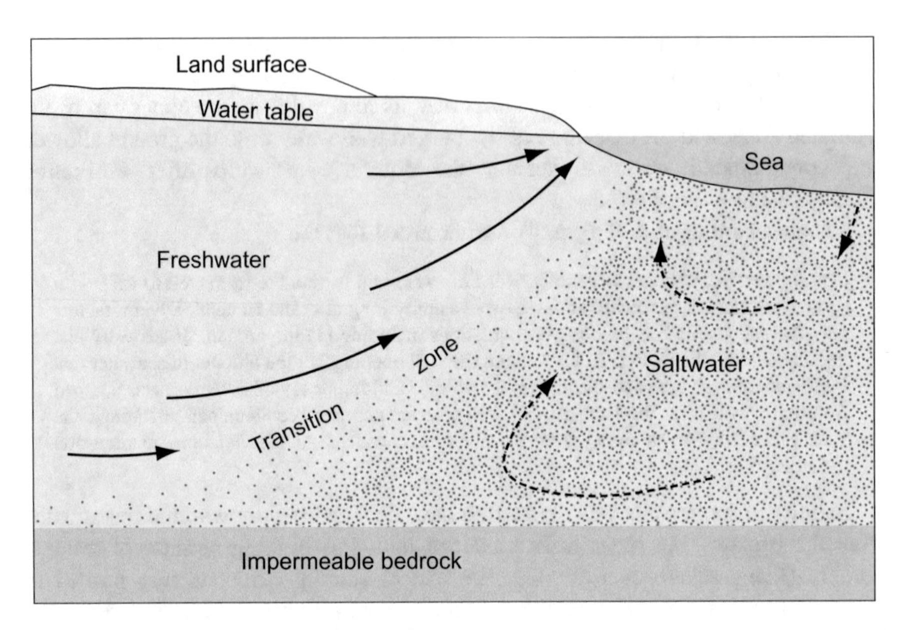

Fig. 6.4 Groundwater flow patterns and the fresh-saline groundwater interface in an idealized coastal aquifer. Modified from Barlow (2003)

aquifers. The interface between fresh and salt water in aquifers is typically a transition (i.e., mixing) zone of varying thickness rather than being a sharp interface (Fig. 6.4).

The depth of a sharp freshwater-salt water interface may be calculated using the well-known Ghyben-Herzberg relation:

$$z = \frac{\rho_f}{(\rho_s - \rho_f)} h \tag{6.1}$$

where,

z = depth of the presumed sharp saline-fresh groundwater interface below sea level (m or ft)

h = height of the freshwater zone above sea level (m or ft)

ρ_s = density of seawater (g/cm^3; lbs/ft^3)

ρ_f = density of freshwater (g/cm^3; lbs/ft^3).

A key consequence of the Ghyben-Herzberg relation is that the depth to the fresh-saline groundwater interface below sea level at a given location is about 40 times the elevation of the top of the fresh groundwater above sea level (z = 40 h) based on the density difference between freshwater and salt water (Fig. 6.5a). The Ghyben-Herzberg relationship is frequently not strictly valid because the underlying assumptions of a sharp interface and the absence of freshwater flow to the sea

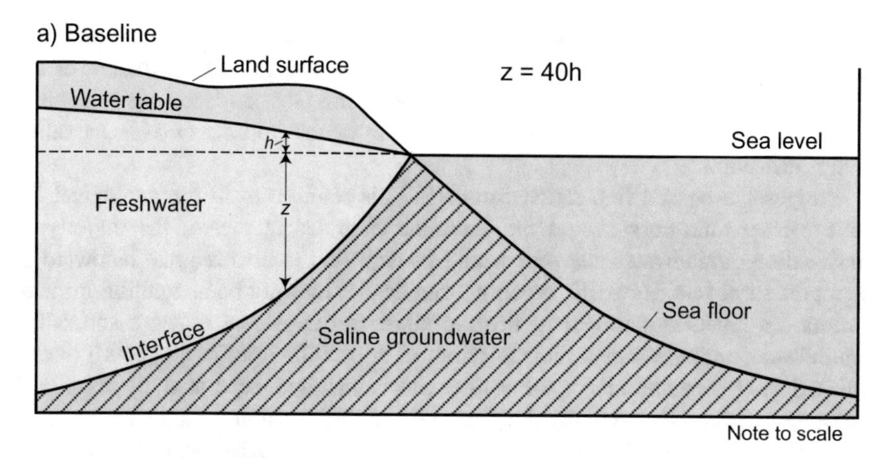

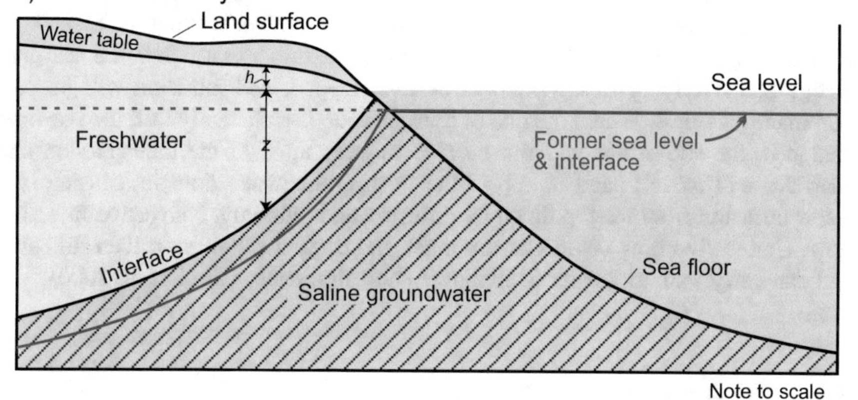

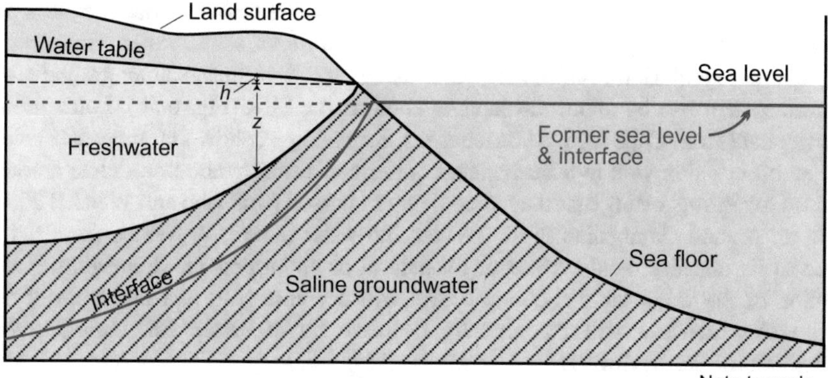

Fig. 6.5 Conceptual diagram of the fresh-saline groundwater interface in an unconfined coastal aquifer. Modified from Barlow (2003). **a** Ghyben-Herzberg relationship, **b** Flux-controlled system in which the water table rises with SLR and the thickness of the freshwater zone is stable, **c** Head-controlled system in which the elevation of the water table is stable and the decreased height of fresh groundwater above sea level results in a decrease in the thickness of the freshwater zone and greater landward migration of the fresh-saline groundwater interface

(i.e., hydrostatic conditions) are not met. Aquifer heterogeneity (vertical and horizontal) can profoundly impact the shape, position, and rate of movement of the fresh-saline groundwater interface. Nevertheless, the Ghyben-Herzberg relation is useful for providing a conceptual understanding of the impacts of SLR on saline water intrusion.

Progressive coastal inundation from RSLR is referred to in the geological literature as a transgression. As the shoreline migrates landward, the underlying fresh-saline groundwater interface would be expected to also migrate landward at about the same rate. However, modern groundwater may not be in equilibrium with current sea level. In some areas, fresh groundwater is present offshore and saline groundwater may be trapped inland (Kooi et al. 2000). Kooi et al. (2000) distinguished two separate modes of saline water intrusion separated by a critical transgression rate. Horizontally dominated seawater intrusion is characterized by a slow rate of transgression in which a quasi-steady state exists with the intruding seawater wedge. As the coastline moves inland, the seawater wedge moves inland at about the same rate and the direction of groundwater flow is essentially unchanged. If the rate of transgression exceeds the critical rate, then the seawater wedge and overlying freshwater will be overstepped, and intrusion will be predominantly vertical. Where the rate of transgression is sufficiently fast, the seawater wedge in the subsurface will not be able to keep up with the transgression and seawater will move inland on top of the fresh groundwater domain, causing seawater intrusion from the top down by diffusion and, possibly, convective fingering flow. Convective flow could cause a much more rapid salinization than diffusion and can carry salts to greater depths over short timescales (Kooi et al. 2000).

6.4.2 Theoretical Modeling

In the case of an unconfined aquifer in which the water table can rise (i.e., it is not controlled by drainage), SLR will tend to result in minor saline water intrusion. At hydraulic equilibrium, both the water table and the fresh-saline groundwater interface will rise by about the same amount as the SLR (Fig. 6.5b). Some monitoring data from Cape Cod, Massachusetts, for example, show that the water levels in an observation well in a freshwater aquifer had risen by about the same amount as sea level, suggesting that they were rising in unison (McCobb and Weiskel 2003; Masterson and Garabedian 2007). Where the water table is prevented from rising due to its reaching land surface or by natural or anthropogenic drainage, the elevation of the water table above sea level will decrease (Δh) and the depth of the salt-water interface will theoretically decrease by 40 times Δh, shrinking the thickness of the freshwater lens or wedge (Fig. 6.5c).

The effects of SLR on the thickness of freshwater lenses in barrier islands, such as Cape Cod and the eastern forks of Long Island, are influenced by rivers, which can either be freshwater rivers that act as constant-stage drains or be tidally influenced and have rising stages and landward migration of saline water with increasing sea level (Masterson and Garabedian 2007). The effects of SLR on a hypothetical freshwater lens were investigated by Masterson and Garabedian (2007) by numerical modeling using the SEAWAT 2000 code (Langevin et al. 2003). Rising sea level was simulated for the period 1929–2050 by specifying an annual average historical SLR rate of 2.65 mm/yr for the constant head and concentration cells at the model boundary. Only advective solute transport was simulated (dispersion was not specified). Simulations were performed for both a freshwater and tidally influenced stream.

The results of both simulations showed increased fresh groundwater levels with rising sea level. The simulations with a constant-stage (gaining) stream predicted increased absolute groundwater levels, but the increases were less than the sea level rise and there was a decline in groundwater levels relative to sea level. The constant-stage stream prevented the nearby water table from rising appreciably above the elevation of the stream bed and the amount of groundwater discharged to the stream increased. The net effect was a thinning of the freshwater lens on the hypothetical coastal aquifer because the rate at which the water table rises is less than the rate of sea level rise (Masterson and Garabedian 2007). The simulated relative water level declines were greatest close to the stream. In the case of the tidal stream, as stream stage rose with rising sea level, there was successively less groundwater discharge to the stream (Masterson and Garabedian 2007). A key conclusion is that the effects of groundwater fed streams on groundwaters levels and the depth to the fresh-saline groundwater interface diminished as the extent of tidal influence in streams propagates inland with rising sea levels (Masterson and Garabedian 2007).

Werner and Simmons (2009) made the distinction between flux-controlled and head-controlled coastal aquifer systems. In a flux-controlled system, the groundwater discharge to the sea is not appreciably changed with SLR; groundwater levels rise, and the hydraulic gradient is maintained. In a head-controlled system, inland freshwater heads remain the same despite the SLR because of various surface controls, such as drains and rivers (Fig. 6.5c). Analytical modeling results indicate that minimal inland movement of the fresh-saline groundwater interface will occur in flux-controlled systems, whereas large changes may occur in head-controlled systems (Werner and Simmons 2009). The migration of the toe of the saline water wedge will be greater with decreasing recharge rates, increasing hydraulic conductivity, and increasing depth of the aquifer below sea level (Werner and Simmons 2009). In the scenarios modelled by Werner and Simmons (2009), saline water intrusion was no greater than 50 m for a flux-controlled system versus hundreds of meters to over a kilometer for a head-controlled system.

The Werner and Simmons (2009) evaluation was based on equilibrium analytical solutions. An unresolved issue is the effects of non-equilibrium transient responses to sea level change. Potential additional saline water intrusion could continue for some time after sea level has stabilized. Webb and Howard (2011) performed numerical simulations of the Werner and Simmons (2009) head-controlled sea-level rise scenario using the SEAWAT 2000 code. Simulations were performed of a 90-yr sea-level rise period, followed by a period in which boundary conditions were stable. Literature values of solute-transport model parameters were used.

A key observation is that the duration of the transient effect is strongly related to the ratio of hydraulic conductivity and recharge (K/W) and effective porosity (n_e). In general, the lower the effective porosity, the closer a system will be to a state of dynamic equilibrium (Webb and Howard 2011). Systems with a low K/W ratio also did not develop a significant degree of disequilibrium and normally stabilized within decades after the cessation of rising sea-level conditions (Webb and Howard 2011). A system with a K/W ratio of 4.6×10^3 was reported to reach equilibrium in about 90 yrs following a sea level rise. High K/W and n_e values resulted in the simulated saline-freshwater interface lagging its equilibrium position by hundreds of meters and that frequently several centuries were required to reach equilibrium conditions (Webb and Howard 2011). The transient response was not sensitive to storativity values. Key lessons of the Webb and Howard (2011) study are that transient effects can greatly impact the position of the fresh-saline groundwater interface in response to sea level rise and the importance of obtaining detailed site-specific data on aquifer hydraulic and solute transport parameters, which can impact the extent of transient responses.

In the case of confined aquifers, SLR can cause a decrease in the seaward hydraulic gradient if the head in the recharge area is unchanged. Chang et al. (2011) investigated the "lifting" effect of SLR on saline water intrusion in confined aquifers in which SLR increases pressure (water levels) throughout the aquifer. The increased water levels on the freshwater side of the fresh-saline groundwater interface would tend to counter the tendency for the SLR to induce saline water intrusion.

Chang et al. (2011) developed a theoretical two-dimensional density-dependent solute transport model of a confined aquifer using the SEAWAT code. A constant-flux scenario was used in which water was applied to the top of the confined aquifer as recharge and at the landward end of the model by injection at a constant rate. A steady-state baseline scenario was first run for the confined aquifer to estimate the initial saltwater wedge position. Subsequent simulations were run with higher sea level elevations. The results showed that the postulated lifting effect would indeed occur, and that it fully offsets the negative impacts of SLR in constant-flow confined aquifer systems. Transient modeling simulations revealed a self-reversal mechanism. With an instantaneous rise in sea level, some landward migration of saltwater wedge is simulated to occur, but the lifting processes was able to subsequently fully reverse the migration. Sensitivity analysis results demonstrated that the rate of the self-reversal process depends on the SLR rate and specific storage value of the aquifer. Simulations of an unconfined aquifer also showed that the lifting effect would be active, but it was difficult to observe any reversal effects due to increases in aquifer transmissivity caused by increases in aquifer saturated thickness (Chang et al. 2011). The key implication is that if ambient recharge remains constant, then sea level rise would have no impact on the steady-state position of the saltwater wedge (Chang et al. 2011). However, transient effects (i.e., self-reversal process) may occur. Chang et al. (2011) cautioned that their results were based on an idealized rectangular aquifer and that climate changes can impact recharge and regional fluxes.

Ataie-Ashtiani et al. (2013) using an analytical model assuming a sharp interface demonstrated that land-surface inundation (LSI) induces significantly more extensive seawater intrusion (SWI) than the effects of pressure changes at the shoreline in unconfined aquifers. In agreement with early work by Kooi et al. (2000), Ataie-Ashtiani et al. (2013, p. 1676) concluded that "If the LSI front advances faster than the SWI wedge, free convection is likely to develop. When free convection occurs, downward salinization occurs at a much faster rate than caused by diffusion, dispersion and advection alone, and it can play a significant role in the spatial and temporal variations of the SWI interface."

Ferguson and Gleason (2012) evaluated the susceptibility of coastal aquifers to SLR using a steady-state analytical solution that assumes a discrete fresh-saline groundwater interface. Only aquifers with a very low hydraulic gradient (<0.001) were found to be vulnerable to saline water intrusion from SLR. These regions generally have low topographic gradients and would, therefore, be impacted by both seawater inundation and saline water intrusion. Low gradient regions include South Florida and many coastal plain and deltaic areas. The impacts of groundwater extraction on saline water intrusion were found to be greater than the impacts of SLR and changes in groundwater recharge. An important conclusion of Ferguson and Gleason (2012) is that human groundwater use is a key driver in the hydrology of coastal aquifers and that efforts that focus on adapting to SLR rather than overall better water management are misguided.

6.4.3 Evaluation of Location of Fresh-Saline Water Interface

Effective management of saline water intrusion requires knowledge of the local position and shape of the fresh-saline groundwater interface. Methods used to locate the position of coastal fresh-saline groundwater interfaces were reviewed by Maliva (2016, 2019). Under natural, pre-development conditions, the seaward flow of fresh groundwater controls the landward migration of the saline water interface. The interface between fresh and saline groundwater is typically roughly wedge-shaped because of density stratification, with the greatest landward extent of saline water occurring at the base of the aquifer, which is commonly referred to as the "toe" of the wedge. The position and shape (slope) of the saline water wedge and thickness of the mixing zone are controlled by the variance in hydraulic conductivity, which in turn controls longitudinal and transverse macro-dispersivities (Kerrou and Renard 2010).

Theoretical modeling results show that the distance of landward penetration of the toe of the saline water wedge increases with increasing aquifer heterogeneity (Kerrou and Renard 2010). The presence of a high-transmissivity flow zone near the base of an aquifer would allow saline water to migrate inland more rapidly and to a greater extent than would occur under more homogeneous conditions. Theoretical methods can provide some rough guidance on the location and shape of the fresh-saline groundwater interface. However, accurate mapping of the interface requires a field investigation, which normally involves borehole and surface geophysical methods.

A multiple-element approach is preferred to determine the position and shape of fresh-saline groundwater interfaces and to detect changes in their positions over time. Aquifer characterization methods were reviewed by Maliva (2016). The main data sources used to map vertical and horizontal differences in groundwater salinity are:

- water samples collected during borehole drilling
- borehole geophysical logging
- water samples collected from monitoring wells
- surface geophysics.

The most accurate salinity data are analyses of actual water samples obtained from discrete depth intervals from boreholes or monitoring wells. There is no adequate substitute for a monitoring well network from which direct data on groundwater salinity can be obtained. Installation of multiple separate monitoring wells can be quite expensive, especially for deep aquifers. Dual- or tri-zone monitoring wells, in which a single well is open to multiple depth intervals, can be less expensive than multiple separate wells. Multilevel sampling (MLS) systems are an alternative to multiple separate or multiple-zone monitoring wells. MLS systems allow for a large number of monitoring points within a single borehole. MLS systems tend to be most economical for deep aquifers (where well drilling costs are high) and where numerous monitoring points are required (Einarson 2006; Cherry et al. 2015). Time-series of samples collected from monitoring wells and MLS systems allow for detection of movement of the fresh-saline groundwater interface over time.

Monitoring well data may be augmented by borehole and surface geophysical data. Resistivity and electromagnetics-based surface geophysical methods can be effective in mapping fresh-saline groundwater interfaces because the interfaces are often marked by a sharp contrast in electrical resistivity. Borehole water quality and geophysical data allow for the calibration and ground truthing of surface geophysical data. Surface geophysical methods can provide a much greater areal coverage than is possible using wells alone but at the expense of lesser vertical resolution.

6.4.4 Saline Water Intrusion Vulnerability Assessments

Methodologies have been developed to map the vulnerability of coastal aquifers to saline water intrusion. Chachadi et al. (2002) and Lobo-Ferreira et al. (2005) presented the GALDIT method for assessing aquifer vulnerability to seawater intrusion. GALDIT is a weighted numerical ranking system that considers the following mappable factors:

- Groundwater occurrence (aquifer type; unconfined, confined and leaky confined)
- Aquifer hydraulic conductivity
- Groundwater Level above sea level
- Distance from the shore
- Impact of existing seawater intrusion in the area
- Thickness of the aquifer.

Werner et al. (2012) proposed a series of analytical equations, based on a sharp interface assumption, to quantify the vulnerability of confined and unconfined aquifers to saline water intrusion. Stresses considered are sea level rise and changes in recharge and seaward discharge. Some of their important observations are:

- SLR impacts in head-controlled unconfined aquifers produce higher rates of saline water intrusion compared to in flux-controlled aquifers
- in unconfined aquifers, flux-controlled boundary conditions produce a larger toe response to changes in recharge than in head-controlled boundary conditions
- SLR is inconsequential in confined aquifers that are flux controlled.

Michael et al. (2013) performed a global assessment of vulnerability to SLR based on whether the coastal groundwater system is either topography-limited (specified head) or recharge-limited (specified flux). The results of their GIS analyses show that under present and future scenarios, over half of the world's coastlines may be topography-limited and thus vulnerable to sea-level rise. Whether a system is topography-limited or recharge-limited was found to depend on a combination of hydrologic and physical features (Michael et al. 2013). Michael et al. (2013, p. 228) concluded that

> Future recharge and sea-level rise scenarios have much less influence on the proportion of vulnerable coastlines than differences in permeability, distance to a hydraulic divide, and recharge, indicating that hydrogeologic properties and setting are more important factors to consider in determining system type than uncertainties in the magnitude of sea-level rise and hydrologic shifts associated with future climate change.

Klassen and Allen (2017) presented the methodology and results of an assessment of the risks of saline water intrusion vulnerability in coastal aquifers on the Gulf Islands in British Columbia. Groundwater in the islands, located near the city of Vancouver, occurs predominantly in fractured rock aquifers. Three main salinization risks applicable in the study area are horizontal intrusion, inundation by SLR and storm surges, and excess pumping. The susceptibility of an aquifer to saline water intrusion was characterize based on distance from coast, topographic slope, and groundwater flux.

6.4.5 Site Specific Modeling of SLR Impacts on Saline Water Intrusion

The current state-of-the-art for evaluating the impact of SLR on saline water intrusion is density-dependent solute-transport modeling using programs such as the U.S. Geological Survey SEAWAT Code (Guo and Langevin 2002; Langevin et al. 2003) or FEFLOW (Diersch 1998) code. Both SEAWAT and FEFLOW can simulate the advective and dispersive transport of saline water and the impacts of differences in water density related to salinity and temperature on groundwater flow. The general procedure is to develop a model that is calibrated to current aquifer heads and salinity, and then perform predictive transient simulations with different SLR rise scenarios, which are simulated by changing boundary conditions (heads assigned to constant or time-variant specified head cells) along the coast.

Density-dependent numerical solute-transport modeling is essential for the simulation of coastal aquifer systems in which there are large vertical and geographic variations in salinity and thus water density. Two-dimensional models oriented perpendiculað to the coast and parallel to the direction of groundwater flow can be used to assess large-scale impacts of SLR and climate change. Three-dimensional models are required for evaluation of the geographic pattern of the migration of the fresh-saline groundwater interface. Three-dimensional modeling is also required to simulate the impacts to and caused by geographically distributed extraction wells. Follows are discussions of a series of modeling studies that examined the potential impacts of SLR on,saline water intrusion, which were chosen to illustrate the diversity of approaches that have been successfully used and the scale of potential impacts.

6.4.5.1 Monterey County, California

Loáiciga et al. (2012) performed numerical simulations of saline water intrusion in response to SLR in the Seaside Area groundwater subbasin of the Salinas Valley Groundwater Basin, Monterey County, California. A four-layered model, representing the four aquifers locally present, was developed using the FEFLOW code (Diersch 1998). The western (coastal) boundary of the model was represented as a time-variant head boundary. Two sea-level rise scenarios were simulated over a one hundred-year period (2006–2106): a 0.5 m sea level rise and a 1.0 m rise. The sea-level rises were incorporated into the simulations assuming a linear increase over time. Groundwater extraction was simulated at the estimated aquifer safe yield (9730 m^3/d) and the average extraction rate for 2002–2006 (15,430 m^3/d).

The model was calibrated to hydraulic heads and groundwater salinity for the period 1956–2006. The model was specifically calibrated so that no appreciable saline water intrusion, as defined by the position of the 10,000 mg/L iso-salinity line, occurred by 2006, consistent with field conditions. Several predictive simulations were then performed for the different sea-level rise and groundwater extraction scenarios. The main result of the modeling is that groundwater extraction is by far the dominant factor in inducing saline water intrusion in the Seaside Area subbasin (Loáiciga et al. 2012).

6.4.5.2 Hilton Head, South Carolina

The confined Upper Florida aquifer is the local water source in the Hilton Head area of coastal South Carolina. The U.S. Geological Survey (Payne 2010) evaluated the impacts of climate change on saline water intrusion in the area using a calibrated three-dimensional density-dependent solute-transport model created using the Sutra 2.1 code (Voss and Provost 2010). Saline water is believed to enter the aquifer through downward leakage where the overlying confining unit is thin or eroded. The simulation results indicate that SLR will have a minor impact on saline water intrusion (except in the Pinckney Island area). A decrease in onshore freshwater recharge, such as might occur by a decrease in precipitation, was found to have the greatest effect on the position of the saline water plume in the aquifer. Changes in recharge may thus be a more important consideration in saline water intrusion management than the estimated rates of SLR (Payne 2010).

6.4.5.3 Shelter Island, New York

Shelter Island is a small (31 km^2) island located at the eastern end of Long Island, New York. Freshwater occurs in a sand-and-gravel aquifer (Upper Glacial aquifer) that is underlain by a marine clay confining unit (Rozell and Wong 2010). Freshwater is present as a lens in the Upper Glacial aquifer and there are no streams on the island that act as drains. Current average groundwater pumping on the island is approximately 1% of the estimated 48,700 m^3/d of submarine groundwater discharge (Rozell and Wong 2010). Rozell and Wong (2010) developed a two-dimensional SEAWAT model of Shelter Island to simulate the impacts of climate change on the freshwater resources of the island. The model was calibrated for March 1995 (a below average precipitation period) and validated using 1996 and 1997 data.

Rozell and Wong (2010) simulated average precipitation conditions and two scenarios based on the 2007 IPCC Report:

- Scenario 2: Precipitation increase of 15% and a SLR of 0.18 m
- Scenario 3: Precipitation decrease of 2% and a SLR of 0.61 m.

The inner edge of the freshwater/saline water interface was defined as 0.25 kg/ m^3 or 250 mg/L of total dissolved solids (50% of the U.S. secondary drinking water standard of 500 mg/L). Scenario 2 resulted in the interface moving seaward by an average of 23 m and a maximum of 60 m near the base of the interface. The average water table rise was 0.27 m. Scenario 3 resulted in the interface moving landward by an average of 16 m and a maximum of 37 m near the base of the interface. The average water table rise was 0.59 m. Under both scenarios, the volume of freshwater in the aquifer increased by 1–3%. The underlying marine clay unit restricted the movement of the bottom of the freshwater lens and the sides of the lens moved only landward and seaward in response to the changed hydrological conditions (Rozell and Wong 2010).

The Shelter Island model did not consider the inundation of low-lying areas. The results of the simulations suggest that the primary challenge of climate change on Shelter Island will not be potable water retention (Rozell and Wong 2010). The Ghyben-Herzberg relationship is not appropriate for Shelter Island due to the underlying bounding marine clay layer. The unexpected fresh groundwater volume increase under unfavorable conditions was best explained by the clay layer under the aquifer restricting the maximum depth of the aquifer (Rozell and Wong 2010).

6.4.5.4 Dutch Delta, The Netherlands

The low-lying Dutch Delta of the Netherlands has long been threatened by RSLR due to land subsidence and can thus serve as a natural laboratory case for other low-lying delta areas of the world (Oude Essink et al. 2010). Climate change is jeopardizing the Dutch Delta by both increased flooding risk and impacts to the fresh water supply (Oude Essink et al. 2010). Nearly one third of the Netherlands lies below mean sea level. The Dutch Delta contains a mosaic of polders, which are areas protected by dikes and have controlled water levels. Brackish and saline groundwater underlie the shallow freshwater aquifer system of the region. The present salinity distribution of the Netherlands is not yet in dynamic equilibrium with hydrological events over the past centuries. The maintenance of water levels below sea level in the polders results in an upward hydraulic gradient and associated seepage of brackish and saline water into the overlying fresh groundwater. Oude Essink et al. (2010) referred to on-going groundwater salinity increases due to past changes as "autonomous salinization."

Oude Essink et al. (2010) modeled the potential impacts of climate change on the groundwater resources of the Dutch Delta using the MOCDENS3D code (Vandenbohede 2007). The model was reported to have cells of 250 m by 250 m and 40 layers. Land subsidence was incorporated into the model by progressively lowering the water levels in the top layer of model cells with land. Four climate change scenarios were evaluated, which differ in rates of sea level rise and precipitation. Future evaporation and precipitation rates were derived from RCMs (developed and run by others) based on GCMs used in the IPCC Fourth Assessment Report (AR4). The modeling results show that groundwater hydraulic heads will rise in response to SLR, but that the effects will be limited to a small zone within 10 km of the coastline and main rivers. A small relative (<1%) decrease in fresh groundwater volume was simulated to occur by 2100 (relative to the autonomous salinization scenario), but the absolute decrease in freshwater volume will be substantial (a decrease of 200–2750 million m^3; Oude Essink et al. 2010).

Two adaptation scenarios were simulated: reclaiming land from the coast offshore (expanding the fresh groundwater zone seaward) and inundating a low-lying polder area (Oude Essink et al. 2010). The increase in hydraulic heads caused by the former option werer found to increase saline water seepage and the salt load into the fresh groundwater. Inundation of the polder was found to eliminate seepage fluxes in the polder, but the inundation increases aquifer heads outside of the inundated area causing increased seepage fluxes and salt load in adjacent areas (Oude Essink et al. 2010). Oude Essink et al. (2010, p. 14) concluded that "It is not easy to stop salinization of the groundwater and surface water system: a combination of different human interventions is possibly needed to decrease the salt load for the future. As such, the Dutch will very likely have to cope with much more saline groundwater in their coastal water system than at present."

6.4.5.5 Island of Faster, Denmark

Rasmussen et al. (2013) assessed the impacts of climate change on the southern part of the island of Faster (Denmark) in the western Baltic Sea. The study area includes a former lagoon that was reclaimed by closing a strait that connected it to the sea and by the construction of a drainage system. The main aquifer is a fractured chalk aquifer that is overlain by Quaternary siliciclastic deposits. The local surface water system is dominated by the artificial drainage canal system.

Groundwater flow and solute transport modeling was performed using the SEAWAT code. An extensive hydrogeological database was developed for the project, which included groundwater chemistry and environmental tracer data, airborne (helicopter) geophysics, borehole geophysical logging, and water level data from numerous wells in the study area. The SEAWAT model had 50 m by 50 m grid cells and 32 layers with thicknesses ranging from 2 to 12 m. The model was calibrated to hydraulic heads and validated against geochemical and geophysical data.

The best estimate climate scenario for the study area is an increase in recharge of 15% (from another study in Denmark) and a relative sea level rise of 0.75 m between 2010 and 2100. The modeling results showed that when sea level and groundwater recharge are increased simultaneously there was almost no effect on the simulated saltwater distribution to the west of the drainage canal. The expected saline water intrusion caused by sea level rise is countered by an increase in groundwater recharge and hence groundwater levels (Rasmussen et al. 2013). On the eastern part of the island, the regional flow system connects the drainage canal to the sea through highly permeable fractured chalk. The rising sea level results in an increasing hydraulic gradient toward the drainage canal where the stage is held constant despite the increase in sea level and recharge, resulting in more pronounced saline water intrusion (Rasmussen et al. 2013).

The results of this study demonstrated how saline water intrusion will be controlled by a combination of changes in groundwater recharge, SLR, groundwater extraction, and canal operation/maintenance (Rasmussen et al. 2013). Rasmussen et al. (2013, p. 441) concluded that

> For the system with flux controlled boundary conditions, only changes in groundwater recharge had an effect of the saltwater distribution, whereas for the system with head-controlled boundary conditions changes in recharge, sea level and the boundary itself (the stage of the canal) were found to be important for saltwater intrusion. For the actual system, changes in recharge were found to be the most important factor, whereas minor sea level rises do not seem to affect the sea water intrusion as much.

6.4.5.6 Borkum, German North Sea

Sulzbacher et al. (2012) performed numerical modeling of the impacts of climate change on fresh groundwater on the German North Sea island of Borkum. This study is noteworthy in its extensive hydrogeological data collection, which included information from boreholes, a seismic survey, a helicopter-borne electromagnetic (HEM) survey, monitoring of the fresh-saline groundwater boundary by vertical electrode chains in two boreholes, measurements of groundwater table elevations, pumping and slug tests, and groundwater sampling. The data were used to determine the hydrostratigraphy, aquifer hydraulic properties, and three-dimensional distribution of salinity on the island, and to use as hydraulic and salinity calibration targets. A three-dimensional FEFLOW model was developed that has 39 horizontal layers and cell sizes that range from 10 to 200 m. The transient calibration period started with the last equilibrium (non-pumping influenced) state of the aquifer in year 1900.

Climate change projections used in the simulations (from the Norddeutsches Klimabüro) were an increase in average annual temperature of 2.9 °C and an increase in annual precipitation of about 10% (summer −5% and winter +25%) by year 2100. Due to the enhanced annual precipitation scenarios used, annual groundwater recharge was simulated to linearly increase to +10% in 2100 (average scenario) and to +5% in 2100 (conservative scenario). Relative sea level was projected to rise by 1 m from 1995 until 2100 or 0.94 m from 2010 to the year 2100, which was simulated using a linear function. The area covered by boundary conditions of this type (time-varying constant head) was extended progressively landward according to the rise in mean sea level (Sulzbacher et al. 2012).

The modeling results for the year 2100 show that SLR through this time will not essentially affect the general shape of the freshwater lens. At greater depths, salinization of the lens is simulated to occur particularly in the area of a waterworks where upconing of sea water and saltwater intrusion will threaten drinking water quality (Sulzbacher et al. 2012). The combination of enhanced recharge and the increased sea level by 2100 will cause the groundwater table to rise in the dune areas by about 0.4 m for the average scenario and 0.3 m for the conservative scenario (Sulzbacher et al. 2012). Drained marshlands will be minimally affected by a rising water table.

6.4.5.7 Broward County, Southeastern Florida

The U.S. Geological Survey developed a variable-density solute-transport groundwater flow model for Broward County, Florida (which includes the city of Fort Lauderdale) using SEAWAT to evaluate the controls over the current and historical distribution of salinity in the Biscayne Aquifer (the primary water source in the county) and to simulate the potential effects of increases in pumping, variable rates of sea level rise, movement of a salinity control structure, and the use of drainage recharge wells on the future distribution of salinity in the aquifer (Hughes et al. 2016). Saline water intrusion was found to be sensitive to changes in sea level, groundwater recharge, and stage of the drainage canals. For the near-coastal extraction wells at risk, chloride concentrations were found to be most sensitive to the stage of the drainage canals and groundwater recharge (Hughes et al. 2016).

The SEAWAT model was calibrated for a 62-yr period (Jan 1950–May 2012) and predictive simulations were run for a 50-yr period. The base case scenario had no SLR, whereas scenarios 2 through 4 had SLRs of 0.77, 1.40, and 2.03 ft (0.23, 0.43, and 0.62 m), respectively, at the end of a 50-yr period. The simulation results show progressive saline water intrusion, mapped as the 1000 mg/L isochlor, with increasing SLR (Fig. 6.6).

6.4.5.8 Laccadive Islands, India

Bobba et al. (2000) in an early study simulated the impacts of SLR on freshwater lenses in the Laccadive Islands, which are coral atolls located in the Arabian Sea, southwest of India. Two-dimension (x, y) single layer models of island aquifers were constructed using the SUTRA code (Voss 1984). The model results showed that even very small SLR due to climate change adversely impacted the thickness of the freshwater lenses on the islands. The simulations show that when the aquifer water budgets are negative, the aquifers will experience seawater intrusion. Bobba et al. (2000) also noted that freshwater aquifers are vulnerable to contamination from abandoned or improperly constructed wells that provide conduits for the migration of contaminated waters.

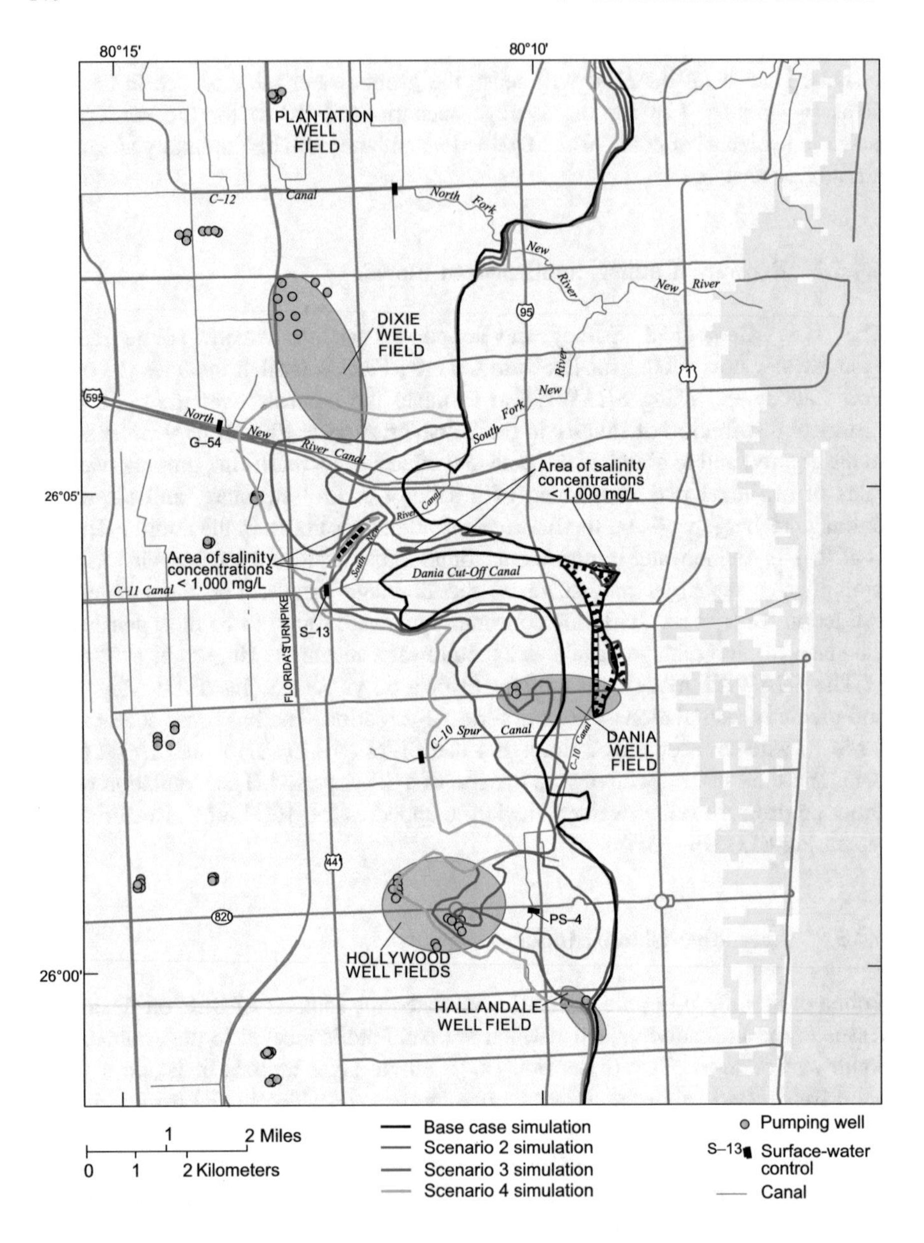

Fig. 6.6 Simulated positions of the 1000 mg/L isochlors in the Biscayne aquifer of Broward County, Florida in 2060 for the base case and three climate change scenarios. Modified from Hughes et al. (2016, Fig. 33)

6.5 Rising Water Tables—Groundwater Inundation

Many urban areas are experiencing rising water tables, which include cities in the Middle East, one of the driest areas of the world (e.g., Riyadh and Jeddah, Saudi Arabia; Kuwait City; Doha, Qatar). The rising water tables are the result of unmanaged and unintentional aquifer recharge, including leakage from water, wastewater, and reclaimed water mains, recharge from on-site sewage treatment and disposal systems, and excessive irrigation of parks and other green areas. Rising water tables can damage building foundations and underground infrastructure, and cause ponding in basements, underground structures (e.g., parking garages and underground vehicle and subway tunnels) and at land surface (Abu-Rizaiza 1999; Alhamid et al. 2007; Al-Sefry and Şen 2006; Allocco et al. 2016). Rising groundwater levels can potentially reduce the supporting capacity of the ground and increase liquefaction (Hay and Mimura 2005). Rising urban groundwater can be controlled by reducing the causes of the rise, such as expanding centralized sewage systems and leak detection and repair, and by the installation of drainage systems.

The water table in unconfined aquifers typically lies above mean sea level, fluctuates with daily tides and other low-frequency sources of ocean energy, and has tidal amplitudes that decrease exponentially with distance from the shoreline (Rotzoll and Fletcher 2013). Rotzoll and Fletcher (2013, p. 477) explained that

> As sea level rises, the water table will rise simultaneously and eventually break out above the land surface creating new wetlands and expanding others, changing surface drainage, saturating the soil, and inundating the land depending on local topography. Flooding will start sporadically but will be especially intense seasonally when high tide coincides with rainfall events.

Increases in water table elevation caused by RSLR depend on multiple factors, including land surface elevation and subsurface permeability (Hay and Mimura 2005). Rising water tables from SLR can occur as far as several tens of kilometers inland resulting in the formation or expansion of standing bodies of fresh and brackish water in low-lying areas (Hay and Mimura 2005). The threat of flooding from SLR will tend to be greater when water tables have also risen in response to heavy rainfall events.

6.5.1 Coastal Groundwater Inundation Vulnerability Mapping Methods

The rate of rise of the water table in response to RSLR is controlled by a number of factors including tidal forcing and other variations in sea level, aquifer hydrogeology, coastline shape, shore slope, and groundwater pumping (Plane et al. 2019). Local SLR-induced groundwater rise may be locally dampened where the water table intercepts low-elevation drainage features that allow for groundwater discharge (Befus et al. 2020).

The basic methodology used to map areas at risk of groundwater inundation is to subtract a future groundwater (water table) elevation surface from a DEM. Areas with negative values are at risk for groundwater flooding. Areas were the water table is expected to rise close to (but not above) land surface are also at risk for adverse impacts, such as flooding of basements and underground infrastructure, and waterlogging of soils. A key technical issue is that DEMs and current and future sea levels must use same datum. For some datums (e.g., NGVD, NAVD), local mean sea level does not equal 0 feet elevation. As is the case for all evaluations of sea level rise impacts, a decision needs to be made as to what RSLR at a given future time to use in the analysis.

Methodologies used to evaluate groundwater inundation risks vary principally in the manner in which future water table elevation GIS layers are generated. The choice of methods will be dictated by the available data (including existing numerical groundwater models of the study area), project timeframes and budgets, and risk assessment objectives. For example, a quicker, less expensive method might be used for a broad area screening to identify geographic locations at high risk, whereas a more detailed, modeling-based approach would be more appropriate for local areas where potential future water table elevations are needed to design local structures or drainage systems. Three main methods that have been used in groundwater inundation studies are described below in order of decreasing technical sophistication and effort required.

6.5.1.1 Three-Dimensional Groundwater Modeling Approach

Future groundwater elevations are obtained using a calibrated three-dimensional groundwater flow model of the area of interest, ideally using a density-dependent solute-transport code (e.g., SEAWAT). Groundwater flow modeling codes, such as MODFLOW, could be used but heads need to be converted to equivalent freshwater heads. RSLR is simulated through boundary conditions, i.e., heads assigned to constant or general head cells along coast. MDCDRER, MDWASD, & FDOD (2018) mapped areas in Miami-Dade County where septic systems are vulnerable to rising groundwater levels using an existing sophisticated model of the county previously developed by the U.S. Geological Survey (Hughes and White 2016). Habel et al. (2017) used a modeling-based approach to assess groundwater inundation risk in Honolulu, Hawaii.

Modeling-based approaches are potentially most accurate, but development of a calibrated groundwater model is labor intensive and requires data on local hydrogeology and sufficient monitoring data (time series of water elevation data from multiple wells) for model calibration and validation.

6.5.1.2 Empirical Water Table Elevation Surface and Hydrostatic Rise Approach

The empirical water table elevation surface with hydrostatic rise approach involves the generation of a current water table elevation surface by interpolation of monitoring well data. Existing water table elevation contour maps might be used but their accuracy should be assessed. Tidal effects may be considered by using the highest water table elevations measured from wells (assumed to represent high tide conditions) or correcting contour maps assumed to reflect mean water table elevations by either (1) applying a constant (MHHW—MSL) across the study area, which is a simple and conservative approach, or (2) multiplying (MHWW—MSL) by the local tidal efficiency (using either a literature function of tidal efficiency versus distance from shore or a calculated site-specific value). A flux-dominated flow system is assumed in which the water table rise near the coast is the same as the selected RSLR scenario.

The main advantages of the empirical water table elevation surface with hydrostatic rise approach are that it involves much less effort than a numerical modeling-based approach and it captures the usual landward increase in water table elevation. Its main disadvantage is that it requires either considerable monitoring data to generate accurate water table elevation contour maps or the use of simplifying assumptions. Varieties of the empirical water table elevation surface with hydrostatic rise approach were used by Rotzoll and Fletcher (2013), Hoover et al. (2017), and Plane et al. (2019).

6.5.1.3 Simple Hydrostatic Rise with No Hydraulic Gradient Approach

The simple hydrostatic rise with no hydraulic gradient approach is a screening method that assumes a flat water table near the coast. Future water table elevation is calculated as the current MHHW plus the relative SLR. The great advantage of this method is that it is a simple technique that requires only a DEM. Habel et al. (2019, p. 9) compared the groundwater modeling and the hydrostatic approach for a study area in Honolulu and concluded that

> Though use of data-assimilating numerical modeling methods are more appropriate in cases where high accuracy simulations are necessary, we find that use of the hydrostatic method (specifically when referencing the local MHHW tide stage) is suitably accurate as a first-cut approach in identifying municipal vulnerabilities to GWI.

The potential error resulting from not including the landward rise in the water table was found to be partially offset by not including the landward attenuation of the tidal signal.

6.5.2 *Coastal Groundwater Inundation Studies*

Fewer investigations have been performed of coastal groundwater inundation vulnerability than for direct seawater inundation. Regional studies tend to use the simple hydrostatic rise approach as a screening tool. More detailed and data intensive methods have been used in local studies.

6.5.2.1 Oahu, Hawaii

Rotzoll and Fletcher (2013) assessed the potential for groundwater inundation as a consequence of RSLR in southern Oahu, Hawaii. Steady-state regional water table elevations at distances perpendicular to the shore were obtained using the equation of Glover (1959), which considers the freshwater flow rate per unit length of shoreline, aquifer hydraulic conductivity, and the density difference between freshwater and seawater. A linear increase in water table elevation with SLR was assumed in this flux-controlled system. Areas vulnerable to inundation from RSLR were mapped by subtracting the DEM elevation from the estimated regional water table elevation in 0.33 m intervals to a 1 m maximum rise. The effect of tides was evaluated by multiplying the offset between current mean sea level and mean higher high water (MHHW) by the tidal efficiency calculated for the distance from the shore and adding the amount to the future mean water table elevation. The results of the analyses are that a 0.6 m RSLR would cause substantial flooding, and 1 m sea-level rise would inundate 10% of a 1 km wide heavily urbanized coastal zone. The flooded area including groundwater inundation is more than twice the area of direct marine inundation alone (Rotzoll and Fletcher 2013).

6.5.2.2 Northern California

Hoover et al. (2017) employed a simplified approach to assess vulnerability to sea level rise from both direct inundation and groundwater inundation at low-lying, coastal sites in California. Water table elevation data were obtained from published groundwater contour maps and well data. Topographic data were obtained from existing NOAA DEMs. Inundation maps were generated for 1 and 2 m RSLR projections. Water table elevations were assumed to rise linearly with local relative SLR.

Hoover et al. (2017) concluded that the extent and degree of SLR-driven groundwater inundation and shoaling are expected to vary from one location to another in California with differences driven by the depth to the water table, local geology, hydrology, and anthropogenic factors (e.g., the extent of groundwater extraction or additions). Coastal communities in central and southern California that are underlain by shallow freshwater aquifers are not expected to have major SLR-driven groundwater emergence issues, even in low-lying areas, primarily because the heavy use of fresh groundwater will keep groundwater levels low (Hoover et al. 2017).

6.5.2.3 Honolulu, Hawaii

Habel et al. (2017) presented a six-component methodology to simulate groundwater inundation induced by sea-level rise and high tides in Honolulu, Hawaii. The main components are:

(1) continuous groundwater-level monitoring
(2) compilation of groundwater-level measurements
(3) estimation of tidal efficiency
(4) development and calibration of a groundwater flow model using the MODFLOW-2005 code
(5) generation of flood maps
(6) performance of a damage assessment.

Groundwater monitoring data were processed to estimate the ocean oscillation-corrected tidal efficiency as a function of distance from the coast. A groundwater-flow model was developed to simulate steady-state groundwater levels at local mean sea level in the unconfined caprock aquifer under current conditions and under future increases in sea level of 0.32, 0.60, and 0.98 m.

Tidal influence was simulated by first applying the analytical solution for tidal efficiency across a raster grid as a function of the distance of each raster cell from the modeled coastline. The tidal efficiency values of each cell were then multiplied by a given tide offset to simulate the tidal surplus. The tidal surplus raster data set was then summed with respective mean water-table elevation raster data sets to obtain the tidally influenced water-table height for the given tide stage (Habel et al. 2017).

A DEM was produced by merging and hydro-flattening 2013 NOAA DEM tiles. The final step was subtracting the water-table raster grid from the DEM on a cell-by-cell basis. Positive cell values represent locations where the simulated water table is situated below the modeled terrain and negative cell values represent areas where the simulated water table is situated above land surface (Fig. 6.7). Areas where the thickness of the unsaturated zone will be significantly reduced were also mapped.

Rising groundwater levels can comprise the operation of on-site sewage disposal systems (OSDSs) resulting in the widespread seepage of untreated sewage directly into coastal groundwaters (Habel et al. 2017). Habel et al. (2017, p. 133) concluded that their results indicate that

> approximately 1.1% of the 13 km^2 study area presently has narrow unsaturated space of less than 0.33 m. SLR of 0.98 m raises this extent to 19.3% and produces GWI over 23.0% of the study area. It follows that under 0.98 m of SLR, nearly half ($\sim$42%) of the region will likely experience either chronic GWI during elevated stages of the tide, or episodic flooding induced by heavy rainfall. The corresponding hazard assessment indicates that the

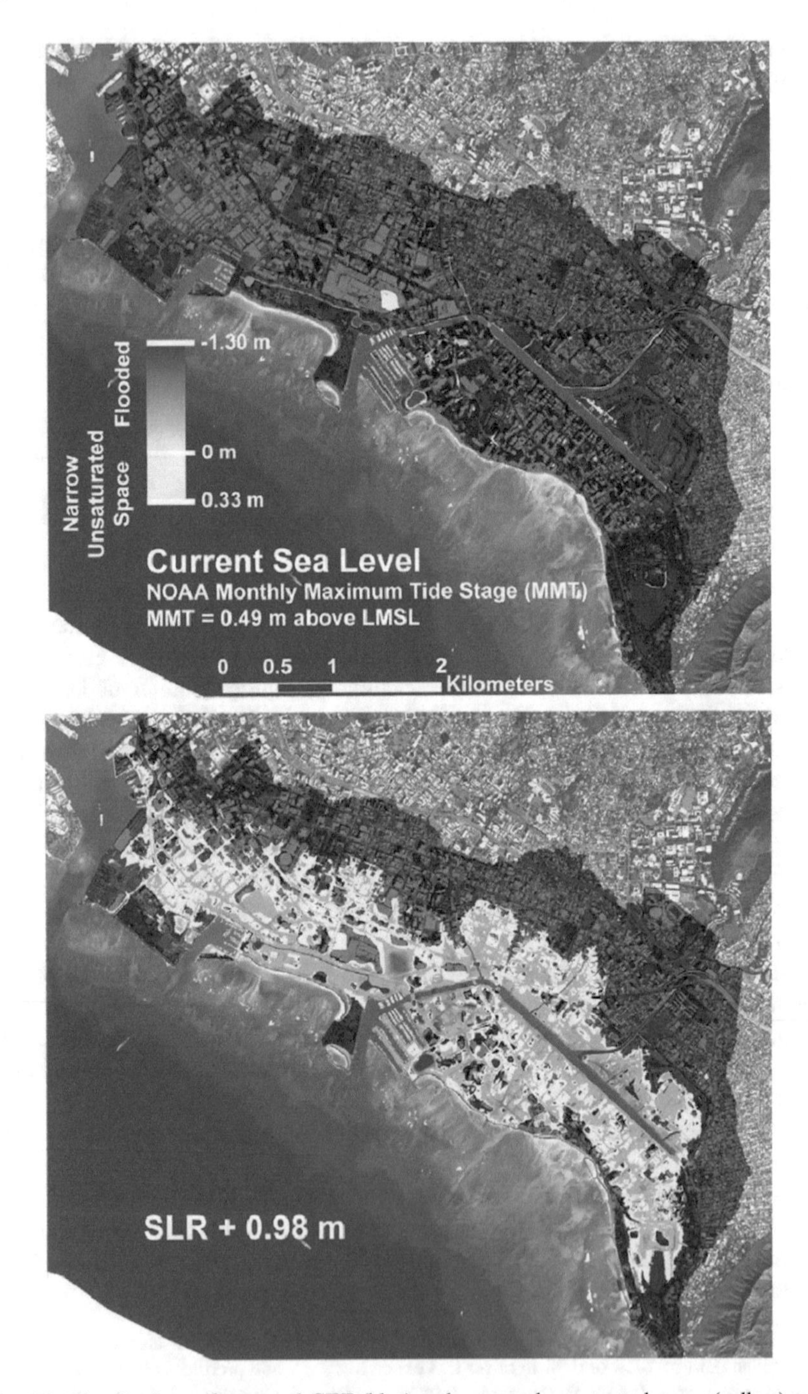

Fig. 6.7 Visualizations of saturated GWI (blue) and narrowed unsaturated space (yellow) areas for simulations of a Honolulu, Hawaii, study area (shaded) representing (top) current conditions and (bottom) a 0.98 m increase in sea level for a tide height representative of the average monthly maximum tidal amplitude (MMT) datum (0.49 m above local mean sea level). The vertical error in the distance between the terrain and groundwater is ±20< cm. *Source* Habel et al. (2017) *Copyright* Elsevier

> currently narrow state of unsaturated space likely causes inundation of 86% of 259 active OSDS sites located across the study area and threatens to fully submerge 39 OSDS sites by the end of the century.

Habel et al. (2019) subsequently compared the results of the numerical groundwater modeling-based evaluation of GWI in the Honolulu area with a simple hydrostatic model in which future sea level was estimated as the current MHHW plus the local RSLR. Errors associated with not including a hydraulic gradient and not considering the landward attenuation of the tidal fluctuations were found to partially offset each other.

6.5.2.4 Coastal New Hampshire

Knott et al. (2019) performed numerical groundwater modeling to assess the potential rise in groundwater levels induced by SLR in the vicinity of the city of Portsmouth, coastal New Hampshire. A five-layer numerical model was developed that was calibrated to average groundwater levels from 1970 to 2014 in 3156 wells (target observations). Water levels were converted to freshwater equivalent heads to simulate the freshwater/saltwater density effects. Knott et al. (2019) cautioned that the accuracy of the model is insufficient to predict the groundwater head at any individual well or location, but the simulations are useful in identifying future changes in regional groundwater-flow patterns and relative groundwater responses to SLR. Areal recharge rates used in the groundwater model were determined by the New Hampshire Geological Survey (NHGS) using the Dripps water balance model (Dripps and Bradbury 2007).

SLRs of 0.3, 0.8, 1.6, and 2 m were simulated, which correspond to the high-emissions scenario in early century (2030), mid-century (2060), and the end of the century (2090 and 2100). The projected rises in sea level were added to the current MSL and then used as the coastal boundary condition for the steady-state simulations. The model considered only the change in hydraulic head at the coast and does not include the inland migration (transgression) of the shoreline.

The modeling results indicate that the ratio of mean groundwater rise to SLR is consistent across the SLR scenarios analyzed and is projected to be approximately 66% between 0 and 1 km, 34% between 1 and 2 km, 18% between 2 and 3 km, 7% between 3 and 4 km, and 3% between 4 and 5 km of the coast (Knott et al. 2019). By the end of the century, the mean groundwater rises ranges from more than 1.3 m between 0 and 1 km from the shore to less than 0.1 m between 4 and 5 km from the coast (Knott et al. 2019). The rising groundwater levels would result in freshwater wetlands expansion from GWI that is projected to begin slowly, with a 3% increase by 2030, and then progresses to a 10% increase by mid-century and a 19–25% increase by the end of century (Knott et al. 2019). The groundwater rise would be dampened near streams as increased gradients would drive more groundwater discharge to streams (Knott et al. 2019). For all scenarios, SLR was found to increase gradients between groundwater piezometric heads and stream stages, increasing groundwater discharge to streams. However, increased tidal-water piezometric head at the coast decreased the coastward hydraulic gradient, reducing groundwater discharge to coastal wetlands (Knott et al. 2019).

6.5.2.5 San Francisco Bay Area

Plane et al. (2019) performed a rapid assessment of areas in the San Francisco Bay area prone to SLR-induced groundwater flooding. A current water table elevation surface was generated by interpolating maximum elevations reported in monitoring wells for the years 1996–2016 and freshwater water table elevations along the coast, which were estimated as the mean tide elevation plus 0.3 m. A sea level rise of 1 m and a simple linear groundwater rise approximation (assuming flux-controlled conditions) were used within 1 km of the bay edge. Depth to water was calculated by subtracting the future water table elevation from a ground surface DEM. The rapid assessment was used to identify locations were more rigorous data collection and dynamic modeling are needed to better evaluate groundwater flooding risks (Plane et al. 2019).

References

Abu-Rizaiza OM (1999) Threats from groundwater table rise in urban areas of developing countries. Water Int 24(1):46–52

Alhamid AA, Alfayzi SA, Hamadto MA (2007) A sustainable water resources management plan for Wadi Hanifa in Saudi Arabia. J King Saud Uni-Eng Sci 19(2):209–222

Allocca V, Coda S, De Vita P, Iorio A, Viola R (2016) Rising groundwater levels and impacts in urban and suburban areas around Naples (southern Italy). Rendiconti Online Soc Geol Ital 41:14–17

Al-Sefry SA, Şen Z (2006) Groundwater rise problem and risk evaluation in major cities of arid lands—Jeddah case in Kingdom of Saudi Arabia. Water Resour Manage 20:91–108

Anderson TR, Fletcher CH, Barbee MM, Romine BM, Lemmo S, Delevaux JM (2018) Modeling multiple sea level rise stresses reveals up to twice the land at risk compared to strictly passive flooding methods. Sci Rep 8(1):1–14

Anderson WP (2002) Aquifer salinization from storm overwash. J Coastal Res 18(3):413–420

Ataie-Ashtiani B, Werner AD, Simmons CT, Morgan LK, Lu C (2013) How important is the impact of land-surface inundation on seawater intrusion caused by sea-level rise? Hydrogeol J 21(7):1673–1677

Bamber JL, Aspinall WP (2013) An expert judgement assessment of future sea level rise from the ice sheets. Nat Clim Change 3(4):424–427

Barlow PM (2003) Ground water in fresh water-salt water environments of the Atlantic Coast. U.S. Geological Survey Circular 1262

Befus KM, Barnard PL, Hoover DJ, Finzi Hart JA, Voss CI (2020) Increasing threat of coastal groundwater hazards from sea-level rise in California. Nat Clim Change. https://doi.org/10.1038/s41558-020-0874-1

Bobba AC, Singh VP, Berndtsson R, Bengtsson L (2000) Numerical simulation of saltwater intrusion into Laccadive Island aquifers due to climate change. J Geol Soc India 55:589–612

Chachadi AG, Lobo Ferreira JP, Noronha L, Choudri BS (2002) Assessing the impact of sea-level rise on salt water intrusion in coastal aquifers using GALDIT model. Coastin: A Coastal Policy Research Newsletter 7:27–32

Chang SW, Clement TP, Simpson MJ, Lee KK (2011) Does sea-level rise have an impact on saltwater intrusion? Adv Water Resoure 34(10):1283–1291

Cherry JA, Parker B, Einarson M, Chapman S, Meyer J (2015) Overview of depth-discrete mutilevel groundwater monitoring technologies: focus on groundwater monitoring in areas of oil and gas well stimulation in California. In: Esser BK (ed) Recommendations on model criteria for groundwater sampling, testing, and monitoring of oil and gas development in California, Final Report (LLNL-TR-669645). California State Water Resources Control Board

Church JA, Clark PU, Cazenave A, Gregory JM, Jevrejeva S, Levermann A, Merrifield MA, Milne GA, Nerem RS, Nunn PD, Payne AJ, Pfeffer WT, Stammer D, Unnikrishnan AS (2013) Sea level change. In: Stocker TF, Qin D, Plattner G-K, Tignor M, Allen SK, Boschung J, Nauels A, Xia Y, Bex V, Midgley PM (eds) Climate change 2013: the physical science basis. Contribution of working group I to the fifth assessment report of the intergovernmental panel on climate change. Cambridge University Press, Cambridge UK, pp 1137–1216

Clark PU, Dyke AS, Shakun JD, Carlson AE, Clark J, Wohlfarth B, Mitrovica JX, Hostetler SW, McCabe AM (2009) The last glacial maximum. Science 325(5941):710–714

Diersch HJ (1998) FEFLOW: interactive, graphics-based finite-element simulation system for modeling groundwater flow, contaminant mass and heat transport processes. In: Getting started; user's manual; reference manual, version 4.6. WASY, Institute for Water Resources Planning and System Research Ltd, Berlin

Dripps WR, Bradbury KR (2007) A simple daily soil–water balance model for estimating the spatial and temporal distribution of groundwater recharge in temperate humid areas. Hydrogeol J 15(3):433–444

Einarson MD (2006) Multilevel ground-water monitoring. In: Nielsen DM (ed) Practical handbook of environmental site characterization and ground-water monitoring, 2nd edn. CRC Press, Boca Raton, pp 808–848

Ericson JP, Vörösmarty CJ, Dingman SL, Ward LG, Meybeck M (2006) Effective sea-level rise and deltas: causes of change and human dimension implications. Glob Planet Change 50(1–2): 63–82

European Environmental Agency (2019) Global and European sea-level rise. https://www.eea.europa.eu/data-and-maps/indicators/sea-level-rise-6/assessment. Accessed 17 Jun 2020

Ferguson G, Gleeson T (2012) Vulnerability of coastal aquifers to groundwater use and climate change. Nat Clim Change 2(5):342–345

Frank T (2019, Apr 11) After a $14-billion upgrade, New Orleans' levees are sinking. Smithsonian American. https://www.scientificamerican.com/article/after-a-14-billion-upgrade-new-orleans-levees-are-sinking/. Accessed 17 Jun 2020

Gallien TW, Schubert JE, Sanders BF (2011) Predicting tidal flooding of urbanized embayments: a modeling framework and data requirements. Coast Eng 58(6):567–577

Gesch DB (2018) Best practices for elevation-based assessments of sea-level rise and coastal flooding exposure. Front Earth Sci 6(article 230)

Glover RE (1959) The pattern of fresh-water flow in a coastal aquifer. J Geophys Res 64(4): 457–459

Guo W, Langevin CD (2002) User's guide to SEAWAT: a computer program for simulation of three-dimensional variable-density ground-water flow. USGS techniques of water-resources investigations book 6, chapter A7, p 77

Habel S, Fletcher CH, Rotzoll K, El-Kadi AI (2017) Development of a model to simulate groundwater inundation induced by sea-level rise and high tides in Honolulu, Hawaii. Water Res 114:122–134

Habel S, Fletcher CH, Rotzoll K, El-Kadi AI, Oki DS (2019) Comparison of a simple hydrostatic and a data-intensive 3D numerical modeling method of simulating sea-level rise induced groundwater inundation for Honolulu, Hawai'i, USA. Environ Res Commun 1(4):041005

Harris A (2020) Feds consider a plan to protect Miami-Dade from storm surge: 10-foot walls by the coast. Miami Herald. https://www.miamiherald.com/news/local/environment/article239967808.html. Accessed 17 Jun 2020

Hay JE, Mimura N (2005) Sea-level rise: implications for water resources management. Mitig Adapt Strat Glob Change 10:717–737

Hoover DJ, Odigie KO, Swarzenski PW, Barnard P (2017) Sea-level rise and coastal groundwater inundation and shoaling at select sites in California, USA. J Hydrol Reg Stud 11:234–249

Hughes JD, White JT (2016) MODFLOW-NWT model used to evaluate the potential effect of groundwater pumpage and increased sea level on canal leakage and regional groundwater flow in Miami-Dade County, Florida. U.S. Geological Survey Data Release. U.S. Geological Survey, Reston

Hughes JD, Sifuentes DF, White JT (2016) Potential effects of alterations to the hydrologic system on the distribution of salinity in the Biscayne aquifer in Broward County, Florida. U.S. Geological Survey Scientific Investigations Report 2016–5022

Illangasekare T, Tyler SW, Clement TP, Villholth KG, Perera APGRL, Obeysekera J, Gunatilaka A, Panabokke CR, Hyndman DW, Cunningham KJ, Kaluarachchi JJ, Yeh WW-G, van Genuchten MT, Jensen K (2006) Impacts of the 2004 tsunami on groundwater resources in Sri Lanka. Water Resoure Res 42(5):W05201

IPCC (2019) Technical summary. In: Pörtner H-O, Roberts DC, Masson-Delmotte V, Zhai P, Tignor M, Poloczanska E, Mintenbeck K, Alegría A, Nicolai M, Okem A, Petzold J, Rama B, Weyer NM (eds) IPCC special report on the ocean and cryosphere in a changing climate. (In press)

Kerrou J, Renard P (2010) A numerical analysis of dimensionality and heterogeneity effects on advective dispersive seawater intrusion processes. Hydrogeol J 18(1):55–72

Klassen J, Allen DM (2017) Assessing the risk of saltwater intrusion in coastal aquifers. J Hydrol 551:730–745

Knott JF, Jacobs JM, Daniel JS, Kirshen P (2019) Modeling groundwater rise caused by sea-level rise in coastal New Hampshire. J Coastal Res 35(1):143–157

Knutson TR, Tuleya RE (2004) Impact of CO_2-induced warming on simulated hurricane intensity and precipitation: sensitivity to the choice of climate model and convective parameterization. J Clim 17(18):3477–3495

Kooi H, Groen J, Leijnse A (2000) Modes of seawater intrusion during transgressions. Water Resoure Res 36(12):3581–3589

Kopp RW, Horton RM, Little CM, Mitrovica JX, Oppenheimer M, Rasmussen DJ, Strauss BH, Tebaldi C (2014) Probabilistic 21st and 22nd century sea-level projections at a global network of tide gauge sites. Earth's Future 2(8):383–406

Kummu M, De Moel H, Salvucci G, Viviroli D, Ward PJ, Varis O (2016) Over the hills and further away from coast: global geospatial patterns of human and environment over the 20th–21st centuries. Environ Res Lett 11(3):034010

Lambeck K, Rouby H, Purcell A, Sun Y, Sambridge M (2014) Sea level and global ice volumes from the last glacial maximum to the Holocene. Proc Natl Acad Sci 111(43):15296–15303

Langevin CD, Shoemaker WB, Guo W (2003) MODFLOW-2000, the U.S. Geological Survey modular groundwater model—Documentation of the SEAWAT-2000 version with the variable-density flow process (VDF) and the integrated MT3DMS transport process (IMT). USGS Open-File Report 03–426

Lin N, Emanuel K, Oppenheimer M, Vanmarcke E (2012) Physically based assessment of hurricane surge threat under climate change. Nat Clim Change 2(6):462–467

Loáiciga HA, Pingel TJ, Garcia ES (2012) Sea water intrusion by sea-level rise: scenarios for the 21st century. Groundwater 50(1):37–47

Lobo-Ferreira JP, Chachadi AG, Diamantino C, Henriques MJ (2005) Assessing aquifer vulnerability to seawater intrusion using GALDIT Method. Part 1: application to the Portuguese aquifer of Monte Gordo. Fourth inter-celtic colloquium on hydrology and management of water resources, Guimarães, Portugal, July 11–14, 2005

Maliva RG (2016) Aquifer characterization techniques. Springer, Cham
Maliva RG (2019) Anthropogenic aquifer recharge. Springer, Cham
Masterson JP, Garabedian SP (2007) Effects of sea-level rise on ground water flow in a coastal aquifer system. Groundwater 45(2):209–217
McCobb TS, Weiskel PK (2003) Long-term hydrologic monitoring protocol for coastal ecosystems. U.S. Geological Survey Open-File Report 02–497
MDCDRER, MDWASD, FDOD (2018) Septic systems vulnerable to sea level rise (November 2018). Miami-Dade County department of regulatory & economic resources, Miami-Dade County water and sewer department & Florida department of health in Miami-Dade County
Michael HA, Russoniello CJ, Byron LA (2013) Global assessment of vulnerability to sea-level rise in topography-limited and recharge-limited coastal groundwater systems. Water Resoure Res 49(4):2228–2240
Mousavi ME, Irish JL, Frey AE, Olivera F, Edge BL (2011) Global warming and hurricanes: the potential impact of hurricane intensification and sea level rise on coastal flooding. Clim Change 104(3–4):575–597
NHC (n.d.) Storm surge overview. National Hurricane Center. https://www.nhc.noaa.gov/surge/. Accessed 28 May 2020
Nicholls RJ, Marinova N, Lowe JA, Brown S, Vellinga P, De Gusmao D, Hinkel J, Tol RS (2011) Sea-level rise and its possible impacts given a 'beyond 4 °C world' in the twenty-first century. Philos Trans R Soc A 369(1934):161–181
Nicholls RJ, Wong PP, Burkett VR, Codignotto JO, Hay JE, McLean RF, Ragoonaden S, Woodroffe CD (2007) Coastal systems and low-lying areas. In: Parry ML, Canziani OF, Palutikof JP, van der Linden PJ, Hanson CE (eds) Climate change 2007: impacts, adaptation and vulnerability. Contribution of working group II to the fourth assessment report of the intergovernmental panel on climate change. Cambridge University Press, Cambridge UK, pp 315–356
NOAA (n.d.a) U.S. sea level trends map. https://tidesandcurrents.noaa.gov/sltrends/slrmap.html. Accessed 29 May 2020
NOAA (n.d.b) Datums for 8723170, Miami Beach FL. https://tidesandcurrents.noaa.gov/datums.html?id=8723170. Accessed 27 May 2020
NOAA (n.d.c) Extreme water levels. Fort Myers, Florida. National Oceanic and Atmospheric Administration. https://tidesandcurrents.noaa.gov/est/curves.shtml?stnid=8725520. Accessed 29 May 2020
NOAA (2017) Detailed method for mapping sea level rise inundation. NOAA Office for Coastal Management. https://coast.noaa.gov/data/digitalcoast/pdf/slr-inundation-methods.pdf
Norheim RA, Mauger GS, Miller IM (2018) Guidelines for mapping sea level rise. Report prepared for the EPA National Estuary Program (NEP), Climate Impacts Group, University of Washington, Seattle
Oostrom M., Hayworth JS, Dane JH, Güven O (1992) Behavior of dense aqueous phase leachate plumes in homogeneous porous media. Water Resources Res 28(8):2123–2134
Oppenheimer M, Glavovic BC, Hinkel J, van de Wal R, Magnan AK, Abd-Elgawad A, Cai R, Cifuentes-Jara M, DeConto RM, Ghosh T, Hay J, Isla F, Marzeion B, Meyssignac B, Sebesvari Z (2019) Sea level rise and implications for low-lying islands, coasts and communities. In: Pörtner H-O, Roberts DC, Masson-Delmotte V, Zhai P, Tignor M, Poloczanska E, Mintenbeck K, Alegría A, Nicolai M, Okem A, Petzold J, Rama B, Weyer NM (eds) IPCC special report on the ocean and cryosphere in a changing climate, pp 321–445. (In press)
Oude Essink GHP, Van Baaren ES, De Louw PG (2010) Effects of climate change on coastal groundwater systems: a modeling study in the Netherlands. Water Resource Res 46(10), W00F04

Park J, Stabenau E, Kotun, K (2016) Sea level rise and inundation projections for Everglades, Biscayne, and Dry Tortugas national park infrastructure. SFNRC Technical Series 2016:1. National Park Service U.S. Department of the Interior. South Florida Natural Resources Center Everglades National Park

Park J, Stabenau E, Kotun K (2017) Sea-level rise and inundation scenarios for national parks in South Florida. Park Sci 33(1):63–73

Payne DF (2010) Effects of climate change on saltwater intrusion at Hilton Head Island, SC, USA. In: SWIM21—21st salt water intrusion meeting proceedings, Azores, Portugal, pp 293–296

Plane E, Hill K, May C (2019) A rapid assessment method to identify potential groundwater flooding hotspots as sea levels rise in coastal cities. Water 11(11):2228

Rasmussen P, Sonnenborg TO, Goncear G, Hinsby K (2013) Assessing impacts of climate change, sea level rise, and drainage canals on saltwater intrusion to coastal aquifer. Hydrol Earth Syst Sci 17(1):421–443

Rohde RA (n.d.) Post-glacial sea level rise. Global Warming Art project. Wikimedia Commons. https://commons.wikimedia.org/wiki/File:Post-Glacial_Sea_Level.png. Accessed 17 June 2020

Rohling EJ, Grant K, Hemleben CH, Siddall M, Hoogakker BAA, Bolshaw M, Kucera M (2008) High rates of sea-level rise during the last interglacial period. Nat Geosci 1(1):38–42

Rotzoll K, Fletcher CH (2013) Assessment of groundwater inundation as a consequence of sea-level rise. Nat Clim Change 3(5):477–481

Rozell DJ, Wong TF (2010) Effects of climate change on groundwater resources at Shelter Island, New York State, USA. Hydrogeol J 18(7):1657–1665

Sanborn (n.d.) Hydro-flattening. https://www.sanborn.com/hydro-flattening/. Accessed 29 May 2020

Sulzbacher H, Wiederhold H, Siemon B, Grinat M, Igel J, Burschil T, Günther T, Hinsby K (2012) Numerical modelling of climate change impacts on freshwater lenses on the North Sea Island of Borkum using hydrological and geophysical methods. Hydrol Earth Syst Sci 16(10)

Terry JP, Falkland AC (2010) Responses of atoll freshwater lenses to storm-surge overwash in the Northern Cook Islands. Hydrogeol J 18(3):749–759

UCAR (n.d.) MAGICC. The climate system in a nutshell. University Corporation for Atmospheric Research. http://www.magicc.org/

United Nations Environment Programme (2005) After the tsunami: rapid environmental assessment. United Nations Environmental Programme, Nairobi. http://wedocs.unep.org/handle/20.500.11822/8372. Accessed 28 July 2020

Van Biersel TP, Carlson DA, Milner LR (2007) Impact of hurricanes storm surges on the groundwater resources. Environ Geol 53(4):813–826

Vandenbohede A (2007) Visual MOCDENS3D: visualization and processing software for MOCDENS3D, a 3D density dependent groundwater flow and solute transport model. User manual. Research Unit Groundwater Modeling, Ghent University

Vithanage M, Engesgaard P, Villholth KG, Jensen KH (2012) The effects of the 2004 tsunami on a coastal aquifer in Sri Lanka. Groundwater 50(5):704–714

Voss CI (1984) SUTRA—a finite-element simulation model for saturated-unsaturated, fluid density-dependent ground-water flow with energy transport or chemically-reactive single-species solute transport. U.S. Geological Survey Water-Resources Investigations Report, 84–4369

Voss CI, Provost AM (2010) SUTRA: a model for saturated-unsaturated, variable-density ground-water flow with solute or energy transport. U.S. Geological Survey Water-Resources Investigations Report, 02–4231

Webb MD, Howard KW (2011) Modeling the transient response of saline intrusion to rising sea-levels. Groundwater 49(4):560–569

Werner AD, Simmons CT (2009) Impact of sea-level rise on sea water intrusion in coastal aquifers. Ground Water 47(2):197–204

Werner AD, Ward JD, Morgan LK, Simmons CT, Robinson NI, Teubner MD (2012) Vulnerability indicators of sea water intrusion. Groundwater 50(1):48–58

Williams SJ (2013) Sea-level rise implications for coastal regions. J Coastal Res 63(sp1):184–196

Wong PP, Losada IJ, Gattuso J-P, Hinkel J, Khattabi A, McInnes KL, Saito Y, Sallenger A (2014) Coastal systems and low-lying areas. In: Field CB, Barros VR, Dokken DJ, Mach KJ, Mastrandrea MD, Bilir TE, Chatterjee M, Ebi KL, Estrada YO, Genova RC, Girma B, Kissel ES, Levy AN, MacCracken S, Mastrandrea PR, White LL (eds) Climate change 2014: impacts, adaptation, and vulnerability. Part A: global and sectoral aspects. Contribution of working group II to the fifth assessment report of the intergovernmental panel on climate change. Cambridge University Press, Cambridge UK, pp 361–409

Chapter 7
Climate Change and Small Islands

7.1 Introduction

Small, low-lying islands include reefal limestone islands (atolls), emergent carbonate sand banks, Pleistocene limestone islands (e.g., Florida Keys), and carbonate and siliciclastic barrier islands. Small island states in the Pacific and Indian Oceans and the Caribbean Sea are particularly vulnerable to global climate change and sea level rise (SLR) with the principal impacts falling into three main categories: shoreline erosion, inundation and flooding, and saline water intrusion into surficial freshwater aquifers (Woodroffe 2008). Virtually an entire island can be considered to be in the coastal zone (Leatherman and Beller-Simms 1997; Nurse et al. 2014). Low elevations make islan1997ds vulnerable to gradual erosion and inundation, and to overtopping and rapid erosion during storms, which will become an even greater threat if storm intensification occurs (Leatherman and Beller-Simms ; Mimura 1999; Khan et al. 2002; Yamano et al. 2007). Some islands face increased coastal erosion from the exploitation of beaches for building materials and other human activities. The construction of sea walls and other infrastructure, and waste dumping on reefs and mangroves undermine the ecological functions on which these island systems depend (Moberg and Folke 1999; Barnett and Adger 2003). Climate change may impact agricultural production on small islands through increased heat stress on plants, changes in precipitation and soil moisture, saline water intrusion from rising sea-levels, and increased damage from extreme weather events (Barnett and Adger 2003).

Fresh groundwater supplies of small islands tend to be fragile because of very high population densities (meaning that large numbers of people are potentially exposed to single events) and their freshwater reserves being restricted to often narrow and thin freshwater lenses that become depleted in times of low rainfall and are easily contaminated by salt water and human and industrial wastes (Barnett and Adger 2003).

R. Maliva, *Climate Change and Groundwater: Planning and Adaptations for a Changing and Uncertain Future*, Springer Hydrogeology,
https://doi.org/10.1007/978-3-030-66813-6_7

Royal and Connell (1991) reiterated an obvious challenge faced by occupants of low-lying small islands. In response to rising sea levels, most coastal dwellers have the option of retreating inland to higher ground. However, small ocean states occupying low coral islands on atolls do not have higher land to which to retreat. As sea level continues to rise, such states may no longer contain habitable land and entire populations will be threatened. The ability of islands to support human habitation is closely tied to the existence of a permanent groundwater system (i.e., freshwater lens), whose volume is proportional to the width and surface area of each island (Royal and Connell 1991). During droughts, the water table elevation falls and the groundwater may become brackish. Limited fresh groundwater resources on small, low-lying islands are vulnerable to saltwater intrusion driven by SLR and overtopping of storm surges.

Leatherman and Beller-Simms (1997, p. 1) observed that

> Many of these countries with limited resource bases are ill equipped to handle existing environmental problems such as explosive population growth, overdevelopment and pollution. These problems will only worsen as the coastal impacts of land submergence, beach erosion, increased storm flooding, high water tables, and reduced fresh water supply take their toll. Such changes make these small land masses, many at or near existing sea level, less habitable for humans, resulting in off-island migration.

Characteristics of small island states that define their vulnerability include (UN General Assembly 1993; Leatherman and Beller-Sims 1997):

- small sizes
- limited ranges of natural resources and fragile resource bases
- susceptibility to natural hazards (tropical cyclones)
- little biological diversity
- relative isolation and great distances to other markets
- extensive land/sea interface per unit area, which makes protective measures very expensive
- great susceptibility to external shocks
- low resilience of a subsistence economy
- narrow range of skills and lack of educated specialists

The IPCC (Nurse et al. 2014, p. 1618) noted that

> it has been suggested that the very existence of some atoll nations is threatened by rising sea levels associated with global warming. Although such scenarios are not applicable to all small island nations, there is no doubt that on the whole the impacts of climate change on small islands will have serious negative effects especially on socioeconomic conditions and biophysical resources—although impacts may be reduced through effective adaptation measures.

It has been reported that island erosion will become so widespread that entire atoll nations will disappear with their inhabitants among the first environmental refugees of climate change (Connell 2003, 2016). Climate change impacts to small island states is also a justice issue as these countries will bear a disproportionate share of the impacts of climate change despite their minimal contribution to its causes (Burkett 2013; Kostakos et al. 2014).

7.2 Small Island Erosion and Inundation

Coral reefs are dynamic systems that have responded in the past (and continue to respond) to changing sea levels. Royal and Connell (1991, p. 1070) observed that "islands are the result of a precarious balance between gains and losses to piles of sediment on atoll rims—a sediment budget that is controlled by a spectrum of physical and biological processes." Atoll island morphology results from a combination of small-scale erosion and accretion, which can be observed on a day-to-day basis, interspersed with catastrophic changes caused by extremely violent tropical storms that occur quite rarely (Royal and Connell 1991).

Similarly, siliciclastic islands are also dynamic features that have changed shape and migrated in response to changes in sea level and sediment supply. Barrier islands are particularly dynamic in some areas (Fig. 7.1). Although islands may persist as dynamic sedimentological features, human communities are typically constructed with the assumption that the land surface upon which they have built is permanent, at least on a human life-time scale.

Stoddart (1990) examined the available data on the accretion rates of coral reefs. Retrospective studies of historical reef growth indicate that the higher rates of reef accretion could match the higher predicted rates of SLR, but the lower rates of accretion cannot keep up with SLR. Stoddart (1990) pointed out that the post-glacial rise in SLR was about 1–2 m/100 years, the contemporary rise is ~0.1 m/100 years, and the projected rise is up to 2 m/100 years. With an estimated mean active reef accretion of 0.7 m/100 years, reefs will drown at higher predicted rates of SLR, as they did during the initial post-glacial transgression, but at lower predicted rates they could maintain themselves (Stoddart 1990). The recent rate of global mean SLR rise of 3.6 mm/year (0.36 m/100 years; IPCC 2019) is still well below the potential rate of active reef growth.

The actual fate of carbonate reef islands in response to sea level is a controversial issue. It is widely presumed that islands would disappear, and natives will be forced to seek refuge elsewhere (Connell 2003). However, it is important to recognize that islands are dynamic features that constantly change due to local influences under current climate conditions (Connell 2003). Connell (2003, p. 104) described that

> Local and media reactions and responses contradict contemporary scientific evidence, while negative environmental changes have explanations other than SLR. Construction of roads between islands (blocking natural lagoon to ocean channels), airport runway sealing, land reclamation, dredging and sea wall construction have all transformed the topography of tiny islands and ensured that the effects of storms and high tides are different to those in earlier times. Sand mining for accelerated construction has had similar effects, eroding the beach and 'giving the impression of rising seas to the casual observer' (Baliunas and Soon 2002, p. 44). Removal of vegetation for various reasons, such as fuel use, has also reduced stability. Building has gone on in areas once considered too hazardous for construction. Removal of coconut trees affects hydrology and groundwater. El Niño fluctuations have resulted in different climatic patterns. Reefs have been harmed by human activity, and by

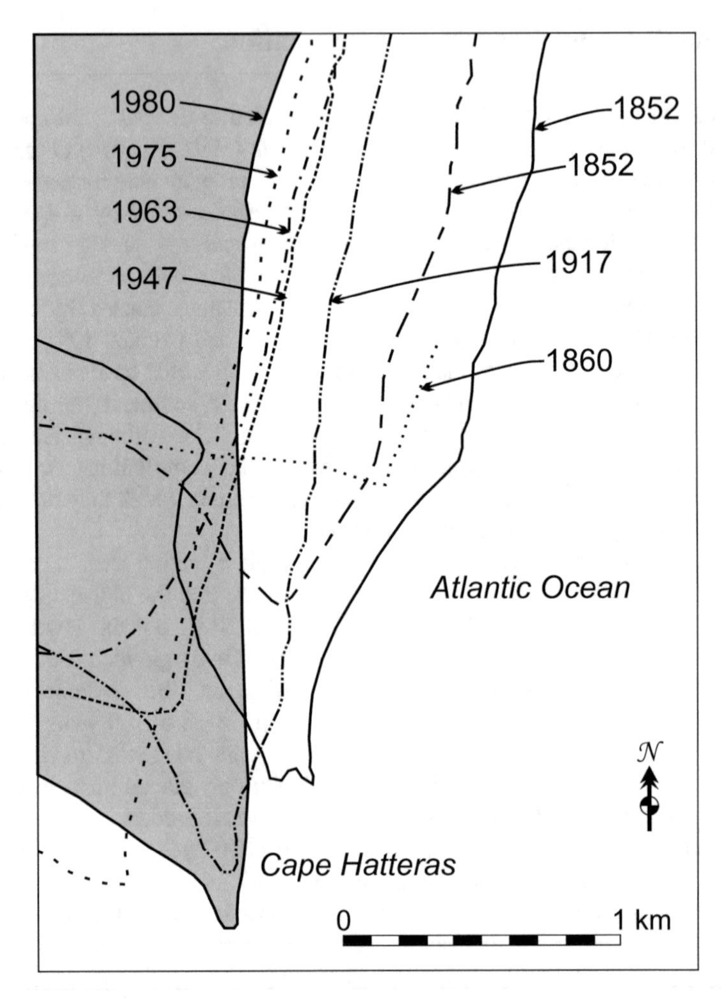

Fig. 7.1 The changing position of the mean high tide line at Cape Hatteras, North Carolina. The island's position in 1980 is shaded gray (after Everts et al. 1983)

> storms, limiting their ability to protect coasts. Spring tides and cyclones have always exacted damage, and storms contribute to mixing of fresh and saline water in lenses. Islands have come and gone, as Bikeman and parts of the Carteret atolls have (Connell 1990), but these changes are localised and unlikely to be the result of long-term climatic change.

Woodroffe (2008, p. 89) similarly argued that the popular generalization of SLR causing shoreline erosion is simplistic and that "although some lagoonal shorelines may be undergoing net erosion, long-term accretion is more likely to outweigh short-term erosion on oceanward beaches." Woodroffe (2008, p. 90) further noted "Erosion of an individual shoreline is far more likely to result from local or proximal causes (such as a particular storm), than to be attributable to the imperceptible gradual subsidence or steric sea-level rise" and that "It seems likely that misrepresentation of local shoreline erosion is all-too-frequently used to support the pre-supposition that the islands will disappear beneath the sea through erosion of exposed shorelines in response to ongoing rise of sea level averaging a few millimetres a year." SLR and climate change are impacting islands but are not the sole cause of all the ecological and economic problems island nations are facing.

Webb and Kench (2010) examined the planform morphological change of 27 atoll islands located in the central Pacific using comparative analysis of historical aerial photography and remotely sensed images. With respect to Central Pacific reef islands, Webb and Kench (2010) reported that the rate of relative SLR during the period of analysis was 2.0 mm/year and that their results show that 86% of islands remained either stable (43%) or increased in area (43%) over the timeframe of analysis. The largest decadal increase in island area was reported to range between 0.1 and 5.6 ha. Only 14% of study islands were found to have exhibited a net reduction in island area. The total change in the area of the reefal islands increased by 63 ha representing 7% of the total land area of all islands studied. Erosion of shorelines facing the ocean reef was detected in 50% of islands examined and accretion of lagoon shorelines was detected in 70% of the islands examined.

The aggregated effect of ocean shoreline displacement and lagoon progradation on the atoll islands was a shift in the nodal position of islands on reef surfaces (Webb and Kench 2010). The movement, small in magnitude, represents a net lagoonward migration of the islands and was observed in 65% of islands studied. Lagoonward migration was most evident on the windward margins of the atolls and in only one case was an island found to have migrated toward the reef edge.

Webb and Kench (2010) concluded that their results contradict widespread perceptions that all reef islands are eroding in response to recent SLR and that instead reef islands are geomorphically resilient landforms that have predominantly remained stable or grown in area over the last 20–60 years. Given this positive trend, Webb and Kench (2010) concluded that reef islands may not disappear from atoll rims and other locations in the near future, as speculated, but will undergo continued geomorphic change.

Changes in reef islands (that are not part of atolls) in the Solomon Islands were investigated using a times series of aerial photographs and high-resolution satellite imagery from 1947 to 2014 (Albert et al. 2016). Five vegetated reef islands (1–5 ha in size) were documented to have recently vanished and a further six islands experienced severe shoreline recession (Fig. 7.2). Shoreline recession was reported at two sites to have destroyed villages that had existed since at least 1935, leading to community relocations. Albert et al. (2016) concluded that the large range of erosion severity observed on the islands highlights the critical need to understand the complex interplay between projected accelerating SLR and other changes in global climate, such as winds and waves, and local tectonics to guide future adaptation planning and minimize social impacts.

Shoreline changes associated with SLR on 41 islands of the Manihi and Manuae atolls, French Polynesia, were investigated by Yates et al. (2013). Only two small islands on Manihi (4%) decreased in surface area between 1961 and 2001, while approximately 29% of the islands were stable, and 67% increased in surface area.

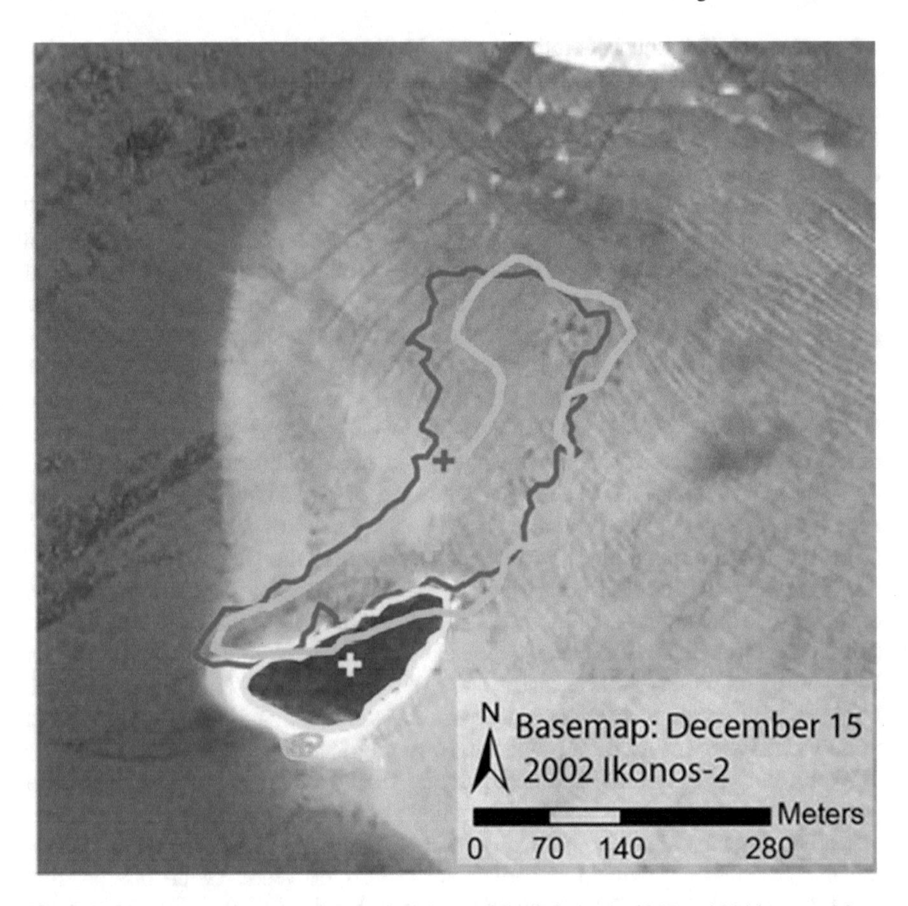

Fig. 7.2 Coastal recession on Kale Island, Solomon Islands between 1947 and 2014 mapped from aerial and satellite imagery. Position of the island in 1947 (blue), 1962 (green), and 2002 (yellow) are shown. Modified from Albert et al. (2016)

The islands on the northwestern rim of the atoll increased in surface area, while the islands on the southeastern atoll rim either increased in surface area or remained in dynamic equilibrium. In contrast to Manihi, the majority of the islands on Manuae (five out of six) decreased in surface area between 1955 and 2008, with erosion having occurred on different sides of the islands depending on their geographic location.

Yates et al. (2013) concluded that their results are agreement with the results of Webb and Kench (2010) in that recent rates of climate induced SLR are not the primary factor controlling shoreline change variability in the studied island environments at this time. Yates et al. (2013) hypothesized that the observed variability in island shoreline change is related to variability in wave exposure. Incoming waves breaking across the reef provide energy that breaks up reefs and transports bioclastic sediments landward (Yates et al. 2013). Human activities, such as sediment dredging, agriculture, pearl farming, tourism, coral and sediment mining, and construction of buildings and protective shoreline structures, may all have direct and indirect impacts on island sediment budgets and shoreline change (Yates et al. 2013). Yates et al. (2013, p. 880) warned that "with predictions of future changes in waves, cyclone intensity, sea-level rise, sea surface temperature, and human-induced pressure, one must be cautious when extrapolating observations of historical shoreline."

In the case of coral islands, other climate change impacts could affect the ability of island accretion to keep pace with SLR. The projected increases in sea-surface temperatures (SSTs) poses the greatest long-term risk to atoll morphology as corals are highly sensitive to sudden changes in SSTs (Barnett and Adger 2003). The impact of higher SSTs is arguably already evident with episodes of mass coral reef mortality through coral bleaching being experienced around the world (Reaser et al. 2000; Barnett and Adger 2003). Rising concentrations of CO_2 in the oceans, and thus lower pH values, may also retard the ability of reefs to grow in step with SLR (Kleypas et al. 1999; Barnett and Adger 2003). Even if the total land area of an island chain is not greatly changed by SLR, accelerated shoreline recession and accretion, combined with more frequent overwash events would still have a significant adverse impact on island water resources and habitability in general.

7.3 Fresh Groundwater on Small Islands

Terry and Chui (2012, p. 76) observed that

> From an outsider's perspective, the natural environment of atoll islands may appear idyllic, resembling exotic notions of 'paradise islands' portrayed by the tourist industry. But appearances are deceptive and the reality is that life on atolls is both harsh and precarious owing to environmental constraints on agricultural productivity and economic development and that probably the greatest impediment for human habitation is the limited availability of freshwater.

Small island nations face some of the most critical water challenges in the world because of expanding populations and very limited freshwater resources, which are vulnerable to over-exploitation, climatic variations, and contamination (White et al. 2007). Small islands obtain their freshwater primarily from fresh groundwater resources, rainwater harvesting (e.g., cisterns), and desalination. The latter can theoretically provide essentially unlimited quantities of water but at a very high cost. Desalination is energy intensive and requires technical sophistication to operate and maintain the systems. Many small island countries have experienced difficulties in maintaining desalination plants due to intermittent power supplies, maintenance and operational costs, and lack of training (White et al. 2007). Freshwater may also be transported to islands via tankers, which may supply small populations with potable water as a response to emergency disruptions in supplies.

Fresh groundwater on small islands occurs primarily as freshwater lenses, floating atop saline groundwater, whose natural size depends on island area, shape, and topography, aquifer properties (transmissivity), and rainfall (Fig. 7.3). Larger islands that are more equant-shaped and have high annual rainfalls can support larger freshwater lenses than smaller or elongate islands in more arid settings. The freshwater lenses, which are critical for small island communities, are extremely vulnerable to both natural climate variations and changes, storm events, and human-caused perturbations and, therefore, require careful assessment, vigilant monitoring, and astute management (White and Falkland 2010).

Freshwater lenses are limited resources that have several important vulnerabilities. First and foremost is that they are vulnerable to salinization from over exploitation. The sustainability of freshwater lenses depends on maintaining a balance between recharge, captured discharge, and pumping. Decreases in recharge during drought periods can result in a profound shrinkage of freshwater lenses. Anthropogenic activities can impact freshwater lenses through induced changes in recharge rates caused by increased impervious covers. Shallow island aquifers are vulnerable to contamination from a variety of sources including sewage disposal, chemical (pesticide and fertilizer) uses, and fuel storage. Shallow aquifers are also vulnerable to salinization from storm overwash. Royal and Connell (1991) noted that preference is usually given to rainwater for potable water supply and that

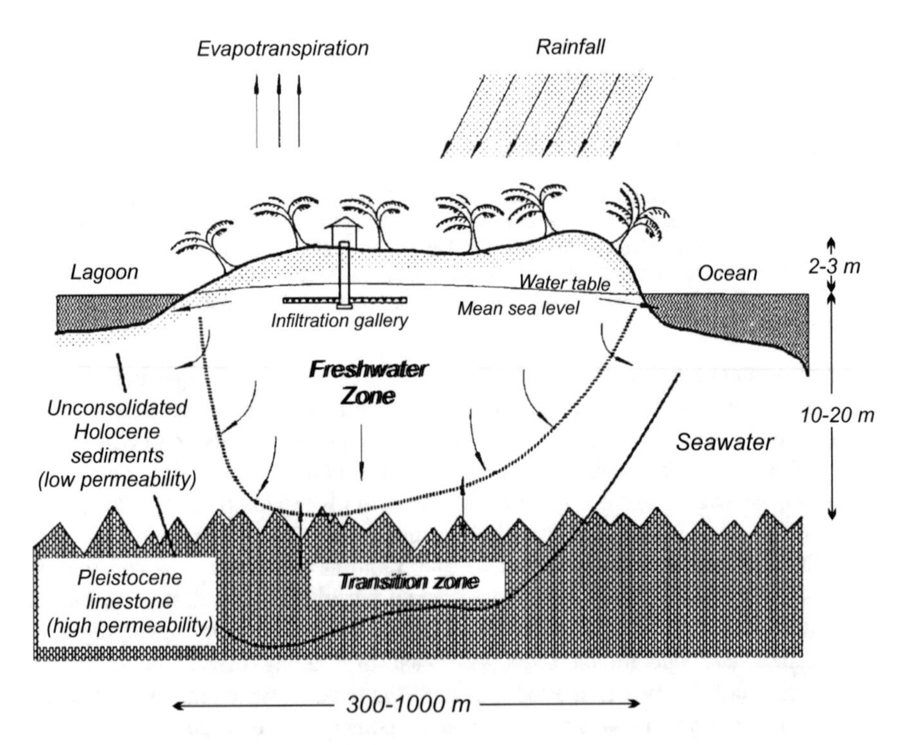

Fig. 7.3 Vertically exaggerated conceptual cross-section through a small coral island showing the main features of a freshwater lens and the location of an infiltration gallery used for groundwater abstraction. Modified from White and Falkland (2010)

construction of better cisterns may enable groundwater use to be minimized or ended. Groundwater is a secondary source but access to it is crucial during drought conditions.

Freshwater lenses have several vulnerabilities to climate change. The most serious threat to fresh groundwater supplies is from marine processes that cause coastal erosion and increase the frequency of storm overwash. Any decline in area can have a very dramatic influence on the availability of freshwater supplies (Royal and Connell 1991). Lagoon shores and low-lying interiors of reef islands that are already inundated during the highest tides and storms events will likely experience inundation conditions exacerbated by sea-level rise (Woodroffe 2008). A transitional to hotter and drier conditions could imperil fresh groundwater resources by increasing evapotranspiration and decreasing recharge. A higher water table associated with sea level rise could increase the vulnerability of aquifers to contamination and cause local groundwater inundation in low-lying areas.

The geology and hydrogeology of carbonate islands were reviewed by Falkland (1999) and Werner et al. (2017), in a series of papers edited by Vacher and Quinn (2004), and in a number of papers on specific islands and island chains. An overriding theme of hydrogeologic studies of carbonate islands is that accurate modeling of the behavior of freshwater lenses is very difficult because the aquifers are highly heterogeneous. On a coarse-scale, many islands have a dual-aquifer system in which an upper aquifer unit of uncemented or poorly cemented Holocene carbonate sediments is underlain by a deeper aquifer unit composed of much more transmissive karstic Pleistocene and older limestones. On a finer-scale, karst processes very commonly result in groundwater flow being dominated by secondary porosity features. Freshwater lenses are underlain by a mixing zone of varying thickness, which is strongly influenced by aquifer heterogeneity and tidal effects. Hence, the Ghyben–Herzberg formula is a poor predictor of the geometry of freshwater lenses.

Rising sea level will cause a rise in the water tables on siliciclastic barrier islands. Depending upon their elevation, barrier islands may have thin vadose zones and even modest SLR can substantially impact their shallow groundwater systems and ecosystems affected by changes in vadose zone thickness (Masterson and Garabedian 2007; Masterson et al. 2014; Manda et al. 2015).

7.4 Field and Modeling Studies of Freshwater Lenses and Their Vulnerability to Climate Change

Freshwater lenses are impacted by groundwater pumping, changes in overlying land use and land cover, natural climatic variation (e.g., droughts and storm events), and global climate change. Key issues are the long-term controls over the volume of stored freshwater and the aquifer response to (and recovery from) storm events (i.e., contamination with saline water from overwash). Follows are summaries of some relevant field and modeling studies of freshwater lenses in small islands.

7.4.1 Theoretical Modeling (Underwood et al. 1992)

Underwood et al. (1992) performed theoretical modeling of a dual-aquifer system that is subjected to laterally and vertically propagated tidal signals. The primary purpose of the study was to investigate relationships among freshwater lens responses, aquifer parameters, and boundary conditions using a general, rather than site-specific, model. The SUTRA code (Voss and Provost 2010) was used for two-dimensional simulations of simplified elongate islands with widths of 250, 500, 750, and 1000 m. A fluctuating boundary condition was added to the original model to simulate oceanic tidal fluctuations. The model was calibrated so that (1) parameter values were within the range of reported literature values, (2) groundwater salinity profiles, represented by the position of the 50% seawater curve, had the same shape and depth as in other studies, and (3) tidal responses were within the range of reported values.

The main results of the simulations were (Underwood et al. 1992):

- The position of the 50% seawater isopleth was controlled primarily by the permeability of the upper aquifer and recharge rates and is largely independent of tidal fluctuations.
- The location of 2.5% isopleth, which defines the limit of potable groundwater, is also sensitive to recharge and the horizontal permeability of the upper aquifer.
- The position of 2.5% isopleth is very sensitive to dispersive processes; hence it is strongly controlled by vertical longitudinal dispersivity and the magnitude the tidal fluctuations.
- Increases in vertical longitudinal dispersivity cause the 2.5% isopleth to move upward and decreases the thickness of the freshwater lens.
- Freshwater volumes depend upon both recharge rates and island area. As recharge rates decrease, greater aquifer widths are required to develop and maintain usable freshwater lenses, which are considered to have minimum thicknesses of 2 to 3 m.
- Tidal response simulations suggest that the tidal pulse will travel through the pathway of least resistance. If the permeability of the upper aquifer is significantly less than that of the lower aquifer, then the tidal pulse observed in the upper aquifer will be propagated from below, and tidal efficiencies and lags will increase and decrease, respectively, with depth.

With respect to technical modeling issues, Underwood et al. (1992) noted that since short-term fluctuating vertical flow is the major dispersive mechanism, the lack of inclusion of tidal dispersion in non-tidal models must be compensated for by increasing the dispersivity value used to simulate the transport process. The tidal dispersion process, which results from actual fluid flow, is replaced by an artificial transport mechanism of increased transverse dispersivity in non-tidal models.

7.4.2 *Home Island, South Keeling Atoll, Indian Ocean*

Accurate estimation of sustainable yield from freshwater lenses and their sensitivity to reduced recharge during drought periods requires accurate numerical simulations. Ghassemi et al. (2000) developed a 3-D model using the SALTFLOW code of a freshwater lens on Home Island, located in the South Keeling Atoll in the Indian Ocean. Difficulties in the transient model calibration arose from (1) numerical instability, (2) large CPU time and memory demands needed for a very fine discretization (which was acknowledged to not be a serious practical limitation with increasing computer power), and, most importantly, (3) inadequately detailed data on the 3-D distribution of key parameters, particularly the hydraulic conductivity and porosity of the karstic Pleistocene limestones and the spatial and temporal variations in recharge and extraction rates.

Mean recharge was estimated to be 44% of precipitation. Consistent with the conclusions of Underwood et al. (1992), very low dispersivity values were needed during the steady-state calibration when tidal fluctuations were simulated, whereas high values were required in non-tidal simulations. Ghassemi et al. (2000, p. 353) concluded that

> Despite an excellent dataset, which was unusually comprehensive by any standards, the quantity and quality of the available data were simply inadequate for a karstic system to meets calibration needs. The requirement for a large amount of quality data represents an immense challenge for the field hydrogeologist, because very few of these data can be reliably generated via the lengthy model-calibration procedure.

7.4.3 *Tarawa, Republic of Kiribata*

White et al. (2007) examined the impacts of ENSO-related droughts on freshwater availability and water quality on Tarawa, a densely populated central Pacific atoll that is part of the Republic of Kiribata. On the island of South Tarawa, water for the public supply (reticulation) system is extracted from the fresh groundwater lens using horizontal infiltration galleries and skimming wells (Fig. 7.4), which are designed to minimize drawdowns and the associated risks of saline water intrusion (up-coning). In response to the drought of 1998–2001, reverse osmosis desalination units were installed in South Tarawa but have subsequently failed and none were reported to be operating by 2007 (White et al. 2007).

High rainfalls occur in Tarawa during El Niño episodes and low rainfalls occur during La Niña episodes. Droughts result in a lowering of the water table and shrinkage of the freshwater lenses (White et al. 2007). Competition often exists between the public water supply system and landowners. Groundwater pumping can lower the water table and impact the health and productivity of traditional crops, such as coconuts and taro. Local governments on some large coral islands have declared lands overlying groundwater source areas to be water reserves and

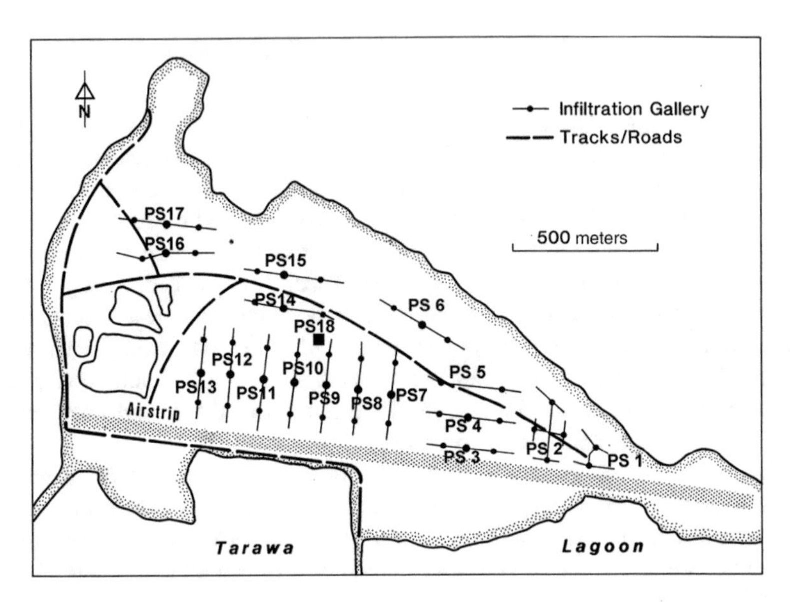

Fig. 7.4 Map of Binriki Island, Tarawa Atoll, showing the distribution of 18 horizontal infiltration galleries and pumping stations (PS) in the water reserve. *Source* White et al. (2007) copyright Soil Science Society of America

restrict land uses, which can adversely impact and create conflicts with landowners (White et al. 2007). Reductions in recharge in ENSO event droughts can imperil local groundwater resources. However, provided that pumping rates are maintained at a sustainable rate, large freshwater lenses can survive through extended droughts with only moderate increases in groundwater salinity (White et al. 2007).

7.4.4 Modeling of Impacts of Storm Overwash and SLR on Pacific Atolls

Terry and Chui (2012) modeled the impacts of sea level rise and storm overwash on the freshwater lenses of Pacific atolls. A two-dimensional model was developed using the SUTRA code of a generic large atoll with a well-developed freshwater lens. The model includes an interior depression (interior swamp) assigned an elevation of 0.5 m Steady-state simulations were run for sea level rises of 10, 20, and 40 cm.

For the 10 cm SLR rise scenario, the shape and thickness of the main body of the freshwater lens (FWL) was largely unaffected. The greatest modeled impacts to the FWL occurred after a 40 cm SLR was imposed on the atoll island. In all the scenarios, freshwater seeps down from the central swamp and recharges the FWL because the hydraulic pressure from the swamp (with its surface 0.5 m above sea level) is always higher than the pressure from the sea. However, in the 40 cm SLR scenario, the downward hydraulic gradient becomes so small that the freshwater seeping downward from the swamp is not enough to work against the seawater intruding from the ocean. As a result, upconing of brackish water was simulated to occur beneath the swamp such that the salt concentration of the groundwater increases to about half that of seawater (Terry and Chui 2012). However, the swamp appears to have been simulated as a constant head feature and the possibility of a lifting effect whereby the water level in the swamp increases with SLR was not evaluated.

Transient simulations of overwash showed the expected increase in salinity near land surface. Over time (several months), a freshening occurs at the top of the aquifer with the resumption of freshwater recharge and the downward migration of more saline water. Salinity rapidly increased with depth and then the groundwater freshens. A deeper zone of undisturbed freshwater is simulated to be present that extends to the base of the FWL. The modeling results show that the saline water plume produced by a washover has the greatest impact following a 40 cm eustatic rise in sea level, which is a consequence of the FWL being thinner at the start of the overwash event.

At shallow depths, a saline plume forms within the FWL and even after a full year of rainfall recharge, the plume does not fully disperse under all SLR scenarios. At greater depths within the aquifer, sandwiched between the overwash introduced saline plume above and the base of the FWL, an undamaged fresher horizon is preserved after a cyclone-driven washover of the atoll islet. The fresher horizon maintains salinity levels below 1.5 g/L (i.e. within usable limits) under the no-SLR and 10 cm SLR scenarios and remains intact for at least a year. In contrast, the diminished FWL configurations after 20 and 40 cm of SLR means that the lenses exhibit far less resilience to subsequent saline damage from storm-wave washover (Terry and Chui 2012).

Chui and Terry (2012) investigated the impacts of washover on FWLs using a 2-D SUTRA model developed of a typical coral atoll islet with dual aquifers (coarse reef-derived carbonate sand and gravel overlying hard, cemented reefal limestone). An initial steady-state simulation was followed by a transient run to simulate FWL damage due to a storm washover event and subsequent recovery by recharge of freshwater. The washover event was simulated to have a duration of 10 h and the tidal range was assumed to be minimal. The results of the simulations and conclusions were:

- rapid infiltration occurs through highly permeable sands and gravels that retain minimal moisture in the unsaturated zone
- greater depths to the water table allow more saline water to infiltrate during a washover event as infiltration slows once the sand and gravel becomes saturated
- longer duration (>1 h) washover events do not significantly worsen the extent of damage

- after the washover event ceases, the saline plume sinks and disperses over time and the high transmissivity limestone basement provides a path for the saline plume to quickly leave the aquifer
- a freshwater horizon may be preserved between the saline plume and seawater salinity groundwater present below the base of the FWL
- topographic depressions (swamps) where seawater can accumulate can be a continuing source of saline water, but the depressional source does not appear to impact the upper part of the aquifer where water is extracted using shallow wells
- impacts of inundations are greatest in dry climates as a far longer time is necessary for a saline plume to disperse and for a freshwater lens to regain pre-disturbance characteristics, given reduced recharge and a lower hydraulic gradient across the atoll islet.

7.4.5 Andros Island, Bahamas

Holding and Allen (2015) performed modeling investigations of the factors that may impact freshwater lenses on small, low-lying islands using Andros Island (Bahamas) as an example. Long-term stressors related to recharge and sea level change and short-term stressors related to overwash events were evaluated. Andros Island contains two main freshwater lens systems (North and South) and the public water supply is obtained from horizontal trench-based collector systems and interconnected shallow boreholes pumped at low rates (Holding and Allen 2015). The objective of the water supply system is to gently skim fresh groundwater from the top of the lens while not inducing significant upconing of saline water.

The Bahamian islands are subject to hurricanes and associated overwash of low-lying areas. The storm surge from Hurricane Frances (September 2005) caused a storm surge on western Andros Island that resulted in extensive salinization of the North Andros wellfield (Holding and Allen 2015). The saline water that entered the impacted water-supply trenches was reported to have been pumped out four days after the storm and the wellfield recovered to normal salinity concentrations between 1 and 2 years after the storm (Holding and Allen 2015). Some parts of the wellfield were unaffected by the overwash.

The impacts of long-term stressors (sea level rise and changes in recharge) were evaluated by numerical density-dependent solute-transport modeling using the SEAWAT code (Guo and Langevin 2002). The SEAWAT model had 44 layers with recharge applied to the top layer with a concentration of 0 g/L of salt. Current annual recharge was estimated using the U.S. Environmental Protection Agency's HELP (Hydrologic Evaluation of Landfill Performance) hydrological modeling software and available climate data from the island and nearest climate station (Miami, Florida). HELP utilizes a storage routing technique based on hydrological water balance principles to estimate recharge, accounting for soil moisture storage, runoff, interception, and evapotranspiration (Holding and Allen 2015).

Published climate change projections for the Bahamas for the A2 emissions scenario, which indicate warmer and drier future conditions, were used to calculate future recharge rates. Recharge in North Andros was predicted to decrease from 877 mm/year (60% of 1442 mm/year average rainfall) to 777 mm/year. Recharge in South Andros was predicted to decrease from 426 mm/year (48% of 889 mm/year average rainfall) to 360 mm/year. The baseline and future recharge rates were applied to the SEAWAT model in each region and SLR rise was simulated by increasing the elevations of the specified head cells in the model domain. Simulations were run for recharge reduction alone, sea level rise alone, and for both stressors. As would be expected, the modeling results indicate that the areal extent and volume of the freshwater lenses will decrease under future drier climate conditions. The reduced recharge rates alone result in the majority of the simulated freshwater lens reduction (Holding and Allen 2015). In the southern wellfield model, a 15% simulated freshwater volume loss was reported to be due to reduced recharge and a 5% loss due to SLR. In the northern wellfield model, the volume losses due to reductions in recharge and SLR were 5% and 0.9%, respectively.

The hydrogeological system on Andros Island is recharge-limited, meaning that the freshwater lenses are able to rise in the subsurface in response to SLR (Holding and Allen 2015). Therefore, the lenses are less vulnerable to SLR because a hydraulic balance is maintained between the fresh and salt water (Michael et al. 2013). This assumption is only valid to a point, as at higher magnitudes of sea level rise, the freshwater lens would likely become topographically limited and, therefore, have a larger response (i.e., loss of lens thickness) to SLR. Although SLR appears to not be a significant threat for saline water intrusion on Andros Island, it may increase the island's vulnerability to other events, such as extreme high tides and storm surge overwash, which have the potential to result in significant impacts to the freshwater lenses.

The HydroGeoSphere (HGS) code was used to model the impacts of overwash on the hydrogeological system, including the inundation of the freshwater supply trenches. Trenches provide direct access for inundating water to enter the freshwater lenses. The entry of salt water into the trenches creates large downward moving plumes of saline water. The simulation results showed that salt concentrations within the trenches will return to levels below the potable water threshold of 0.3 g/L within 149 days following the surge. If draining of the trenches as a remedial action occurs within 1 day of the storm surge, potable water was simulated to return in the trenches within 120 days. After a delay of 3 days, drainage of the trench provides no benefits as far as expediting recovery to potable standards (Holding and Allen 2015). The modeling results thus indicate that the timing of remedial action, specifically the rapid draining of the salt water from the trenches, is critical for expediting the restoration of the aquifer (Holding and Allen 2015).

7.4.6 Modeling of Effects of SLR on Waves and Overwash

Most analyses of the inundation of atolls by SLR are passive "bathtub" models that do not incorporate the interaction between SLR and waves (Storlazzi et al. 2015). Storlazzi et al. (2015) noted that waves in most atoll reef crests and reef flats are water depth limited, which suggests that as water depth increases with SLR, larger waves will propagate inland. With increasing water depths, there is reduced hydrodynamic roughness and associated wave energy dissipation, which will cause larger waves to impact the coastline (Storlazzi et al. 2015). Numerical modeling results indicate that waves will synergistically interact with SLR causing twice as much land to be flooded then predicted by passive models (Storlazzi et al. 2015). The implication of the dynamic analysis is that low-lying atoll islands will be annually flooded by seawater sooner in the future than predicted by passive models (Storlazzi et al. 2015). Frequent flooding could result in salinization of freshwater lenses, potentially forcing inhabitants to abandon their islands in decades, not centuries, as previously thought (Storlazzi et al. 2015).

7.4.7 Roi-Namur Island, Kwajalein Atoll, Republic of the Marshall Island, Overwash

Roi-Namur Island, located at the northeast tip of Kwajalein Atoll, Republic of the Marshall Islands, was flooded by an extra-tropical storm related to the El Niño Southern Oscillation on December 7–9, 2008 (Gingerich et al. 2017). The island's water source is a rainwater-catchment system (involving three large storage tanks) and a freshwater lens. Ponded seawater from the storm was reported to have infiltrated into the subsurface within a few days and impacted a horizontal skimming well used as part of the island's water supply. It took approximately 22 months for the chloride concentrations in all the horizontal galleries to decline to potable values measured prior to the flooding event (Gingerich et al. 2017).

Gingerich et al. (2017) performed solute-transport modeling using the SUTRA code to evaluate the controlling factors and basic management and mitigation options for salinization caused by overwash. Managed aquifer recharge is performed on the island by diverting captured rainwater into a swale above the skimming wells. Freshwater lenses are restored by the recharge of freshwater; hence restoration will occur more rapidly if a wet period follows a flooding event. Gingerich et al. (2017) concluded that artificial freshwater recharge to the freshwater lens is an obvious choice for mitigating the effects of seawater flooding after seawater has infiltrated into the freshwater lens.

The modeling results also indicate that avoidance of post-flooding groundwater withdrawals generally increases the speed of recovery of water quality in the lens, irrespective of other hydrological factors. Pumping results in greater mixing and the capture of deeper, more saline water from below. Artificial recharge of captured rainwater provides a clear benefit, lowering the salinity of the pumped groundwater no matter when and under what conditions it is applied (Gingerich et al. 2017). Gingerich et al. (2017) noted that even for an island water supply that is not impacted by a seawater-flooding event, artificial recharge to enhance FWL water quality provides a significant benefit in decreasing the salinity of pumped groundwater.

Gingerich et al. (2017, p. 688) concluded that

> the key to rapid recovery of the freshwater lens as a potable water supply on low-elevation sandy islands of the Roi-Namur type (atoll and barrier islands) is the total post-flood recharge; more recharge, whether natural or artificial, decreases the length of resource salinization. Furthermore, in situations where rainfall recharge is limited (either occasionally or on a permanent basis) following seawater overwash events, recharge enhancement by artificial recharge of freshwater is the most effective management/mitigation intervention possible for increasing freshwater supply and for shortening the post-flood length of time needed before pumped groundwater again becomes potable.

7.4.8 Supertyphoon Haiyan, Samar Island, Philippines

Cardenas et al. (2015) investigated the impacts of Supertyphoon Haiyan (STH) at the coastal village of San Antonio, Basey, Samar Island, Philippines (November 8, 2013). The storm surge was reported to have reached 7 m above sea level. Domestic water is reported to be obtained exclusively from hundreds of tube wells that tap either a shallow (0–5 m) unconfined aquifer or a deeper (>10 m) confined aquifer. Water samples collected 2 months after STH from 34 wells revealed widespread contamination with seawater, with concentrations ranging from no contamination (two wells) to up to 17.6% seawater, with a median of 4.5% based on chloride concentrations.

Modeling results indicate that two months after the surge from STH, seawater would percolate through the unsaturated zone to the water table and eventually mix with the shallow groundwater. By July, density fingers enhanced the spreading of contaminated groundwater below the water table by 2–5 m, but dilution and dispersion decreased concentrations. According to the model, the shallow portion of the surficial aquifer will require 1–2 years to become potable, but its deeper portions will become more saline. Most of the surficial aquifer will be potable again within 5–10 years (Cardenas et al. 2015).

Cardenas et al. (2015) concluded that it is highly unlikely that most of the wells in the village, screened typically at depths of 7–10 m within the confined aquifer, could have been contaminated by seawater percolating through the surficial aquifer. Residents reported contamination of the wells within a few hours. The groundwater modeling results also showed that it should take months for seawater to reach the deep aquifer. Cardenas et al. (2015), therefore, concluded that the seawater contamination occurred through the wells. The tube wells in San Antonio use simple piston hand pumps with old and inadequate seals that readily leak water as evidenced by the need to prime the pumps before using. During the storm surge, these poorly sealed wells would allow pure seawater to flow directly into the aquifer (Cardenas et al. 2015).

Cardenas et al. (2015) also hypothesized that the rapid recovery of the wells in San Antonio is accelerated by large groundwater flow rates through the very permeable and productive confined aquifer. In addition to the contamination through the wells in San Antonio, seawater that had infiltrated into the ground could eventually percolate to the deep wells and is an ongoing threat (Cardenas et al. 2015).

7.4.9 Distant Source Waves

Hoeke et al. (2013, p. 128) noted that it "has long been recognized that long-wavelength wind-waves (swells) produced by mid-latitude storms can propagate across entire ocean basins, sometimes to distances greater than 20,000 km" and that the arrival of such swells may trigger inundation events along such coasts. Hoeke et al. (2013) documented the environmental context and impacts surrounding a series of major inundation events that occurred in the western Pacific during December 2008. The reviewed reports indicate that widespread and severe damage to infrastructure and key natural resources (e.g., soils and freshwater) occurred on islands in Micronesia, the Marshall Islands, Kiribati, Papua New Guinea, and the Solomon Islands.

The analysis of Hoeke et al. (2013) shows that the December 2008 inundation event was caused primarily by remotely generated swell waves and that their severity was greatly increased by anomalously high regional sea levels linked with ENSO and SLR. The impacts of this event would have been greater had other sea level drivers also been more extreme (e.g. if the high wave energy had arrived during the peak of spring tides and stronger La Niña conditions). Hoeke et al. (2013, p. 137) cautioned that more frequent and severe wave-driven inundation events will have multiple longer-term impacts on coastal areas and that "the limits to adaptation and possibly habitability on many low-lying islands will be defined by the changes in the occurrence of these episodic events more than progressively rising sea levels."

7.5 Small Island Climate Change Adaptation Options

The choice of adaptation strategies to SLR on small islands will depend upon environmental, economic, and social factors (Stoddart 1990). Developing countries have the option of protecting their coasts through beach renourishment, sea walls, and other techniques, which may not be affordable for less wealthy small island states (Stoddart 1990). Key structures can be constructed on stilts or mounds to avoid flooding in storm events. A more novel approach is floating and amphibious architecture in which buildings normally rest on the ground but can float when necessary (Baggaley 2018; Lin et al. 2019). However, the habitability of small islands depends on much more than protecting structures. An alternative of an orderly retreat would require a lead time of several decades and presumes that there is higher land available somewhere to which to retreat (Stoddart 1990).

The combination of changes in mean conditions and the frequency and intensity of extreme events may ultimately cause some atoll environments to become unable to sustain human habitation, a possibility with even a moderate amount of climate change over the next century (Barnett and Adger 2003). Historically migration has been an adaptation strategy that contributes to the resilience of societies facing stresses (Barnett and Adger 2003). While island-to-island migration has historically occurred, wholesale abandonment of regions to other countries that are willing to accept climate change refugees would have tragic consequences in the loss of cultures.

With respect to water resources, desalination remains a viable option for potable supply, particularly low-maintenance units that can operate off the electrical grid. Solar-powered desalination units can either use solar energy for heating of water or for the generation of electricity used to power membrane treatment systems (Cheah 2004; Kabeel et al. 2019). Commercial modular photoelectric desalination systems have been developed and marketed (e.g., Elemental Water Makers 2020). Wind powered desalination systems are also viable options for supplying remote communities (Tzen 2008). Successful implementation of desalination requires technical capacity building for the local population to operate and maintain the system or perhaps operation and maintenance could be provided by national agencies, NGOs, or private contractors. Expanded household and communal rainwater harvesting can contribute toward increased resilience of water supplies but is vulnerable to prolonged droughts (White et al. 2007).

Optimization of the use of the freshwater lenses remains a great challenge. White et al. (2007) proposed that adaptation strategies for small islands can be grouped under three main themes:

(1) capacity strengthening
(2) demand management and refurbishment
(3) protection and supplementation of freshwater resources

White et al. (2007) emphasized that a key issue is provision of appropriate knowledge and that on many small islands, meteorological services and water supply agencies are under resourced. More knowledge is needed on the amount and quality of available freshwater resources, the demands on the resources, and sustainable extraction rates. Measures need to be taken to protect groundwater resources from contamination from surface activities.

The common approach to the optimization of the use of aquifers involves groundwater modeling to assess the potential impacts from extractions and climate change and to evaluate different management options. However, the carbonate aquifers on small islands are hydrogeologically complex and large amounts of site-specific data are required for accurate modeling, which is typically not now available and expensive to obtain (Underwood et al. 1992). The sheer numbers of small islands vulnerable to SLR preclude detailed aquifer characterization and modeling of all freshwater lenses that are supporting local populations.

Some impacts to FWLs are practically impossible to completely avoid such as salinization from overwash events. Overwash events are normal, low-frequency, extreme events under current climate conditions and their water quality impacts to FWLs dissipate over several years. Potable water can be provided immediately after events by tankering freshwater or installation of temporary desalination plants. The impacts from overwash events can be reduced somewhat by installing water-tight seals on wells to reduce the threat of the wells serving as direct conduits for saltwater to enter FWLs.

Management of freshwater lenses can be improved through improved well and wellfield design, monitoring of aquifer water levels and salinity, controls on groundwater pumping, implementation of land use practices to reduce the risk of anthropogenic contamination, and managed aquifer recharge. Horizontal wells and galleries are preferred over vertical wells as they allow for freshwater to be skimmed off the top of FWLs and the shallow depths of horizontal wells effectively reduce the potential for over-exploitation (i.e., groundwater levels cannot be lowered below the screen elevation). A key issue is maximizing the recharge from rainfall events, which could include steps to increase the perviousness of the land surface over recharge zones and to reduce runoff of freshwater to tide.

References

Albert S, Leon JX, Grinham AR, Church JA, Gibbes BR, Woodroffe CD (2016) Interactions between sea-level rise and wave exposure on reef island dynamics in the Solomon Islands. Environ Res Lett 11(5):054011

Baggaley K (2018) How floating architecture could help save cities from rising seas (April 9, 2018). https://www.nbcnews.com/mach/science/how-floating-architecture-could-help-save-cities-rising-seas-ncna863976 Accessed 18 June 2020

Baliunas S, Soon W (2002) Viewpoint: is Tuvalu really sinking? Pac Mag 28(2):44–45

Barnett J, Adger WN (2003) Climate dangers and atoll countries. Clim Change 61(3):321–337

Burkett M (2013) A justice paradox: on climate change, small Island developing states, and the quest for effective legal remedy. Univ Hawaii Legal Rev 35:633–670

Cardenas MB, Bennett PC, Zamora PB, Befus KM, Rodolfo RS, Cabria HB, Lapus MR (2015) Devastation of aquifers from tsunami-like storm surge by Supertyphoon Haiyan. Geophys Res Lett 42(8):2844–2851

Cheah S-F (2004) Photovoltaic reverse osmosis. Desalination and water purification research and development report 104. Denver, Bureau of Reclamation

Chui TFM, Terry JP (2012) Modeling fresh water lens damage and recovery on atolls after storm-wave washover. Groundwater 50(3):412–420

Connell J (1990) Modernity and its discontents. In: Connell J (ed) Migration and development in the South Pacific, Pacific Research Monograph 24. Canberra, National Centre for Development Studies

Connell J (2003) Losing ground? Tuvalu, the greenhouse effect and the garbage can. Asia Pac Viewpoint 44(2):89–107

Connell J (2016) Last days in the Carteret Islands? Climate change, livelihoods and migration on coral atolls. Asia Pac Viewpoint 57(1):3–15

Elemental Water Makers (2020) Solar desalination. Plug and play solution. https://www.elementalwatermakers.com/solutions/plug-play-solar-desalination/ Accessed 17 June 2020

Everts CH, Battley JP Jr, Gibson PN (1983) Shoreline movements. Report 1, Cape Henry Virginia to Cape Hatteras, North Carolina, 1849–1960. Technical Report CERC-83-1. Washington, DC: U.S. Army Corps of Engineers and National Oceanic and Atmospheric Administration

Falkland T (1999) Water resources issues of small island developing states. Nat Resour Forum 23(3):245–260

Ghassemi F, Alam K, Howard K (2000) Fresh-water lenses and practical limitations of their three-dimensional simulation. Hydrogeol J 8(5):521–537

Gingerich SB, Voss CI, Johnson AG (2017) Seawater-flooding events and impact on freshwater lenses of low-lying islands: controlling factors, basic management and mitigation. J Hydrol 551:676–688

Guo W, Langevin CD (2002) User's guide to SEAWAT: a computer program for simulation of three-dimensional variable-density ground-water flow. USGS Techniques of Water-Resources Investigations Book 6, Chapter A7

Hoeke RK, McInnes KL, Kruger JC, McNaught RJ, Hunter JR, Smithers SG (2013) Widespread inundation of Pacific islands triggered by distant-source wind-waves. Global Planet Change 108:128–138

Holding S, Allen DM (2015) From days to decades: numerical modelling of freshwater lens response to climate change stressors on small low-lying islands. Hydrol Earth Syst Sci 19(2): 933–949

IPCC (2019). Technical summary. In: Pörtner H-O, Roberts DC, Masson-Delmotte V, Zhai P, Tignor M, Poloczanska E, Mintenbeck K, Alegría A, Nicolai M, Okem A, Petzold J, Rama B, Weyer NM (eds) IPCC special report on the ocean and cryosphere in a changing climate. (In press)

Kabeel AE, El-Said EM, Dafea SA (2019) Design considerations and their effects on the operation and maintenance cost in solar-powered desalination plants. Heat Transfer—Asian Res 48(5): 1722–1736

Khan TMA, Quadir DA, Murty TS, Kabir A, Aktar F, Sarker MA (2002) Relative sea level changes in Maldives and vulnerability of land due to abnormal coastal inundation. Mar Geodesy 25(1–2):133–143

Kleypas JA, Buddemeier RW, Archer D, Gattuso JP, Langdon C, Opdyke BN (1999) Geochemical consequences of increased atmospheric carbon dioxide on coral reefs. Science 284(5411):118–120

Kostakos G, Zhang T, Veening W (2014) Climate security and justice for small island developing states. The Hague Institute for Global Justice

Leatherman SP, Beller-Simms N (1997) Sea-level rise and small island states: an overview. J Coastal Res 24(Special Issue):1–16

Lin YH, Chih Lin Y, Tan HS (2019) Design and functions of floating architecture—a review. Mar Georesour Geotechnol 37(7):880–889

Manda AK, Sisco MS, Mallinson DJ, Griffin MT (2015) Relative role and extent of marine and groundwater inundation on a dune-dominated barrier island under sea-level rise scenarios. Hydrol Process 29(8):1894–1904

Masterson JP, Garabedian SP (2007) Effects of sea-level rise on ground water flow in a coastal aquifer system. Groundwater 45(2):209–217

Masterson JP, Fienen MN, Thieler ER, Gesch DB, Gutierrez BT, Plant NG (2014) Effects of sea-level rise on barrier island groundwater system dynamics—ecohydrological implications. Ecohydrology 7(3):1064–1071

Michael HA, Russoniello CJ, Byron LA (2013) Global assessment of vulnerability to sea-level rise in topography-limited and recharge-limited coastal groundwater systems. Water Resour Res 49(4):2228–2240

Mimura N (1999) Vulnerability of island countries in the South Pacific to sea level rise and climate change. Climate Res 12(2–3):137–143

Moberg F, Folke C (1999) Ecological goods and services of coral reef ecosystems. Ecol Econ 29(2):215-233.

Nurse LA, McLean RF, Agard J, Briguglio LP, Duvat-Magnan V, Pelesikoti N, Tompkins E, Webb A (2014) Small islands. In: Barros VR, Field CB, Dokken DJ, Mastrandrea MD, Mach KJ, Bilir TE, Chatterjee M, Ebi KL, Estrada YO, Genova RC, Girma B, Kissel ES, Levy AN, MacCracken S, Mastrandrea PR, White LL (eds) Climate change 2014: impacts, adaptation, and vulnerability. Part B: regional aspects. Contribution of working group II to the fifth assessment report of the intergovernmental panel on climate change (pp. 1613–1654). Cambridge, UK: Cambridge University Press

Reaser JK, Pomerance R, Thomas PO (2000) Coral bleaching and global climate change: Scientific findings and policy recommendations. Conserv Biol 14:1500–1511

Roy P, Connell J (1991) Climatic change and the future of atoll states. J Coastal Res 7(4):1057–1075

Stoddart DR (1990) Coral reefs and islands and predicted sea-level rise. Prog Phys Geogr 14(4): 521–536

Storlazzi CD, Elias EP, Berkowitz P (2015) Many atolls may be uninhabitable within decades due to climate change. Sci Rep 5:14546

Terry JP, Chui TFM (2012) Evaluating the fate of freshwater lenses on atoll islands after eustatic sea-level rise and cyclone-driven inundation: A modelling approach. Glob Planet Change 88:76–84

Tzen E (2008) Wind-powered desalination-principles, configurations, design, and implementation. In: Gude VG (ed) Renewable energy powered desalination handbook: application and thermodynamics, pp 91–139. Oxford: Elsevier

Underwood MR, Peterson FL, Voss CI (1992) Groundwater lens dynamics of atoll islands. Water Resour Res 28(11):2889–2902

UN General Assembly (1993) Global Conference on the Sustainable Development of Small Island Developing States: resolution (A/RES/47/189)

Vacher LH, Quinn TM (eds) (2004) Geology and hydrogeology of carbonate islands. Elsevier, Amsterdam

Voss CI, Provost AM (2010) SUTRA: a model for saturated-unsaturated, variable-density ground-water flow with solute or energy transport. U.S. Geological Survey Water-Resources Investigations Report, 02-4231

Webb AP, Kench PS (2010) The dynamic response of reef islands to sea-level rise: Evidence from multi-decadal analysis of island change in the Central Pacific. Glob Planet Change 72(3):234–246

Werner AD, Sharp HK, Galvis SC, Post VE, Sinclair P (2017) Hydrogeology and management of freshwater lenses on atoll islands: review of current knowledge and research needs. J Hydrol 551:819–844

White I, Falkland T (2010) Management of freshwater lenses on small Pacific islands. Hydrogeol J 18(1):227–246

White I, Falkland T, Metutera T, Metai E, Overmars M, Perez P, Dray A (2007) Climatic and human influences on groundwater in low atolls. Vadose Zone J 6(3):581–590

Woodroffe CD (2008) Reef-island topography and the vulnerability of atolls to sea-level rise. Glob Planet Change 62(1–2):77–96

Yamano H, Kayanne H, Yamaguchi T, Kuwahara Y, Yokoki H, Shimazaki H, Chikamori M (2007) Atoll island vulnerability to flooding and inundation revealed by historical reconstruction: Fongafale Islet, Funafuti Atoll, Tuvalu. Glob Planet Change 57(3–4):407–416

Yates ML, Le Cozannet G, Garcin M, Salaï E, Walker P (2013) Multidecadal atoll shoreline change on Manihi and Manuae, French Polynesia. J Coastal Res 29(4):870–882

Chapter 8
Groundwater Related Impacts of Climate Change on Infrastructure

8.1 Introduction

Climate change will impact physical infrastructure, including water, wastewater, transportation, energy, residential, commercial, and industrial facilities, in many manners. Most basically, communities and facilities located in low-lying coastal areas will become increasingly vulnerable to inundation from rising sea levels, initially during storm and extreme tidal events, and eventually permanently as mean sea level approaches land surface. Heavy precipitation events and associated flooding will also impact infrastructure in low-lying coastal and riverine (flood plain) areas. Coastal flooding can be caused by direct ingress of seawater, through rising groundwater levels (i.e., groundwater inundation; GWI), and in some instances backflow through stormwater drain systems (Habel et al. 2020). Serious impacts to infrastructure may occur even if the water table does not rise to land surface. Rising groundwater levels decrease the thickness of the unsaturated zone, reducing the storage capacity of the soil, and thus increase runoff and the frequency and severity of flooding from rainfalls. GWI can be more difficult to manage than direct marine flooding because conventional physical barriers (e.g., sea walls, dikes, and levees), depending on their construction and local hydrogeology, may be largely ineffective. Climate change impacts could accelerate physical degradation of the built environment, and increase operation, maintenance, engineering design, and construction costs to support long-lived public water, wastewater, road, and other systems.

Wastewater treatment plants are normally sited in low-lying coastal or river areas to facilitate operation of gravity collection systems and because offshore and river discharge is (or was) used for treated wastewater disposal. Hence, wastewater treatment (aka water reclamation) plants are vulnerable to both direct and groundwater inundation. Inland wastewater treatment plants constructed in floodplains may experience more frequent and extreme flooding if an intensification of the hydrological cycle occurs.

R. Maliva, *Climate Change and Groundwater: Planning and Adaptations for a Changing and Uncertain Future*, Springer Hydrogeology,
https://doi.org/10.1007/978-3-030-66813-6_8

Port facilities and many major airports are located along the coast and are vulnerable to periodic flooding with progressive sea level rise (SLR). Vulnerable airports in the United States include San Francisco International, Oakland International, Honolulu International, New Orleans Louis Armstrong International, Tampa International, Miami International, Ft. Lauderdale International, Ronald Reagan Washington National, Newark Liberty International, LaGuardia (New York City), Philadelphia International, and John F. Kennedy International (New York City), all of which have runways within 12 feet (3.6 m) of current sea level (Agravante 2019). Where direct flooding is of concern, protective measures can be implemented, such as the building of sea walls around the perimeter of airports (Alves 2019) and elevating structures (Azevedo de Almeida and Mostafavi 2016).

Energy and transportation infrastructure in low-lying coastal areas are also subject to direct and groundwater inundation. Vulnerable infrastructure includes tunnels and low-lying roads and rail lines. Flooding of some roads can disrupt traffic flow and cause significant congestion throughout traffic management systems as traffic is rerouted (Azevedo de Almeida and Mostafavi 2016). The usual solution to increased flooding risk to physical infrastructure is some type of hardening, which may include construction of barriers, elevating facilities, and otherwise enhancing or upgrading facilities so that they are less susceptible to damage. Retreat involves moving infrastructure to less vulnerable areas.

The impacts of rising groundwater extend beyond flooding of aboveground facilities. Rising water tables can impact underground utilities, building foundations, and the performance of on-site sewage treatment and disposal systems (OSTDSs). Wetter conditions may be beneficial to agriculture by improving rainfed crop production and decreasing irrigation water requirements. On the contrary, rising water tables can result in waterlogged soil conditions and soil salinization, decreasing crop production. Declining groundwater levels adversely impact water availability and water quality through saline water intrusion. Falling groundwater levels can also trigger land subsidence, which can damage both above and underground structures and impact local drainage.

8.2 Urban Rising Groundwater Levels

Rising groundwater levels are already causing serious impacts in many cities and ironically in some cities located in the driest areas of the world, such as Kuwait City, Doha, Jeddah, Riyadh, and Madinah (Hamdan and Mukhopadhyay 1991; George 1992; Rushton and Al-Othman 1994; Al-Otaibi and Mukhopadhyay 2005; Kreibich and Thieken 2008; Al-Senafy et al. 2015; Maliva and Missimer 2012; Bob et al. 2016). The adverse impacts of rising groundwater levels in urban areas include (George 1992; Al-Sefry and Şen 2006; Maliva 2019):

- flooding of house basements
- flooding of underground parking garages, elevator shafts, train tunnels and other utilities
- deterioration of roads and highways
- damage to building foundations
- differential settlement in waterlogged soils
- damage to walls by capillary rise
- contamination of soils
- public health impacts from ponded waters
- offensive odors
- breeding of mosquitos
- contamination of shallow aquifers
- dewatering required during construction

George (1992, p. 171) observed that "Much damage has occurred because the potential for rising groundwater and associated problems were not recognised prior to development, the effect of a higher water table not often being considered in the design of deep basement buildings and buried services." Site-specific remediation (e.g., installation of drainage systems) is generally costly as considerable damage has often already been done prior to its implementation and its construction and operation may adversely affect adjacent structures or services (George 1992). Processes contributing to urban recharge and rising shallow groundwater levels include (Lerner 1986, 1990; Brassington and Rushton 1987; Foster 1990; Wakode et al. 2018):

- leakage from utility mains and lines
- return flows from over irrigation of parks, gardens, landscaped areas, and residential lawns
- OSTDSs in unsewered areas
- impervious covers that are more pervious than expected
- stormwater retention/detention/infiltration ponds
- discharges to losing (ephemeral) streams
- reductions in groundwater pumping after very long periods of pumping
- change from a vegetated land cover to impervious cover with an associated reduction in evapotranspiration (ET)
- sea level rise

Areas most likely to experience rising groundwater levels are low-lying areas with relatively low permeability aquifers, especially if the aquifers are not being exploited for groundwater supply (Al-Sefry and Şen 2006). Health hazards and nuisances may occur where the water table rises to land surface and results in standing water. Health impacts depend on the presence or absence of contamination in the recharged water (Kreibich and Thieken 2008). In situ sewage disposal is a proven vector of pathogen transmission and a source of increasing groundwater nitrate and dissolved organic carbon (including trace organic compounds) concentrations (Foster 1990).

Rising groundwater levels can damage water-tight buildings and underground structures through buoyancy effects. Empty (air-filled) underground storage systems can be damaged because of their great buoyancy when they become submerged below the water table. To avoid structural damage, the lifting effect can be countered by intentionally flooding basements or ensuring that buildings or structures are heavy enough to not be subject to buoyancy lifting (Kreibich and Thieken 2008).

Methods used to identify the cause of rising urban groundwater levels include (Lerner 2002; Maliva 2019):

- Piezometry–mapping of local rises in the water table
- inorganic tracers (major cations and anions) and trace elements
- organic chemical tracers, including trace organic compounds related to specific sources (e.g., some pharmaceutical compounds may be indicative of domestic sewage sources)
- microbiological tracers (e.g., fecal coliform bacteria)
- stable isotopes tracers (^{2}H, ^{15}N, ^{18}O, ^{35}S)
- water budget analysis
- unaccounted for water analysis
- minimum night flow analysis
- inverse groundwater modeling

The piezometry method involves comparing the geographic pattern of historical water table rise with land uses and the distribution of utility infrastructure. For example, if water table rise tends to be focused in areas using OSTDSs, then it may be inferred that recharge from such systems is likely at least a contributing cause of the rise. Chemical or microbiological tracers may confirm this interpretation. Similarly, higher groundwater levels in irrigated parks and landscaped areas would suggest that return flows are a cause of rising water tables.

Sources of groundwater rise may be identified using chemical, isotopic, or microbiological tracers that are indicative of specific source waters. Tracer-based methods require data on the concentrations of the tracers in each potential source water, which is complicated by spatial and temporal variations in the concentrations in each source water type. Mixing models are used to estimate the contribution of different sources. Where water use is metered, leakage from potable water mains can be estimated from the unaccounted-for water and by minimum night flow analysis.

Mitigation and adaptation responses to rising urban groundwater levels fall into three main categories: source elimination, hardening, and drainage. Source elimination involves investigation of the specific causes of local groundwater rise and the implementation of amelioration measures, which might include leak detection and repair, extension of centralized sewage service to areas using OSTDSs, and reduction of irrigation rates. Impacted underground structures and facilities may be hardened, such as by waterproofing basements and raising structures above ground. Groundwater levels are being lowered in some impacted cities using drainage systems.

A common drainage system design consists of horizontal drains (buried perforated pipes) that collect groundwater through gravity action (Zamfirescu et al. 2006; Al-Othman 2011), which is similar to the subsurface drainage system design widely used for farm fields (Van der Molen et al. 2007). Drainage is also performed by pumping vertical wells, which can be the preferred solution in already heavily developed urban areas where construction of horizontal drains may be too disruptive. Drainage water is a potential source of water for some uses depending on its quality (Al-Otaibi and Mukhopadhyay 2005). Typically drainage water is of too poor a quality for potable and most-non potable uses without advanced treatment (membrane treatment/desalination). However, in regions facing great water scarcity, treating urban drainage water may be an economically viable option. The lower salinity of drainage water theoretically makes it less expense to treat than seawater. However, other factors affect the economics of desalination including variability in drainage water quality, the presence of contaminants and foulants (and thus pretreatment requirements), and a much lesser economy of scale compared to very large seawater desalination plants.

Rising sea levels will exacerbate the existing rising groundwater level problems currently being experienced in some coastal cities. There are few realistic approaches to address accelerating groundwater rise from sea level rise in already developed urban areas. The main viable option will be to install or upgrade drainage systems. Gravity-driven systems may have to be upgraded to pumped systems if SLR reduces the hydraulic gradients required for system operation.

8.3 Stormwater Management Systems

Climate change will impact stormwater management systems primarily through changes in the frequency and intensity of precipitation events. Intensification of rainfall events, in terms of rate and duration, may result in exceedances of the capacity of stormwater management systems to convey water out of urban and suburban areas, causing local flooding. Rising groundwater elevations will decrease the available storage capacity of the unsaturated zone, decreasing infiltration and increasing runoff, which can also threaten to overwhelm stormwater management systems and cause local flooding.

Sea level rise can increase heads at stormwater outfalls, decreasing the hydraulic gradient in gravity systems, which can reduce flow rates and cause back-ups of water to land surface that contribute to flooding (Habel et al. 2020). Options for maintaining stormwater system capacity and avoiding back-ups include the installation of one-way valves and installing or enhancing pumping systems (Azevedo de Almeida and Mostafavi 2016; Habel et al. 2020). Various forms of green infrastructure, such as bioswales and bioretention systems, can reduce the impacts of climate change on stormwater drainage systems by increasing infiltration rates and decreasing run off from heavy rains (Azevedo de Almeida and Mostafavi 2016).

8.4 Centralized Sewage Systems

Wastewater treatment plants sited in low-lying coastal or river areas are vulnerable to both direct and groundwater inundation (Hummel et al. 2018). Inland wastewater treatment plants constructed in floodplains may experience more frequent and extreme flooding. Recent flooding in the midwestern United States resulted in "wet weather bypass" from inundated or overwhelmed wastewater treatment plants, releasing untreated wastewater into waterways (Howard 2019). Combined sewer systems (CSSs), which collect rainwater runoff, domestic sewage, and industrial wastewater in single pipes, have a greater vulnerability to climate change as induced increases in runoff can cause exceedances of the treatment capacity of plants. Exceedances of plant capacities result in direct discharges of untreated stormwater and wastewater into nearby streams, rivers, and other water bodies.

Water mains are operated under pressure (typically 30–80 psi) to provide all users in the distribution system with sufficient water (volume and pressure), meet fire flow requirements, and maintain water quality (i.e., prevent infiltration into the mains). Most sewer systems, on the contrary, are gravity systems because they provide reliable water movement with no energy costs. Lift stations are used to either lift sewage for entry into higher elevation gravity sewer mains or into wastewater treatment plants.

In some older areas of cities, gravity sewer mains are over a century old and may be in questionable condition. Gravity mains are normally not metered and receive less monitoring and maintenance than potable water mains. Factors affecting the integrity of sewer pipes include age, poor, outdated, and/or insufficient maintenance, inadequate funding relative to the high costs of replacement and rehabilitation, and geological conditions (Ellis 2001). A closed-circuit television (CCTV) inspection of the Rastatt, Germany, sewer system revealed 31,006 defects within the 208 km sewer system with the most common type of defects being damaged or improperly installed house connections and joint displacements (Wolf et al. 2006). The Rastatt sewer system would be considered a rather well-maintained system (Wolf et al. 2006), so greater defect rates would be expected to occur in other older and less-well maintained systems.

Inasmuch as gravity sewer mains are not operated under pressure, widespread leaks individually and collectively go largely unnoticed. Where the mains are located above the water table, leakage from sewer mains (exfiltration) has been shown to be an important contributor to groundwater recharge. In low-lying coastal areas, such as South Florida, sanitary sewer pipes are commonly installed below the water table and are susceptible to infiltration of groundwater through cracked pipes or poorly constructed pipes systems, thereby consuming capacity in treatment plants (Bloetscher et al. 2017). Infiltration increases the volume of water requiring treatment and thus treatment costs. Infiltration of saline water can impact biological treatment processes and impair the quality of reclaimed water and thus its suitability for reuse. When the flow in a sewer pipe network exceeds the capacity of the pipes to transmit water or the capacity of the plant to treat the water, surcharging can occur, which is essentially a backflow of water onto streets and into buildings. Sanitary sewer overflows can result in fines assessed against utilities and property damage (Bloetscher et al. 2017).

Rising groundwater levels associated with SLR will increase infiltration into submerged gravity sewer mains (Flood and Cahoon 2011). More frequent inundation from rainstorms, extreme (king) tides, and storm surges is expected to increase inflows into sanitary sewer systems through unsealed manholes, open cleanouts, and problem surface connections (Bloetscher et al. 2017).

Infiltration can be reduced through a leak detection and repair program. Sewer mains can be replaced using open cuts (excavations). A usually more cost-effective and less disruptive option is trenchless repairs. A variety of trenchless repair techniques have been developed including pipe bursting (in-line expansion), slip lining, cured-in-place pipe, and modified cross-section liners (USEPA 1999a). Pipe bursting is a technique whereby an existing pipe is forced outward and opened by a bursting tool, which is dragged ahead of the new pipeline. The existing pipe is expanded radially outward until it cracks, allowing a new pipe of the approximately the same diameter to be installed. After the pipe bursting is completed and new pipe installed, laterals are re-connected, typically with robotic cutting devices (USEPA 1999a). Slip lining, cured-in-place pipe, and modified cross section liners involve the installation of a smaller diameter pipe or liner inside an existing pipe (USEPA 1999a). Lift stations may also require upgrading to handle increased flows. The stations may also require hardening if they are located in areas that are becoming increasingly subject to flooding.

8.5 On-Site Sewage Treatment and Disposal Systems

OSTDSs are used primarily in rural areas and some rapidly growing urban areas that are not yet served by centralized sanitary sewer systems. The most common type of OSTDS is septic systems, but in some areas cesspools, seepage pits, and dry wells are used. In the United States, approximately 25% of households rely on septic systems to treat and dispose of sanitary waste, which includes wastewater from kitchens, clothes washing machines, and bathrooms (USEPA 1990, 2001).

Household septic systems usually consists of a septic tank, a distribution box, and a drain field. The septic tank is a rectangular or cylindrical container made of concrete, fiberglass, or polyethylene into which wastewater flows (Fig. 8.1). The wastewater is retained in the septic tank for a period of time to allow suspended solids to settle out to the bottom of the tank where they are partially decomposed by microbial activity. Grease, oil, and fat, along with some digested solids, float to the surface to form a scum layer (USEPA 1990). The partially clarified wastewater between the scum layer and bottom sludge flows to the distribution box from which it is distributed to the drain field.

Drains fields (also referred to a leach fields and soil treatment areas) are networks of perforated pipes laid horizontally in gravel-filled trenches or beds. Depending

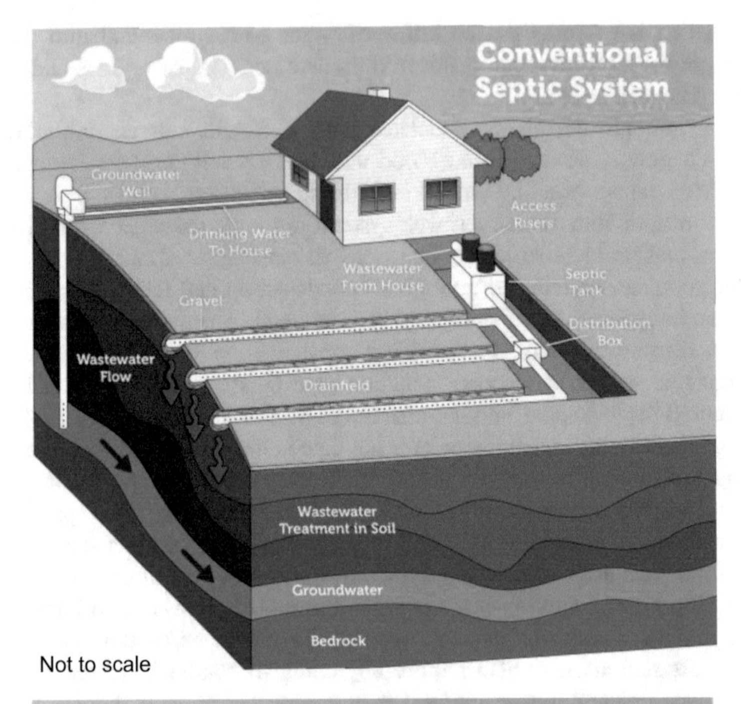

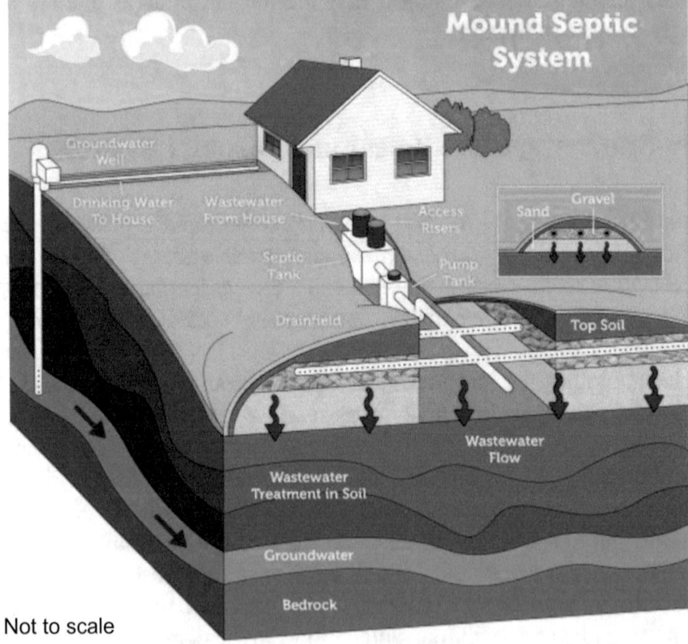

Fig. 8.1 Conceptual diagrams of conventional and mound-type septic systems. *Source* USEPA (n.d.)

upon jurisdiction, septic system drain fields are required to have from 0.3 to 0.9 m (1–3 ft) minimum vertical separation from the wet season high water table. Most jurisdictions have adopted minimum horizontal setback distances for septic systems from buildings and drinking water wells (USEPA 1990).

Septic systems can be a significant source of ground water contamination leading to waterborne disease outbreaks and other adverse health effects. Bacteria, protozoa, and viruses found in sanitary wastewater can cause numerous diseases, including gastrointestinal illness, cholera, hepatitis A, and typhoid (USEPA 2001; Cooper et al. 2016). Nitrogen present in sanitary wastewater, primarily from urine, feces, food waste, and cleaning compounds, can impact local groundwater quality (USEPA 2001). Chemical and biological processes in the unsaturated (soil) zone play an important role in the removal of microbial and chemical contaminants present in wastewater entering drainfields. The unsaturated zone is the site of aerobic contamination attenuation processes, and it is important that water from the drainfield be retained in the aerobic zone for a sufficiently long time for contaminant attenuation processes to occur. Highly permeable soils are avoided because they allow for too rapid flow through the unsaturated zone.

Rising water levels can impair the operation of septic systems and can cause waterlogging of soils. Rising water tables caused by climate change would reduce vadose zone thicknesses, resulting in reduced aerobic treatment and greater transport of pathogens and nutrients to shallow aquifers (Cooper et al. 2016). Reduced treatment in septic systems will pose a threat to the quality of shallow groundwater and hydrologically connected surface water bodies. Increasing temperatures can also impact the performance of septic systems. Cooper et al. (2016) noted that microbial activity increases with warmer temperatures, which may enhance removal of biochemical oxygen demand (BOD5; i.e., organic compounds). However, lower levels of BOD5 may limit heterotrophic processes such as nitrate removal by denitrification. Higher temperatures will also reduce O_2 solubility and increase microbial dissolved oxygen (DO) consumption, resulting in less DO available for aerobic treatment processes. A reduction in DO can lead to lowered redox conditions, resulting in the geochemical reduction of metals (and their associated mobilization) and a diminished phosphorous removal capacity of soils.

Septic systems in low-lying coastal areas are particularly vulnerable to being compromised by rising water tables caused by SLR. Septic systems that are vulnerable to SLR can be identified by mapping areas where the water table is projected to rise above a prescribed minimum distance below land surface (Habel et al. 2017; MDCDRER, MDWASD, & FDOD 2018). For example, in Miami-Dade County, Florida, there are approximately 105,000 parcels served by septic systems, approximately 100,000 of which are within the Urban Development Boundary (UDB). The results of a modeling study of the impacts of SLR on groundwater levels indicated that within the next 25 years, Miami-Dade County can expect the number of residential septic systems that may be periodically compromised during storms or wet years to increase from approximately 56% today (58,349 parcels) to more than 64% by 2040 (67,234 parcels; MDCDRER, MDWASD, & FDOD 2018). A 2016 study indicated that the estimated cost to connect the residential pockets with septic systems within Miami-Dade Water and Sewer District's service area to the regional wastewater system was approximately US$ 3.3 billion (MDCDRER, MDWASD, & FDOD 2018).

On an individual household level, Cooper et al. (2016) noted that shallow narrow soil treatment areas (STAs, drainfields) that infiltrates pre-treated wastewater higher in the soil profile may be more resilient to climate change. Mound systems (Fig. 8.1) are also an option in areas with shallow water tables but require a substantial amount of space and periodic maintenance and are more expensive to construct (USEPA 1999b).

8.6 Agricultural and Changing Groundwater Levels

Groundwater levels in agricultural areas may be impacted by climate changes through changes in precipitation (and thus recharge), changes in extraction rates required to meet irrigation water demands, and sea level rise in coastal regions. Rising groundwater levels can have either beneficial or adverse impacts on agriculture. Potential agricultural benefits of rising groundwater water levels include reductions in water stress and irrigation requirements and increases in crop yields (Zipper et al. 2015). On the contrary, rising water tables can adversely impact crops when waterlogging of the root zone results in a dissolved oxygen deficiency (Zipper et al. 2015) and soil salinization. Whether shallower water table depths are net beneficial or harmful to crops depends upon the amount of groundwater rise, soil texture, and growing season weather conditions (Zipper et al. 2015).

Soil salinization occurs in over 100 countries of the world to varying extents, adversely impacting agricultural production, environmental health, and economic welfare (Rengasamy 2006). Saline water in plant root zones inhibit plant growth by an osmotic effect, which reduces the ability of plants to take up water, and by ion excesses (Rengasamy 2006). Three main types of soil salinization occur: groundwater associated, non-groundwater associated, and irrigation associated (Rengasamy 2006). Groundwater associated salinization occurs when the water table rises closes to land surface and salts within the groundwater are concentrated by evapotranspiration. Non-groundwater associated salinization occurs when poor hydraulic conditions result in the accumulation of salts introduced from rain, weathering of minerals in the soil, and aeolian deposition (Rengasamy 2006). Irrigation associated salinization is the evapotranspirative concentration in the root zone of crops of salts present in irrigation water. Irrigation associated salinization is usually managed by applying excess irrigation water to leach and transport the salts to below the root zone and by drainage.

Groundwater associated salinization occurs most frequently (but not exclusively) in lower-lying areas of semiarid and arid lands (e.g., valley or basin floors). The direct evaporation of water from the capillary zone and transpiration by plants remove essentially pure water, while salts are left behind. Soil salinization tends to occur when the following conditions occur together: the presence of soluble salts in the soil and groundwater, a high water table, a high rate of evaporation, and a low annual rainfall (USDA 1998). Groundwater associated soil salinization generally occurs where water tables are within 2 m of land surface (Rengasamy 2006). Rising groundwater levels can also cause contamination of shallow groundwater by mobilizing salts, nutrients, and chemicals that accumulated in the vadose zone (Greene et al. 2016).

Groundwater associated salinization has been caused by human activities that resulted in a more positive water balance and thus a net increase in water storage in the water table aquifer. A number of studies have documented how changes in land cover have resulted in a pronounced decrease in evapotranspiration and increase in aquifer recharge. A change from deep-rooted native vegetation to more shallowly rooted agricultural crops and grasses has resulted in dramatic increases in recharge in some areas (e.g., Gee et al. 1992; Petheram et al. 2002; Keese et al. 2005; Scanlon et al. 2005, 2006; Kim and Jackson 2012). Recharge rates have increased by one or two orders of magnitude in some areas with a natural or human-caused change in vegetative type (Scanlon et al. 2006).

Climate change can contribute to soil salinization in several manners:

- warmer temperatures may increase ET rates and thus the accumulation of salts in the root zone
- SLR raise may cause higher groundwater levels in coastal areas
- increases in precipitation may contribute to higher groundwater levels in lower-lying areas
- climate change could cause changes in vegetation that impact ET and recharge rates
- low-lying area may be subject to more frequent and greater magnitude flooding with saline water during storm events.

Chen and Mueller (2018) reported that in coastal Bangladesh, flood inundation alone has had negligible effects on population migration and agricultural production. However, gradual increases in soil salinity were found to have a direct effect on internal and international population migration through the adverse impacts of salinization on crop production and associated income losses.

Reducing the severity and extent of soil salinization is primarily a problem of water management (USDA 1998). Where rising groundwater levels have an anthropogenic origin, management options include reducing recharge (e.g., by decreasing irrigation and lining canals) and planting deep-rooted vegetation (USDA 1998). Subsurface drainage systems can be effective in lowering the water table beneath farm fields (Hanson et al. 2006). The recovered water is often too saline for direct use and its disposal can cause environmental damage. There is considerable interest in some water scarce regions in the desalination and reuse of drainage water (e.g., Molseed et al. 1987; Sorour et al. 1992; Lee et al. 2003; Talaat and Ahmed 2007; McCool et al. 2010). Cost-effective disposal of the residual salts (concentrate) can be a key feasibility issue for desalination systems. Another adaptation strategy to salinization is to transition to more salt tolerant crops.

8.7 Land Subsidence

Climate change can contribute to land subsidence primarily through its negative impacts on aquifer water budgets and associated declines in groundwater levels (pressures). A hotter and drier climate could prompt additional groundwater pumping to meet increased demands or to augment declining surface water supplies. Hotter and drier climate conditions might also reduce aquifer recharge rates. Conversely, a transition to wetter conditions could reduce the rate of groundwater use and associated land subsidence.

Declining groundwater levels can cause land subsidence and associated impacts (e.g., fissuring). Land subsidence induced by groundwater pumping was reviewed by Galloway et al. (1999), Holzer and Galloway (2005), and Galloway and Burbey (2011). Groundwater pumping induces land subsidence by decreasing the hydrostatic pressure in strata and, in turn, increasing effective stress. The amount of land subsidence that an area experiences is a function of both the increase in effective stress and the compressibility and thickness of the impacted strata. Silt and clay beds are usually much more compressible than relatively clean sands and lithified aquifer strata. Areas underlain by thick sequences of fine-grained sediments with high compressibilities are most vulnerable to subsidence.

Land subsidence has both reversible and irreversible components. Some elastic deformation is inherent in the production of groundwater from confined aquifers. Water is produced by the expansion of the water and compression of the aquifer material. Elastic responses are evidenced by a land surface rebound occurring during periods of increased aquifer water levels (heads). Inelastic compaction of sediments is the result of the reorientation and repacking of grains and matrix into a tighter, less porous fabric. Land subsidence caused by inelastic compaction is irreversible. The stress history of sedimentary deposits plays an important role in their vulnerability to compaction. Deposits exposed to greater stresses in the past (e.g., due to previous loading or lower aquifer water levels) may become over-consolidated (preconsolidated) and thus less vulnerable to further consolidation resulting from decreases in hydrostatic pressure. Inelastic compaction may resume if aquifer water levels drop below previous historical lows.

The area globally impacted by land subsidence is enormous. Within the United States alone, an area of more than 26,000 km^2 (10,039 mile^2) has been permanently lowered as the result of withdrawal of underground fluids (National Research Council 1991; Holzer and Galloway 2005). The land subsidence in the San Joaquin Valley, California (Fig. 8.2) caused largely by groundwater pumping for irrigation, was observed by Galloway and Riley (1999, p. 23) to be "one of the single largest alterations of the land surface attributed to mankind."

Land subsidence can cause a wide variety of adverse impacts to infrastructure including (Galloway et al. 1999; Leake 2004; Holzer and Galloway 2005):

Fig. 8.2 Photograph of the approximate location of maximum land subsidence in the United States (San Joaquin Valley, California). Signs on the pole show the approximate position of the land surface in 1925, 1955, and 1977. *Source* Galloway and Riley (1999)

- permanent inundation and increased vulnerability to flooding and storms
- changes in the elevation and slope of streams, canals, and drains
- damage to bridges, roads, railroad tracks, storm drains, sanitary sewers, water mains, canals and sewers
- damage to private and public buildings
- failure of well casings and screens
- impacts to the operation of gravity sewer and storm drains
- opening of earth fissures
- modification of aquifer hydraulic properties
- the need to raise levees and other flood control structures.

Subsidence is evident in some impacted cities (e.g., Mexico City) by the leaning of older buildings and cracks in sidewalks, roads, and foundations. In recent years, several studies examined the impacts of land subsidence on high-speed railroads (e.g., Ge et al. 2010; Ye et al. 2015). Declining water levels can also trigger sinkhole development in karstic terrains (Foose and Humphreville 1979; Sinclair 1982; Lamoreaux and Newton 1986).

Where land subsidence occurs over large areas at the same rate, its effects may not be noticeable by casual observations. Land subsidence may be evident by well casings, and other features anchored below the compacting strata, extending upward increasing heights above land surface (i.e., acting as crude extensometers). Land subsidence caused by differential compaction can cause angular distortions that can be more damaging. The effects of land subsidence due to groundwater pumping on roadways, highways, and bridges are usually minor and the associated damage has been generally repairable. Economic consequences of subsidence, while large, are generally factored into yearly maintenance budgets rather than accounted for as unique natural disasters at a single point in time (Cabral-Cano et al. 2008).

Land subsidence is measured through time series of land surface elevation measurements. The most accurate method to measure land surface elevation changes is borehole extensometers, which are typically either a steel pipe or cable that is cemented in place at the bottom of a well. The extensometer establishes a subsurface benchmark at the bottom of the borehole and compaction (or expansion) is measured as the change in the distance between the bottom of the well and land surface. Extensometers are expensive to construct and thus few can usually be constructed, providing only a small number of series of high-resolution measurements. Time series of conventional or global positioning system (GPS) surveys allow for a greater spatial density of point measurement but at the cost of a lesser vertical resolution.

Satellite interferometric synthetic aperture radar (InSAR) involves the reflection of radar signals transmitted by a satellite off the ground. InSAR was reviewed by Massonnet and Feigl (1998), Bürgmann et al. (2000), and Galloway and Hoffmann (2007). Time series of measurements are made of changes in the distance between a satellite and the ground and thus changes in land surface elevation. The InSAR process measures the phase shift between synthetic aperture radar (SAR) images obtained from two satellite passes. InSAR provide millions of data points over a large region (about 10,000 km^2, 3,860 mile^2), which is not practical using land-based techniques. Pixels on an InSAR displacement map are usually 30–90 m^2 (323–969 ft^2) on the ground (Bawden et al. 2003). The use of InSAR to map changes in land surface elevation has increased dramatically over the past two decades because of the ability of technique to efficiently produce spatially complete detailed maps of elevation changes that cover broad areas, such as an entire groundwater basin or metropolitan area (Fig. 8.3).

The predominant inelastic compaction component of subsidence is irreversible. Future land subsidence rates can be reduced by restoring aquifer water budgets (eliminating overdraft) through decreased pumping or managed aquifer recharge. However, subsidence may continue for some time after an overdraft is terminated.

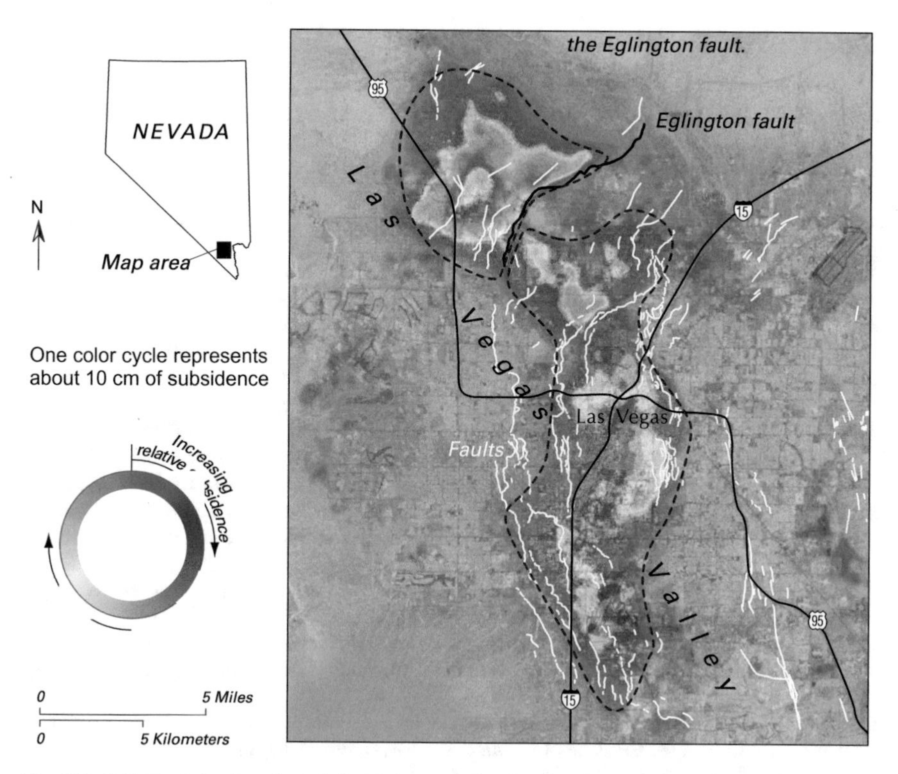

Fig. 8.3 InSAR derived surface deformation map for the Las Vegas Valley showing subsidence between April 1992 and December 1997. *Source* Galloway et al. (2000)

A time delay may occur that is caused by the low hydraulic conductivity of fine-grained sediments. Subsidence can occur only as fast as water can migrate out of the strata undergoing compaction. It may take decades or centuries for pore pressures in thick aquitards (semiconfining units) to equilibrate with pressure changes in aquifers. Compaction and associated land subsidence will, therefore, tend to continue even if groundwater levels have stabilized or even partially recovered (Hoffmann et al. 2003).

Impacts of land subsidence can reduced by employing subsidence resistant construction that is able to accommodate some subsidence without damage and avoiding areas in which differential subsidence is occurring or may occur. In karst areas, geotechnical surveys for buried features subject to collapse are routinely performed, especially for major construction projects. Compression sections are used for wells constructed in subsiding areas to avoid casing damage during subsidence.

References

Agravante M (2019) Climate change is forcing major airports to future-proof against rising sea levels and floods. Inhabitat (October 14, 2019). https://inhabitat.com/climate-change-is-forcing-major-airports-to-future-proof-against-rising-sea-levels-and-floods/. Accessed 24 May 2020

Al-Otaibi M, Mukhopadhyay A (2005) Options for managing water resources in Kuwait. Arabian J Sci Eng 30:55–68

Al-Othman AA (2011) Implication of gravity drainage plan on shallow rising groundwater conditions in parts of ArRiyadh City, Saudi Arabia. Int J Phys Sci 6(7):1611–1619

Al-Sefry SA, Şen Z (2006) Groundwater rise problem and risk evaluation in major cities of arid lands–Jedddah Case in Kingdom of Saudi Arabia. Water Resour Manage 20(1):91–108

Al-Senafy M, Hadi K, Fadlelmawla A, Al-Fahad K, Al-Khalid A, Bhandary H (2015) Causes of groundwater rise at Al-Qurain residential area, Kuwait. Procedia Environ Sci 25:4–10

Alves L (2019) San Francisco Airport Building Wall to Contain Sea Level Rise (October 11, 2019). Miami Beach Times. https://miamibeachtimes.com/real-estate/national-and-global-real-estate/san-francisco-airport-building-wall-to-contain-sea-level-rise/. Accessed 18 June 2020

Azevedo de Almeida B, Mostafavi A (2016) Resilience of infrastructure systems to sea-level rise in coastal areas: Impacts, adaptation measures, and implementation challenges. Sustainability 8 (11):1115

Bawden GW, Sneed N, Stork SV, Galloway DL (2003) Measuring human-induced land subsidence from space. U.S. Geological Survey Fact Sheet 069-03

Bob M, Rahman N, Elamin A, Taher S (2016) Rising groundwater levels problem in urban areas: a case study from the Central Area of Madinah City, Saudi Arabia. Arabian J Sci Eng 41(4): 1461–1472

Bloetscher F, Hoermann S, Berry L (2017) Adaptation to Florida's urban infrastructure to climate change. In: Chassignet EP, Jones JW, Misra V, Obeysekera J (eds) Florida's climate: changes, variations, & impacts. Florida Climate Institute, Gainesville, pp 311–338

Brassington FC, Rushton KR (1987) A rising water table in central Liverpool. Q J Eng Geol Hydrogeol 20(2):151–158

Bürgmann R, Rosen PA, Fielding EJ (2000) Synthetic aperture radar interferometry to measure Earth's surface topography and its deformation. Annu Rev Earth Planet Sci 28:169–209

Cabral-Cano E, Dixon TH, Miralles-Wilhelm F, Díaz-Molina O, Sánchez-Zamora O, Carande RE (2008) Space geodetic imaging of rapid ground subsidence in Mexico City. Geol Soc Am Bull 120:1556–1566

Chen J, Mueller V (2018) Coastal climate change, soil salinity and human migration in Bangladesh. Nat Climate Change 8(11):981–985

Cooper JA, Loomis GW, Amador JA (2016) Hell and high water: diminished septic system performance in coastal regions due to climate change. PloS One 11(9)

Ellis JB (2001) Sewer infiltration/exfiltration and interactions with sewer flows and groundwater quality. In: 2nd international conference interactions between sewers, treatment plants and receiving waters in urban areas–Interurba II, pp 19–22

Flood JF, Cahoon LB (2011) Risks to coastal wastewater collection systems from sea-level rise and climate change. J Coastal Res 27(4):652–660

Foose RM, Humphreville JA (1979) Engineering geological approaches to foundations in the karst terrain of the Hershey Valley. Bull Assoc Eng Geol 16(3):355–381

Foster SSD (1990) Impacts of urbanization on groundwater. In: Massing H, Packman J, Zuidema F (eds) Hydrological processes and water management in urban areas. Papers from urban water '88 symposium at Duisburg, Germany, April 1988, IAHS Publ. 198, pp 187–207. International Association of Hydrological Sciences, Wallingford

Galloway D, Jones DR, Ingebritsen SE (1999) Land subsidence in the United States. Circular 1182. U. S. Geological Survey, Washington, DC

Galloway DL, Burbey TJ (2011) Regional land subsidence accompanying groundwater extraction. Hydrogeol J 19(8):1459–1486

Galloway DL, Hoffmann J (2007) The application of satellite differential SAR interferometry-derived groundwater displacements in hydrogeology. Hydrogeol J 15:133–154

Galloway D, Riley FS (1999) San Joaquin Valley, California. In Galloway D, Jones DR, Ingebritsen SE (eds) Land subsidence in the United States. Circular 1182, pp 23–34. U. S. Geological Survey, Washington, DC

Galloway D, Jones DR, Ingebritsen SE (2000) Land Subsidence in the United States. Fact Sheet 165-00. U. S. Geological Survey, Washington, DC

Ge L, Li X, Chang HC, Ng AH, Zhang K, Hu Z (2010) Impact of ground subsidence on the Beijing-Tianjin high-speed railway as mapped by radar interferometry. Ann GIS 16(2):91–102

Gee GW, Fayer MJ, Rockhold ML, Campbell MD (1992) Variations in recharge at the Hanford Site. Northwest Science 66:237–250

George DI (1992) Rising groundwater: a problem of development in some urban areas of the Middle East. In: McCallD GJH, Laming JC, Scott SC (eds) Geohazards. Springer, Dordrecht, pp 171–182

Greene R, Timms W, Rengasamy P, Arshad M, Cresswell R (2016) Soil and aquifer salinization: Toward an integrated approach for salinity management of groundwater. In: Jakeman AJ, Barreteau O, Hunt RJ, Rinaudo J-D, Ross A (eds) Integrated groundwater management. Springer, Cham, pp 377–412

Habel S, Fletcher CH, Rotzoll K, El-Kadi AI (2017) Development of a model to simulate groundwater inundation induced by sea-level rise and high tides in Honolulu, Hawaii. Water Res 114:122–134

Habel S, Fletcher CH, Anderson TR, Thompson PR (2020) Sea-level rise induced multi-mechanism flooding and contribution to urban infrastructure failure. Sci Rep 10:3796

Hamdan L, Mukhopadhyay A (1991) Numerical simulation of subsurface-water rise in Kuwait City. Groundwater 29(1):93–104

Hanson BR, Grattan SR, Fulton A (2006) Agricultural salinity and drainage. UC Davis Publication 3375

Hoffmann J, Galloway DL, Zebker HA (2003) Inverse modeling of interbed storage parameters using land subsidence observations, Antelope Valley, California. Water Resour Res 39(2):1031

Holzer TL, Galloway DL (2005). Impacts of land subsidence caused by withdrawal of underground fluids in the United States. In: Ehlen J, Haneberg WC, Larson RA (eds) Humans as geologic agents. Reviews in Engineering Geology v. XVI, pp 87–99. Geological Society of America, Boulder

Howard F (2019) Spring floods send raw sewage into nation's waterways (May 8, 2019) Agri-Pulse. https://www.agri-pulse.com/articles/12163-spring-floods-send-raw-sewage-into-nations-waterways. Accessed 18 June 2020

Hummel MA, Berry MS, Stacey MT (2018) Sea level rise impacts on wastewater treatment systems along the US coasts. Earth's Future 6(4):622–633

Keese KE, Scanlon BR, Reedy RC (2005) Assessing controls on diffuse groundwater recharge using unsaturated flow modeling. Water Resour Res 41:W06010

Kim JH, Jackson RB (2012) A global analysis of groundwater recharge for vegetation, climate, and soils. Vadose Zone J 11(1). https://doi.org/10.2136/vzj2011.0021ra

Kreibich H, Thieken AH (2008) Assessment of damage caused by high groundwater inundation. Water Resour Res 44(9):W09409

Lamoreaux PE, Newton JG (1986) Catastrophic subsidence: an environmental hazard, Shelby County, Alabama. Environ Geol Water Sci 8(1–2):25–40

Leake SA (2004) Land subsidence from ground-water pumping: U.S. Geological Survey. http://geochange.er.usgs.gov/sw/changes/anthropogenic/subside/. Accessed 2 June 2020

Lee RW, Glater J, Cohen Y, Martin C, Kovac K, Milobar MN, Bartel DW (2003) Low-pressure RO membrane desalination of agricultural drainage water. Desalination 155(2):109–120

Lerner DN (1986) Leaking pipes recharge ground water. Ground Water 24:654–662

Lerner DN (1990) Recharge due to urbanization. In: Lerner DN, Issar AS, Simmers I (eds) Groundwater recharge, a guide to understanding and estimating natural recharge, Contributions to Hydrogeology 8. International Associations of Hydrogeologists, Kenilworth, pp 210–214

Lerner DN (2002) Identifying and quantifying urban recharge: a review. Hydrogeol J 10(1):143–152

Maliva RG (2019) Anthropogenic aquifer recharge. Springer, Cham

Maliva RG, Missimer TM (2012) Arid lands water management. Springer, Berlin

Massonnet D, Feigl KL (1998) Radar interferometry and its application to changes in the earth's surface. Rev Geophys 36(4):441–500

McCool BC, Rahardianto A, Faria J, Kovac K, Lara D, Cohen Y (2010) Feasibility of reverse osmosis desalination of brackish agricultural drainage water in the San Joaquin Valley. Desalination 261(3):240–250

MDCDRER, MDWASD, & FDOD (2018). Septic systems vulnerable to sea level rise (November 2018). Miami-Dade County Department of Regulatory & Economic Resources, Miami-Dade County Water and Sewer Department & Florida Department of Health in Miami-Dade County

Molseed AC, Hunt JR, Cowin MW (1987) Desalination of agricultural drainage return water. Part I: Operational experiences with conventional and nonconventional pretreatment methods. Desalination 61(3):249–262

National Research Council (1991) Mitigation losses from land subsidence in the United States. National Academy Press, Washington, DC

Petheram C, Walker G, Grayson R, Thierfelder T, Zhang L (2002) Towards a framework for predicting impacts of land-use on recharge: 1. A review of recharge studies in Australia. Soil Res 40(3):397–417

Rengasamy P (2006) World salinization with emphasis on Australia. J Exp Bot 57(5):1017–1023

Rushton KR, Al-Othman AAR (1994) Control of rising groundwater levels in Riyadh, Saudi Arabia. In: Groundwater problems in urban areas: proceedings of the international conference organized by the Institution of Civil Engineers and held in London, 2–3 June 1993, pp 299–309. Thomas Telford Publishing, London

Scanlon BR, Keese KE, Flint AL, Flint LE, Gaye CB, Edmunds WM, Simmers I (2006) Global synthesis of groundwater recharge in semiarid and arid regions. Hydrol Process 20:3335–3379

Scanlon BR, Reedy RC, Stonestrom DA, Prudic DE, Dennehy KF (2005) Impact of land use and land cover change on groundwater recharge and quality in the southwestern US. Glob Change Biol 11:1577–1593

Sinclair WC (1982). Sinkhole development resulting from ground-water withdrawal in the Tampa area, Florida. U.S. Geological Survey Water-Resources Investigations, 81–50

Sorour MH, Abulnour AG, Talaat HA (1992) Desalination of agricultural drainage water. Desalination 86(1):63–75

Talaat HA, Ahmed SR (2007) Treatment of agricultural drainage water: technological schemes and financial indicators. Desalination 204(1–3):102–112

USDA (1998) Soil quality resource concerns: salinization. USDA Natural Resources Conservation Service, Washington, DC

USEPA (1990) Onsite wastewater treatment and disposal systems, design manual (EPA 625/1-80-012). U.S. Environmental Protection Agency Office of Water Program Operations, Washington, DC

USEPA (1999a) Collection systems. O&M fact sheet trenchless sewer rehabilitation (EPA 832-F-99-032). Washington, DC: United States Environmental Protection Agency Office of Water

USEPA (1999b) Decentralized systems technology fact sheet mound systems (EPA 832-F 99-074). United States Environmental Protection Office of Water, Washington, DC

USEPA (2001) Managing septic systems to prevent contamination of drinking water (EPA 816-F-01-021). United States Environmental Protection Office of Water, Washington, DC

USEPA (n.d.) Types of septic systems. United States Environmental Protection Office of Water. https://www.epa.gov/septic/types-septic-systems. Accessed June 2, 2020

Van der Molen WH, Beltrán JM, Ochs WJ (2007) Guidelines and computer programs for the planning and design of land drainage systems. FAO Irrigation and Drainage Paper 62. Food and Agriculture Organization of the United Nations, Rome

Wakode HB, Baier K, Jha R, Azzam R (2018) Impact of urbanization on groundwater recharge and urban water balance for the city of Hyderabad, India. Int Soil Water Conserv Res 6(1):51–62

Wolf LEIF, Eiswirth M, Hotzl H (2006) Assessing sewer-groundwater interaction at the city scale based on individual sewer defects and marker species distributions. Environ Geol 49:849–857

Ye C, Yang Y, Tian F, Luo Y, Zhou Y (2015) Numerical analysis to determine the impact of land subsidence on high-speed railway routes in Beijing, China. Proc Int Assoc Hydrol Sci 372:493–497

Zamfirescu F, Popa I, Danchiv A (2006) Gravitational drainage systems in urban areas: case study (Galaţi town, Romania). Environ Geol 49(6):887–896

Zipper SC, Soylu ME, Booth EG, Loheide SP (2015) Untangling the effects of shallow groundwater and soil texture as drivers of subfield-scale yield variability. Water Resour Res 51 (8):6338–6358

Chapter 9
Adaptation and Resilience Concepts

9.1 Introduction

There is now an increasingly large literature on all aspects of vulnerability and adaptation to climate change. The "AR5 Climate Change 2014: Impacts, Adaptation, and Vulnerability" reports (IPCC 2014a, b) provide an admirable overview of these issues with extensive references to key papers. Adaptation was defined by the IPCC (2014c, p. 1758) as "the process of adjustment to actual or expected climate and its effects, in order to moderate harm or exploit beneficial opportunities." Resilience was defined as "the ability of a system and its component parts to anticipate, absorb, accommodate, or recover from the effects of a hazardous event in a timely and efficient manner, including through ensuring the preservation, restoration, or improvement of its essential basic structures and functions" (IPCC 2014c, p. 1772). Resilience is thus a subset of adaptation. Adger et al. (2005) described the three cornerstones of adaptation as reduction of the sensitivity of a system to climate change, altering the exposure of a system to climate change, and increasing the resilience of a system to cope with changes.

Mitigation was defined by the IPCC (2014c, p. 1769) as the "lessening of the potential adverse impacts of physical hazards (including those that are human-induced) through actions that reduce hazard, exposure, and vulnerability." The critical element of climate change mitigation is reducing the emissions of greenhouse gases (GHGs). Both mitigation and adaption will be clearly be necessary to address anthropogenic climate change. In the 1990s, adaptation came out of favor as it was viewed as an alternative that allowed for the avoidance of mitigation (Pielke et al. 2007). However, the taboo against adaptation has been abandoned as it was recognized that there is a timescale mismatch—even with decarbonization of the global energy system—climate change is still unavoidable (Pielke et al. 2007). The most stringent mitigation efforts cannot avoid additional impacts of climate change over the next few decades, which makes adaptation unavoidable (Klein et al. 2007, 2014).

R. Maliva, *Climate Change and Groundwater: Planning and Adaptations for a Changing and Uncertain Future*, Springer Hydrogeology,
https://doi.org/10.1007/978-3-030-66813-6_9

Successful mitigation will necessarily involve an effective international commitment to reduce greenhouse gas emissions. Mitigation, to the extent that it is implemented, will be primarily driven by international agreements and ensuing national public policies, complemented by unilateral and voluntary actions (Klein et al. 2007, 2014). Individual, community, and national efforts to reduce their carbon footprints can incrementally contribute to mitigation, and certainly should not be discounted, but they will not solve the problem if other countries are increasing their emissions to support their population and economic growth.

Adaptation, on the contrary, "typically works on the scale of an impacted system, which is regional at best, but mostly local" (Klein et al. 2007, p. 750). Klein et al. (2007, p. 750) also observed that "the bulk of adaptation actions have historically been motivated by the self-interest of affected private actors and communities, possibly facilitated by public policies". Most adaptation involves private actions of affected entities, public arrangements of impacted communities, and national policies. Mitigation and adaptation operate on different time scales. The effects of emissions reductions may take several decades to fully manifest, whereas most adaptation measures have more immediate benefits (Füssel and Klein 2006). Mitigation generally applies the polluter-pays principle, whereas the need for adaptation measures will tend be greatest and capacity to adapt the least in developing countries whose historical contribution to climate change has been small (Füssel 2007).

Vulnerability to climate change is increasing due to factors other than greenhouse gas emissions, such as increases in population in vulnerable coastal areas (Pielke et al. 2007). Pielke et al. (2007, p. 598) observed that "Virtually every climate impact projected to result from increasing greenhouse-gas concentrations—from rising storm damage to declining biodiversity—already exists as a major concern". Adaptation measures also have the benefit of reducing risks associated with current climate variability (Füssel and Klein 2006). For example, adaptations to anthropogenic-driven future hotter and drier local climate conditions will reduce the risks from naturally occurring droughts. As long as adaptation is discussed in terms of its marginal effects on anthropogenic climate change, its real importance for society is obscured (Füssel and Klein 2006).

Adaptation can be either anticipatory or reactive (Adger et al. 2005, 2007). Anticipatory adaptation is the act of adapting to a potentially harmful condition before actually confronting the problem. Reactive adaptation is triggered by past or current events, but it is also anticipatory in the sense that it is based on some assessment of conditions in the future (Adger et al. 2005). For example, the City of Miami Beach's efforts to improve its flood protection by installing pumping systems and elevating critical roads are both a reaction to current tidal flooding and an anticipatory response to projected future more frequent and greater magnitude flooding from sea level rise (SLR).

Anticipatory responses are significantly more likely to be undertaken by higher levels of government, particularly at the national level, whereas individual or household adaptive responses are more likely to arise from a reactive response to existing stimuli (Berrang-Ford et al. 2011). A gradient in anticipatory adaptive capacity appears to occur with institutions and governments showing potentially higher capacity or resources to proactively engage in adaptive initiatives (Berrang-Ford et al. 2011).

Adaptation may be either autonomous or planned. Autonomous adaptation is defined by the IPCC (2014c, p. 1579) as "Adaptation in response to experienced climate and its effects, without planning explicitly or consciously focused on addressing climate change." Autonomous adaptation has also been described as the "innate ability to adapt to one's environment" (Engle 2011, p. 648). Engle (2011) cautioned that autonomous adaptation does not always end well, and, in some circumstances, it does not moderate harm, but instead exacerbates it. Individuals and communities can make what turns out to be wrong choices.

Planned adaptation are measures taken specifically to address climate change. The distinction between autonomous and planned adaptation is blurred as actions that contribute to the sustainable development goal of making societies more resilient to current climate conditions and changes due to non-climate factors (e.g., population growth) are often consciously intended to also provide resilience to climate change resulting from anthropogenic greenhouse gas emissions. With respect to water management, decision makers endeavor to plan water supply systems to meet current and future water demands under the climate conditions that might occur within a planning period. Planning for and implementing measures to address droughts that might occur under current climate variability provides some resilience to an anthropogenic transition to drier conditions with the general caveat that future climate conditions may be more extreme than recorded historical conditions.

Füssel (2007, p. 268) observed that

> Planned adaptation to climate change means the use of information about present and future climate change to review the suitability of current and planned practices, policies, and infrastructure. Adaptation planning involves addressing questions such as: How will future climatic and non-climatic conditions differ from those of the past? Do the expected changes matter to current decisions? What is a suitable balance between the risks of acting (too) early and those of acting (too) late?

Prerequisites for planned adaptation include awareness of the problem, availability of resources for implementing adaptation measures, cultural acceptability of the measures, and incentives for implementing the measures (Füssel 2007).

Future climate conditions inherently have considerable uncertainty because it is unknown which future GHG emissions scenario will come to pass. Future atmospheric GHG concentrations are poorly constrained and there are uncertainties as to how GHG concentrations will actually impact global and local temperatures and precipitation (i.e., there are errors inherent in climate modeling). Weaver et al. (2013, p. 42) aptly explained that

> Once we start considering timescales beyond a decade or two, the changing human influence on climate cannot be ignored. Societies will likely react in complex ways to climate change, ways that will alter the trajectory of subsequent changes (e.g., in greenhouse gas emissions, land use, and geoengineering), with further feedbacks on human behavior, and these changes will likely be unpredictable.

Adaptation to climate change is (and will continue to be) undertaken privately or publicly, and by individuals, informal groups of people, governmental agencies, and businesses. Adaptation can be undertaken by individuals for their own benefit or it can be made up of actions by governments and public bodies to protect their citizens (Adger et al. 2005). The degree to which individuals, companies, communities, and governments take measures to adapt to climate change depends on numerous factors. Perhaps most important are their actual and perceived exposure to climate change and thus their urgency to act. Climate change is an insidious problem in that the rates of change tend to be small relative to the normal human decision-making time frame. For example, the current global sea level rise rate of about 3.6 mm/yr (Lindsey 2019) is barely perceptible to coastal dwellers on an annual basis and is thus often insufficient to warrant immediate actions considering other pressing demands on personal and societal resources. However, climate change impacts can be profound on decadal and longer time scales, especially as the rates of change are accelerating. The low ranking of climate change amongst the general public reflects a widespread perception that the issue is generally removed in space and time and that whilst it is considered socially relevant, most individuals do not feel it poses a prominent personal threat (Lorenzoni et al. 2007).

Effective adaptation to climate change is contingent on both the availability of information on what changes to adapt to and how to adapt, and the resources needed to implement adaptation measures (Füssel and Klein 2006; Smit and Wandel 2006). Adaptation thus has external and internal dimensions. The external dimensions are the exposures of a system to climate variations. The internal dimensions are the sensitivity of a system to the variations and the capacity of the impacted parties to responds to the variations (Füssel and Klein 2006).

Adaptive capacity is defined as the "ability of systems, institutions, humans, and other organisms to adjust to potential damage, to take advantage of opportunities, or to respond to consequences" (IPCC 2014c, p. 1758). Adaptive limits, as defined by Moser and Ekstrom (2010, p. 22026), refers to "obstacles that tend to be absolute in a real sense: they constitute thresholds beyond which existing activities, land uses, ecosystems, species, sustenance, or system states cannot be maintained, not even in a modified fashion." Adaptive barriers are defined as "obstacles that can be overcome with concerted effort, creative management, change of thinking, prioritization, and related shifts in resources, land uses, institutions, etc." (Moser and Ekstrom 2010, p. 22027). Adaptative barriers can make adaptation less efficient or effective or may require costly changes that lead to missed opportunities or higher costs (Moser and Ekstrom 2010).

Individuals and communities vary greatly in their adaptive capacity with respect to climate change. Adaptive capacity depends to a large degree on available economic and technological resources, but sociocultural factors can also come into play. The local sociopolitical institutions must be willing and capable of making, implementing, and enforcing adaptation decisions. Reactive adaptation decisions can be forced by circumstances. Once the proverbial "well runs dry" individuals and communities have no choice but to adapt in some manner.

9.2 Vulnerability Assessments

Planned adaptation to climate change starts with a recognition of vulnerabilities to climate change. The IPCC (2014c, p. 1775) defines vulnerability as "The propensity or predisposition to be adversely affected. Vulnerability encompasses a variety of concepts and elements including sensitivity or susceptibility to harm and lack of capacity to cope and adapt." A quick search of the internet reveals numerous publications on climate change vulnerability assessments performed for specific hazards, geographic areas, economic sectors, environments, and/or species, and documenting various vulnerability assessment tools. There is also a long history of vulnerability assessments developed in other contexts, such as for natural disasters and risk management in general (Füssel and Klein 2006).

A commonly used model for climate change vulnerability assessments is the risk-hazard framework used in the risk and disaster management context (Füssel and Klein 2006). Risk assessments can be defined broadly as the processes of estimating the probability of occurrence of events and then the probable magnitude of adverse impacts of the events on safely, health, and ecology over a specified time period (Asano et al. 2007). Vulnerability assessments typically involve a synthesis of the currently available scientific information on projected local climate change that is used to evaluate the degree to which key assets, resources, ecosystems, or other features of interest may be affected (adversely or beneficially) by the variability of current climate and potential future local changes in climate and sea level (Joyce and Janowiak 2011).

The U.S. Climate Resilience Tool Kit (NOAA 2019) provides a practical tool for assessing vulnerability and risks from climate change using the risk-hazard methodology. The initial step in the assessment process is an evaluation of past weather events and possible future trends. A determination is then made as to how future climate and sea level conditions could impact various assets, which could include various elements of water treatment and distribution systems and aquifers or surface water sources used for water supply. Assets that are potentially exposed to harm are identified. An assessment of the risk to each exposed asset is then performed, which involves evaluation of both the probability of adverse impacts to each asset and the magnitude of the consequences. A decision is made as to whether the risk climate change poses to each identified vulnerable asset is acceptable or not. The next step in the process is consideration of various strategies that could be employed to reduce the risks to vulnerable assets. Inasmuch as resources to respond (adapt) to climate changes are limited, the final step of the methodology is the ranking of the expected value of each adaptation action and integrating the highest-value actions into a stepwise action plan (NOAA 2019). The USEPA (2015) "Adaptation strategies guide for water utilities" includes a similar sequential approach for utilities to plan for climate change.

With respect to water supply systems, climate change vulnerability assessments involve the following basic steps:

- A general evaluation of potential local climate and relative sea level changes
- assessment of how each component of the water supply, treatment, and distribution system might be impacted by the potential local climate and relative sea level changes
- a risk assessment of identified vulnerabilities—an evaluation of the likelihood, magnitude, and timing of impacts
- identification and evaluation of adaptation options.

Vulnerability assessments can most basically start with a review of IPCC and regional and national climate assessment reports for projections of local climate changes and relative SLR. The publicly available climate modeling results will give a range of values reflecting the uncertainties in future GHG emissions and inherent in climate modeling. Projections for most areas will indicate increasing temperatures. Climate modeling results may provide some insights into whether precipitation will likely locally increase or decrease.

The second step is a systematic evaluation of the potential vulnerability of each element of water supply systems to projected climate changes. It is, in essence, asking and answering a series of "what if" questions. Many of the answers are obvious. Clearly a pronounced decrease in rainfall in a watershed could threaten surface water supplies. Increases in rainfall might overload gravity-driven stormwater conveyance and wastewater mains systems or impair the quality of surface water bodies used for water supply. Vulnerabilities can be identified by examining historical impacts from current climate variations. For example, insights on vulnerabilities to future drier conditions can be obtained through examination of the impacts of past droughts, bearing in mind that future climate conditions may be even more severe (i.e., deviate to a greater degree from the historical mean) than recent historical conditions.

Risk assessments are more complex in that they involve consideration of the probabilities and actual impacts of climate changes on systems. Climate change evaluations can be performed using a "top-down" modeling-based approach or a "bottom-up" impact scenario approach (Sects. 5.2 and 5.7). The top-down approach starts with GCM projections using selected emissions scenarios and proceeds to downscaling and then hydrological and groundwater modeling. The bottom-up approach involves consideration of the impacts of a series of plausible climate changes and then proceeds to an evaluation of the probability of impacts that are considered serious threats to systems and system elements.

Consider for example, a projected future decrease in regional precipitation. The top-down approach would attempt to simulate the potential magnitude of the precipitation change and then simulate how the decrease in precipitation might impact, for example, surface water flows and groundwater recharge and aquifer water levels and, in turn, water utility operations. The bottom-up approach to evaluating the impacts of potential decreases in local precipitation might, for example, start by evaluating how utility operations would be impacted by 5, 10, 20, and 30% reductions in the flows of a surface water source or groundwater recharge. If a 20% or greater reduction was found to likely result in a major disruption in water supplies, then climate modeling data would be considered (or possibly performed) to evaluate the likelihood of such a future reduction and the timeframe in which such a reduction might occur. Finally, if such a reduction in available water is determined to be likely, then various options would be evaluated to respond to the water shortfall including conservation measures and obtaining alternative water supplies.

Freas et al. (2008, 2010) described a variation of the bottom-up approach referred to as the "threshold risk assessment approach." The threshold risk assessment approach starts with an evaluation of the threshold or performance criteria for each water system component, which are the range of conditions for normal operation. Conditions outside of the range are considered to disrupt the function of the component. The next steps in the process are to explore climate changes (e.g., in precipitation amounts and temperature) that could result in a violation of threshold or performance criteria and then whether the identified climate sensitivities fall within the range of anticipated local changes in climate variables. Adaptations strategies are then developed for identified vulnerable components.

From a practical perspective, the bottom-up approach to climate change risk assessment is much preferred. The top-down approach, while technically very rigorous, suffers from the "chain of uncertainties" resulting from unknowns and inaccuracies in each step of the process (Foley 2010; Falloon et al. 2014). "The range (or envelope) of uncertainty expands at each step of the process to the extent that potential impacts and their implied adaptation responses span such a wide range as to be practically unhelpful" (Wilby and Dessai 2010, p. 181).

The hope is that increased sophistication and higher-resolution climate models will better characterize and constrain uncertainty in the regional climate projections offered to decision makers (Wilby and Dessai 2010). The ability to downscale to finer time and space scales does not imply that confidence will be any greater in the resulting scenarios. Downscaling has serious practical limitations, especially where the historical meteorological data required may be of dubious quality or patchy, the links between regional and local climate are poorly understood or resolved, and where technical capacity is not in place (Wilby and Dessai 2010). Wilby and Dessai (2010) cautioned that high-resolution downscaling can be misconstrued as accurate downscaling (Dessai et al. 2009; Wilby and Dessai 2010).

The final step in climate change vulnerability assessments is the identification and evaluation of adaptation options for elements of systems that are identified as being vulnerable to projected climate change. The adaptation assessment part of climate change vulnerability assessments is little different from the normal water supply planning process with the exception being that decision making is being made under less certain future conditions. It must be stressed that there is a great variety in the approaches and detail involved in climate change vulnerability assessments and certainly one-size does not fit all.

Gober and Kirkwood (2010) provided a good example of a vulnerability assessment for water supply in an area already facing water scarcity, Phoenix,

Arizona. Phoenix is located in a desert environment in which water is now provided from overdrafted aquifers and externally sourced river systems; the Salt and Verde Rivers and the Colorado River through the Central Arizona Project. Previous modeling studies projected future runoffs of 19–123% of the historical mean for the Salt and Verde Rivers system and 61–118% for the Colorado River. Groundwater was assumed to be used to compensate for any surface water deficits. Groundwater use in Arizona is regulated under the Groundwater Management Act of 1980, which mandates that Active Management Areas (AMAs), including the Phoenix AMA, achieve a safe yield (withdrawals = recharge) by 2025.

The integrated simulation model WaterSim was used to investigate the consequences of policies to manage groundwater, water use, and urban development in Phoenix (Gober and Kirkwood 2010; Gober et al. 2011). The metrics used to evaluate combinations of scenarios and policies were per capita water availability and cumulative groundwater deficit. The results of the simulations were that under current water use rates and projected population increase, it is not possible to achieve groundwater sustainability under any scenario (Gober and Kirkwood 2010). Gober and Kirkwood (2010, p. 21286) concluded that "designing a system to supply enough water for business as usual in the most pessimistic climate-change scenarios would be very expensive and perhaps, physically impossible" and that "Ignoring these scenarios and designing for a best guess case could leave Phoenix vulnerable to water shortages with little time to adapt." Policy options identified that would allow for sustainable water use under most climate scenarios are limiting population growth to 50% of projected levels and eliminating irrigated outdoors landscaping and private backyard pools (Gober and Kirkwood 2010). Other water supply options not included in the analyses were purchasing Native American water rights, conversion of agricultural lands to urban purposes, and desalination (Gober and Kirkwood 2010). The Phoenix vulnerability assessment is noteworthy in that it illustrates the difficult challenges some areas will face if climate change exacerbates already existing water scarcity and unchecked population growth results in ever-increasing water demands.

Water and other utilities own and operate infrastructure that is both concentrated at specific sites (e.g., water treatment plants) and distributed throughout their service areas. Determination of potential climate change impacts requires a systematic vulnerability assessment approach to evaluate the needs for operational changes and the installation new infrastructure or hardening of existing infrastructure to make systems more robust (Bloetscher et al. 2014, 2017). A suggested risk assessment approach for utilities to quantify the risks of a utility's current system to potential climate change impacts involves nine steps (Freas et al. 2008; Bloetscher et al. 2014, 2017):

1. selection of a range of climate change scenarios based on commonly accepted models
2. translation of the large-scale scenarios into local climate change scenarios
3. identification of climate change variables of importance (e.g. rainfall frequency, rainfall volume, temperature, sea level)
4. determination of local system responses to climate changes (e.g., incorporate rainfall changes into surface flow models)

5. development of adaptation strategies
6. evaluation of the robustness of the adaptive strategies
7. identification of hardening measures
8. evaluation of overall system performance
9. implementation of the decisions needed to adapt to the changes.

The conventional anticipatory approach should be replaced or supplemented with one that recognizes the importance of building resiliency because vulnerability can never be estimated with 100% accuracy (Bloetscher et al. 2017).

9.3 Adaptation Planning Under Uncertainty

How future local climate conditions will differ from present conditions has considerable uncertainty other than that it will generally get warmer and sea level will continue to rise. The challenging question that is faced is "how can we ensure that adaptation measures realize societal benefits now, and over coming decades, despite uncertainty about climate variability and change?" (Wilby and Dessai 2010, p. 180). Uncertainty has been defined as lacking confidence in one's knowledge relating to a concrete question (Sigel et al. 2010). Uncertainty can be either phenomenological or epistemological (Sigel et al. 2010). Phenomenological uncertainty relates to inherent unpredictability in the phenomena themselves. With respect to climate change, future GHG concentrations at different times in the future cannot be accurately predicted because they will depend largely on the actions that nations will take in the future. Epistemological uncertainty relates to the state of knowledge of a person (e.g., models are not good enough to precisely simulate relevant processes; Sigel et al. 2010).

In standard design, the goal is usually to identify the optimal solution, which may consider a diversity of factors including function, cost, social acceptability, and environmental impacts. In light of the uncertainties in local future climate conditions, it is now widely accepted that the preferred adaptation strategies are ones that are robust and that perform well (although not necessarily optimally) over a wide range of conditions faced now and potentially under future climate scenarios (Wilby and Dessai 2010). Some performance is sacrificed in order to reduce sensitivity to broken assumptions and thus minimize regret under some particular futures (Weaver et al. 2013).

The decision-making process is also impacted by the risk aversion of the stakeholders involved, particularly with respect to low-probability and severe outcome events. Individuals and organizations vary in their risk tolerance and some are very averse to bad outcomes even when they have very low probabilities. With respect of climate change, decision makers are faced with the decision as to which range of possible future climate conditions and associated adverse impacts (particularly at the high-impact, low-probability end of the spectrum) to plan for and invest in infrastructure to address.

A frequently expressed goal for robust adaptation is "no regret" strategies that will yield benefits regardless of climate change (Heltberg et al. 2009). Wilby and Dessai (2010) noted that in practice, there are opportunity costs, trade-offs, and externalities associated with adaptation actions, so it is better to refer to such interventions as "low regret". Robust adaptation measures tend to be low regret (or reversible), incorporate safety margins, employ "soft" solutions, and are flexible and mindful of actions being taken by others to either mitigate or adapt to climate change (Lempert and Schlesinger 2000; Hallegatte 2009; Wilby and Dessai 2010). Examples of no or low regret adaptations are protecting water sources from salinization and other contamination as they are sound strategies under any climate context (Wilby and Dessai 2010). Conservation is also a low regret adaptation to present and future water scarcity.

No or low regrets actions are usually sound water management practices. Refsgaard et al. (2013, p. 344) noted that

> No-regrets actions are those that one would wish to take for other reasons and that also have benefits in terms of reducing the impacts of possible climate change. If these actions are truly justified on other grounds than their climate benefits, then the uncertainty of their climate benefits is of no consequence as it does not affect the decision.

Robust adaptation may also incorporate adaptive management. Adaptive management is frequently referred to as "learning by doing" but it is much more. It is an iterative structured process in which learning is derived from the deliberate formal processes of inquiry (Stankey et al. 2005). Initial actions are taken based on available information, data are collected, hypotheses tested, and plans and actions adjusted based on experiences. Adaptative management "does not postpone action until 'enough' is known but acknowledges that time and resources are too short to defer some action" (Stankey et al. 2005, p. 9). Adaptations that incorporate flexibility, monitoring, and review differ very little from what are considered best water management practices (Wilby and Dessai 2010).

A variety of strategies are available for low or no regret decision making. A basic method outlined by Wilby and Dessai (2010) is to first construct an inventory of all adaptation options for a given situation, which could include hard engineering solutions and soft solutions involving reallocation of resources, behaviour change, institutional and sectoral reform or restructuring, awareness-raising, or risk spreading via financial instruments. The set of adaptation options is then screened and appraised to identify a subset of preferred adaptation measures that would reduce vulnerability under present and future climate regimes, whilst being socially acceptable, and technically and economically feasible given the prevailing regulatory environment.

Cost is an important consideration in adaptation decision making. Some form of cost-benefit analysis (CBA) is usually performed to evaluate various options to achieve a specific goal. Risk and uncertainty can be incorporated into CBAs through an expected value analysis (Boardman et al. 1996). In expected value analysis, the future is characterized in terms of a number of distinct contingencies that are mutually exclusive and capture the full range of likely variations in the costs and benefits of a project or policy. For each adaptation option considered, probabilities are assigned to the occurrence of each possible contingency (e.g., specific climate change). The expected net benefits (ENB) of each adaptation option is probability weighted sum of the net present value for each contingency.

Robust decision making (RDM) has been specifically defined as "a decision analytic framework based on the concept of identifying strategies robust over a wide range of often poorly-characterized uncertainties" (Groves and Lempert 2007, p. 75). RDM rests on three key concepts: (1) multiple views of the future, (2) a robustness criterion, and (3) an iterative vulnerability-and-response-option analysis (Groves and Lempert 2007; Lempert and Groves 2010; Weaver et al. 2013). The central idea of RDM is to perform a large number of runs of computer simulation models to identify those scenarios most important to the choices facing decision makers (Groves and Lempert 2007). RDM uses computer simulation models, not to predict the future, but to create large ensembles of hundreds to millions of plausible future states that are used to identify candidate robust strategies and systematically assess their performance (Groves and Lempert 2007). The main goals of RDM are to (1) identify robust strategies whose satisfactory performance is largely independent of the eventually revealed true values of most unknowns (e.g., future climate conditions) and (2) characterize the few deep uncertainties most important to the choice among strategies.

As described by Groves and Lempert (2007, p. 76),

> RDM is an iterative process that begins with decision options and then runs the expected utility machinery many times in order to identify potential vulnerabilities of these candidate strategies, that is, combinations of model formulations and input parameters where the strategy performs relatively poorly compared to the alternatives. The analysis then suggests new or modified strategies that might perform better in these vulnerable futures and characterizes the tradeoffs involved in choosing among these decision alternatives.

An important characteristic of RDM is that it is independent of any assessment of the likelihood of the considered future states of the world.

The RDM process is intended to identify both strategies that are robust under a wide range of future conditions and vulnerabilities, which are cases where the candidate strategy fails to meet policymakers' (or other stakeholders') performance standards under some combinations of key uncertainty factors (Groves and Lempert 2007; Lempert and Groves 2010). At the end of the process, the plausibility of vulnerabilities may be assessed, for example, using climate modeling results.

9.4 Adaptative Capacity

The prerequisites for planned adaptation include (Füssel 2007):

- awareness of the problem
- availability of effective adaptation measures
- information about potential adaptation measures

- availability of resources for implementing the measures
- cultural acceptability of the measures
- incentives for implementing the measures

The capacity to adapt to climate change is thus a function of societal, economical, and technical resources and the ability and willingness of societies to effectively respond. Clearly wealthy developed countries have much greater financial and technical resources to address current and future climate changes than do poorer countries. High-income countries are characterized by more proactive or anticipatory adaptations to climatic changes and the adaptations in developed countries are more likely to include governmental participation (Berrang-Ford et al. 2011).

Climate change adaptations in low-income countries, on the contrary, are characterized by reactive adaptations in response to short-term motivations and most adaptations are occurring at the individual level with weak involvement of government stakeholders. Adaptation mechanisms tend to include community-level mobilization rather than institutional, governmental or policy tools (Berrang-Ford et al. 2011). Developing nations and poorer communities tend to be the least capable of adapting because they lack the resources and institutions to mobilize available resources (Engle 2011).

Adaptative capacity is more complex than a simple function of national or more local-level incomes. Political organization patterns also impact local adaptative capacity. The organization of community water supply systems in the United States (outside of major urban areas) and in other countries tends to be fragmented with small utilities serving local populations (Mullin 2020). Decentralization is often preferred by communities because it gives them greater autonomy over their water supplies. However, fragmentation of community water supplies can be an impediment to drought and climate change-related water security as small utilities often lack the technical and financial resources to develop more resilient water supplies. Consolidation of and collaboration between community water supply systems are means for expanding water security (i.e., increasing adaptive capacity) in the face of droughts and other hazards under current and future climate conditions (Mullin 2020).

Adger (2003, p. 388) observed that societies have inherent capacities to adapt to climate change, which are bound up in their ability to act collectively. More specifically he observed that

> Decisions on adaptation are made by individuals, groups within society, organizations and governments on behalf of society. But all decisions privilege one sets of interests over another and create winners and losers.

and that

> effectiveness of strategies for adapting to climate change depend on the social acceptability of options on adaptation, the institutional constraints on adaptation, and the place of adaptation in the wide landscape of economic development and social evolution.

Adaptation depends upon social capital, which is made up of the norms and networks that enable people to act collectively (Adger 2003).

Factors, such as perception of risk, habit, social status, and age, operate at individual decision-making levels but can constrain collective action (Adger et al. 2009). Individual adaptation hinges on whether an impact (anticipated or experienced) is perceived as a risk that should (and could) be acted upon. Adger et al. (2009) further opined that at the policy level, adaptation policies are constrained by inertia, cultures of risk denial, and other phenomena.

Discussions of social capital and adaptative capacity tend to focus on poorer developing countries. However, it can be argued that adaptive capacity in the United States has been eroded through the decline of social capital manifested by hyper political partisanship. Planned adaptation to climate changes is not possible where there is denial that anthropogenic climate change is occurring. The denial is driven by political contrarianism (i.e., if an opposing political party advocates something, then it must be wrong) and an unwillingness to accept the repercussions and responsibilities of anthropogenic climate change. A lack of leadership from elected officials and within organizations has been cited in surveys as a key hurdle to climate change adaptation (Kay et al. 2018).

The availability of adaptive capacity in itself does not ensure the absence of adaptation barriers (Azhoni et al. 2018). Organizations and communities may have considerable adaptive capacity that is untapped due to ineffective leadership. Effective utilization of adaptive capacity is reduced by barriers, such as cultural and normative behaviors, rigid and outdated laws, bureaucratic procedures, and inaccessibility to information, among others causes (Azhoni et al. 2018). Where the management or senior technical experts of a water utility or wholesale water supplier are engaged in climate change issues, then there is a great likelihood that vulnerabilities to climate change will be given serious consideration. Some of the strongest aids in advancing adaptation are strong leadership and increased public awareness of and interest in climate change (Moser and Ekstrom 2012; Kay et al. 2018).

Crabbé and Robin (2006) in a study of institutional adaptation of water resource infrastructures to climate change in Eastern Ontario identified that municipal adaptive capacity is a function of various factors:

- the range of available technological options
- available resources and their distribution across the municipal population
- structure of critical institutions and the criteria for decision making
- human and social infrastructures
- access to risk-spreading mechanisms
- the ability of decision makers to manage credible information and their own credibility
- the public's perception of both the source of the impact and its significance to its local manifestations.

Although relative differences in adaptive capacity are apparent between countries and regions, it cannot be practically quantified. Engle (2011) highlighted two of the greatest limitations of the concept of adaptive capacity. First is that adaptive capacity is difficult to gauge because of its latent nature, meaning that it is very difficult to measure it until after its realization or mobilization within a system. For example, the ability of a society to adapt to drier conditions cannot be fully gauged until it faces such conditions. Second is that it is possible that future climate change conditions could far exceed what systems have experienced in the past. Lessons from historical experiences may not be a guide to the ability of a society to adapt to future more extreme conditions. However, it is more likely that most climate changes will occur gradually, stretching the boundaries of previous extremes, and with societies and institutions adapting along the way (Engle 2011).

9.5 Effectiveness of Adaptation

Effectiveness relates to the capacity of an adaptation action to achieve its expressed objectives and can be gauged either through reducing impacts from adverse conditions or in terms of reducing risk and avoiding danger and promoting security (Adger et al. 2005). Defining success simply in terms of the effectiveness of meeting objectives, however, is insufficient for two reasons. First, the ability of an adaptation to reduce impacts will not be known until the exposure occurs, which could be a way in the future. For example, the effectiveness of measures to reduce vulnerability to drought would not be known for certain until a severe drought occurs, at which time there may not be sufficient time to explore other options.

Whilst an action may be successful in terms of one stated objective, it may impose externalities at other spatial and temporal scales. What appears successful in the short term may turn out to be less successful in the longer term. For example, the installation of domestic and commercial air conditioning in western Europe following summer heat waves represents an effective adaptation for its adopters, but it is based on energy- and emissions-intensive technologies and thus may not be sustainable in the long term (Adger et al. 2005).

Hence, a key part of successful adaptation is to avoid complete commitment to any one strategy, but rather to retain flexibility to alter course as new data become available (i.e., to employ adaptive management). Incremental approaches can provide benefits while still retaining flexibility. For example, in a study of adaptations to SLR, it was found that incremental adaptation costs were less than the economic damage in almost all emissions scenarios, providing an incentive to take more immediate actions to respond to climate change (Tamura et al. 2019).

References

Adger WN (2003) Social capital, collective action, and adaptation to climate change. Econ Geogr 79(4):387–404

Adger WN, Arnell NW, Tompkins EL (2005) Successful adaptation to climate change across scales. Glob Environ Change 15(2):77–86

Adger WN, Agrawala S, Mirza MMQ, Conde C, O'Brien K, Pulhin J, Pulwarty R, Smit B, Takahashi K (2007) Assessment of adaptation practices, options, constraints and capacity. In: Parry ML, Canziani OF, Palutikof JP, van der Linden PJ, Hanson CE (eds) Climate change 2007: impacts, adaptation and vulnerability. Contribution of Working Group II to the Fourth Assessment Report of the Intergovernmental Panel on Climate Change. Cambridge University Press, Cambridge, UK, pp 717–743

Adger WN, Dessai S, Goulden M, Hulme M, Lorenzoni I, Nelson DR, Naess LO, Wolf J, Wreford A (2009) Are there social limits to adaptation to climate change? Clim Change 93:335–354

Asano T, Burton FL, Leverenz HL, Tsuchihashi R, Tchobanoglous G (2007) *Wastewater reuse. Issues, technologies and applications*. McGraw-Hill, New York

Azhoni A, Jude S, Holman I (2018) Adapting to climate change by water management organisations: Enablers and barriers. J Hydrol 559:736–748

Berrang-Ford L, Ford JD, Paterson J (2011) Are we adapting to climate change? Glob Environ Change 21(1):25–33

Bloetscher F, Hammer NH, Berry L (2014) How climate change will affect water utilities. J Am Water Works Assoc 106(8):176–192

Bloetscher F, Hoermann S, Berry L (2017) Adaptation to Florida's urban infrastructure to climate change. In: Chassignet EP, Jones JW, Misra V, Obeysekera J (eds) Florida's climate: Changes, variations, & impacts. Florida Climate Institute, Gainesville, pp 311–338

Boardman A, Greenberg D, Vining A, Weimer D (1996) Cost-benefit analysis: Concepts and practice. Prentice Hall, Upper Saddle River

Crabbé P, Robin M (2006) Institutional adaptation of water resource infrastructures to climate change in Eastern Ontario. Clim Change 78(1):103–133

Dessai S, Hulme M, Lempert R, Pielke R (2009) Climate prediction: A limit to adaptation. In: Adger N, Lorenzoni I, O'Brien K (eds) Adapting to climate change: Thresholds, values, governance. Cambridge University Press, Cambridge, UK, pp 64–78

Engle NL (2011) Adaptive capacity and its assessment. Glob Environ Change 21(2):647–656

Falloon P, Challinor A, Dessai S, Hoang L, Johnson J, Koehler AK (2014) Ensembles and uncertainty in climate change impacts. Front Environ Sci 2:33

Foley AM (2010) Uncertainty in regional climate modelling: A review. Prog Phys Geogr 34 (5):647–670

Freas K, Bailey R, Muneavar A, Butler S (2008) Incorporating climate change in water planning. J Am Water Works Assoc 100:93–99

Freas K, Bailey R, Muneavar A, Butler S (2010) Incorporating climate change in water planning. In: Howe C, Smith JB, Henderson J (eds) Climate change and water. International perspectives on mitigation and adaptation. Denver and London, American Water Works Association and IWA Publishing, pp 173–182

Füssel HM (2007) Adaptation planning for climate change: Concepts, assessment approaches, and key lessons. Sustain Sci 2(2):265–275

Füssel HM, Klein RJ (2006) Climate change vulnerability assessments: An evolution of conceptual thinking. Clim Change 75(3):301–329

Gober P, Kirkwood CW (2010) Vulnerability assessment of climate-induced water shortage in Phoenix. Proc Natl Acad Sci 107(50):21295–21299

Gober P, Wentz EA, Lant T, Tschudi MK, Kirkwood CW (2011) WaterSim: A simulation model for urban water planning in Phoenix, Arizona, USA. Environ Plan 38(2):197–215

Groves DG, Lempert RJ (2007) A new analytical method for finding policy-relevant scenarios. Glob Environ Change 17:73–85

Hallegatte S (2009) Strategies to adapt to an uncertain climate change. Glob Environ Change 19 (2):240–247

Heltberg R, Siegel PB, Jorgensen SL (2009) Addressing human vulnerability to climate change: toward a 'no-regrets' approach. Glob Environ Change 19(1):89–99

IPCC (2014a) Climate change 2014: Impacts, adaptation, and vulnerability. Part A: Global and sectoral aspects. Contribution of Working Group II to the Fifth Assessment Report of the Intergovernmental Panel on Climate Change. In: Field CB, Barros VR, Dokken DJ, Mach KJ, Mastrandrea MD, Bilir TE, Chatterjee M, Ebi KL, Estrada YO, Genova RC, Girma B, Kissel ES, Levy AN, MacCracken S, Mastrandrea PR, White LL (eds). Cambridge University, Cambridge UK

IPCC (2014b) Climate change 2014: Impacts, adaptation, and vulnerability. Part B: Regional aspects. Contribution of Working Group II to the Fifth Assessment Report of the Intergovernmental Panel on Climate Change. In: Field CB, Barros VR, Dokken DJ, Mach KJ, Mastrandrea MD, Bilir TE, Chatterjee M, Ebi KL, Estrada YO, Genova RC, Girma B, Kissel ES, Levy AN, MacCracken S, Mastrandrea PR, White LL (eds). Cambridge University, Cambridge UK

IPCC (2014c) Annex II: Glossary (Agard J, Schipper ELF, Birkmann J, Campos M, Dubeux C, Nojiri Y, Olsson L, Osman-Elasha B, Pelling M, Prather MJ, Rivera-Ferre MG, Ruppel OC, Sallenger A, Smith KR, St. Clair AL, Mach KJ, Mastrandrea MD, Bilir TE (eds)). In: Field CB, Barros VR, Dokken DJ, Mach KJ, Mastrandrea MD, Bilir TE, Chatterjee M, Ebi KL, Estrada YO, Genova RC, Girma B, Kissel ES, Levy AN, MacCracken S, Mastrandrea PR, White LL (eds). Climate change 2014: Impacts, adaptation, and vulnerability. Part B: Regional Aspects. Contribution of Working Group II to the Fifth Assessment Report of the Intergovernmental Panel on Climate Change. Cambridge University Press, Cambridge UK, pp 1757–1776

Joyce LA, Janowiak MK (2011) Climate change vulnerability assessments. U.S. Department of Agriculture, Forest Service, Climate Change Resource Center. www.fs.usda.gov/ccrc/topics/assessments/vulnerability-assessments. Accessed 3 June 2020

Kay R, Scheuer K, Dix B, Bruguera M, Wong A, Kim J (2018) Overcoming organizational barriers to implementing local government adaptation strategies, CNRA-CCA4-2018-005, California's Fourth Climate Change Assessment. California Natural Resources Agency, Sacramento

Klein RJT, Huq S, Denton F, Downing TE, Richels RG, Robinson JB, and Toth FL (2007) Inter-relationships between adaptation and mitigation. In: Parry ML, Canziani OF, Palutikof JP, van der Linden PJ, Hanson CE (eds.) Climate change 2007: Impacts, adaptation and vulnerability. Contribution of Working Group II to the Fourth Assessment Report of the Intergovernmental Panel on Climate Change. Cambridge University Press, Cambridge UK, pp 745–777

Klein RJT, Midgley GF, Preston BL, Alam M, Berkhout FGH, Dow K, Shaw MR (2014) Adaptation opportunities, constraints, and limits. In: Field CB, Barros VR, Dokken DJ, Mach KJ, Mastrandrea MD, Bilir TE, Cshatterjee M, Ebi KL, Estrada YO, Genova RC, Girma B, Kissel ES, Levy AN, MacCracken S, Mastrandrea PR, White LL (eds) Climate change 2014: Impacts, adaptation, and vulnerability. Part A: Global and sectoral aspects. Contribution of Working Group II to the Fifth Assessment Report of the Intergovernmental Panel on Climate Change. Cambridge University Press, Cambridge, UK, pp 899–994

Lempert RJ, Groves DG (2010) Identifying and evaluating robust adaptive policy responses to climate change for water management agencies in the American west. Technol Forecast Soc Chang 77(6):960–974

Lempert RJ, Schlesinger ME (2000) Robust strategies for abating climate change. Clim Change 45 (3–4):387–401

Lindsey R (2019) Climate change: global sea level. https://www.climate.gov/news-features/understanding-climate/climate-change-global-sea-level. Accessed 3 June 2020

Lorenzoni I, Nicholson-Cole S, Whitmarsh L (2007) Barriers perceived to engaging with climate change among the UK public and their policy implications. Glob Environ Change 17(3–4):445–459

Moser SC, Ekstrom JA (2010) A framework to diagnose barriers to climate change adaptation. Proc Natl Acad Sci 107(51):22026–22031

Moser SC, Ekstrom JA (2012) Identifying and overcoming barriers to climate change adaptation in San Francisco Bay. Results from case studies. White paper for the California Energy Commission's California Climate Change Center, July, Sacramento

Mullin M (2020) The effects of drinking water service fragmentation on drought-related water security. Science 368:274–277

NOAA (2019) U.S. climate resilience toolkit. NOAA Climate Program Office. https://toolkit.climate.gov/. Accessed 4 June 2020

Pielke R Jr, Prins G, Rayner S, Sarewitz D (2007) Climate change 2007: lifting the taboo on adaptation. Nature 445(7128):597–598

Refsgaard JC, Arnbjerg-Nielsen K, Drews M, Halsnæs K, Jeppesen E, Madsen H, Markandya A, Eivind J, Porter JR, Christensen JH (2013) The role of uncertainty in climate change adaptation strategies—A Danish water management example. Mitig Adapt Strat Glob Change 18(3):337–359

Sigel K, Klauer B, Pahl-Wostl C (2010) Conceptualising uncertainty in environmental decision-making: The example of the EU water framework directive. Ecol Econ 69(3):502–510

Smit B, Wandel J (2006) Adaptation, adaptive capacity and vulnerability. Glob Environ Change 16(3):282–292

Stankey GH, Clark RN, Bormann BY (2005) Adaptive management of natural resources: theory, concepts and management institutions.General Technical Report PNW-GTR-654. USDA Forest Service, Pacific Northwest Research Station, Portland

Tamura M, Kumano N, Yotsukuri M, Yokoki H (2019) Global assessment of the effectiveness of adaptation in coastal areas based on RCP/SSP scenarios. Clim Change 152(3–4):363–377

USEPA (2015) Adaptation strategies guide for water utilities. U.S. Environmental Protection Agency, Office of Water, Washington DC, February

Weaver CP, Lempert RJ, Brown C, Hall JA, Revell D, Sarewitz D (2013) Improving the contribution of climate model information to decision making: The value and demands of robust decision frameworks. Wiley Interdiscip Rev: Clim Chang 4(1):39–60

Wilby RL, Dessai S (2010) Robust adaptation to climate change. Weather 65(7):180–185

Chapter 10
Adaptation Options

10.1 Introduction

Countries and communities exposed to climate change will have to adapt in some manner. Adaptation options for coastal communities and inland communities exposed to increased flooding and other direct physical threats fall into three categories: retreat, accommodation, and protection (Dronkers et al. 1990). Retreat involves abandoning a location and moving to a less exposed area. Protection is the employment of structural means to reduce the likelihood of a hazard. For example, the potential for local flooding is commonly reduced by the construction of walls, levees, and other hydraulic barriers. Accommodation involves measures to reduce vulnerability to hazards and includes hardening of infrastructure to withstand extreme events and elevating buildings and other infrastructure on mounds or pilings.

Climate change can also be net beneficial in some areas; a warmer climate and more plentiful rainfall might be advantageous for local agriculture. Global adaptation to climate change will include taking advantage of benefits resulting from positive local changes to offset negative impacts elsewhere. Nevertheless, the primary focus of climate change adaptation is to reduce the negative local consequences of climate change.

The principal threats to groundwater supplies from climate change are:

- increased aridity decreasing recharge and thus the amount of water that can be sustainably extracted
- deceases in annual average precipitation could decrease surface water supplies causing additional demands on groundwater to make up the difference
- changes in the timing and form of precipitation (e.g., decreases in annual snowpacks) could decrease recharge
- increased temperatures and decreases in precipitation may drive additional demands for water in general

R. Maliva, *Climate Change and Groundwater: Planning and Adaptations for a Changing and Uncertain Future*, Springer Hydrogeology,
https://doi.org/10.1007/978-3-030-66813-6_10

- sea level rise could contribute to salinization of coastal aquifers through horizontal intrusion and more frequent and extensive overwash events
- climate change-induced deterioration of surface waters could adversely impact the quality of recharged water and thus shallow groundwater.

Climate change can also have adverse impacts on communities through rising groundwater levels, which can be caused by sea level rise or increased precipitation. Groundwater rise can be managed through drainage and accommodated by elevating or waterproofing of vulnerable structures.

Adaptations to the negative consequences of climate change on groundwater supplies are essentially the same options available to address water scarcity in general. Decreasing groundwater supplies and reliability may be addressed by some combination of:

- demand management
- development of alterative water supplies (e.g., desalination, wastewater reuse, rainwater harvesting)
- optimization of existing supplies (e.g., managed aquifer recharge, conjunctive use, and community water system interconnections, coordination, and consolidations).

Saline water intrusion in coastal aquifers caused by sea level rise (SLR) can be managed by reducing groundwater extractions near the coast and the construction of physical and hydraulic barriers that either block the landward flow of saline water or maintain a seaward hydraulic gradient at the fresh-saline groundwater interface.

It is important to emphasize that in many areas the impacts of climate change on groundwater will be less than those from overexploitation, which will likely increase with population growth and economic development. Hence adaptations to climate change include reductions in the multiple non-climate-related pressures on freshwater resources (e.g., pollution and large water withdrawals) as well as improvement of water supply and sanitation infrastructure in developing countries (Kundzewicz et al. 2008).

Techniques available for groundwater management in arid lands were reviewed by Maliva and Missimer (2012) and Maliva (2019). Follows are summaries (with references to key publications) of some of the main adaptation approaches for responding to the impacts of climate change on groundwater resources and water resources in general.

10.2 Demand Management

10.2.1 Demand Management Basics

The fundament objective of water management is to bring water supplies and demands into a long-term balance. The balance can be achieved by some combination of decreasing the demand for water (or at least stabilizing demand) and obtaining new supplies. As far as obtaining new water supplies, the "low hanging fruit" has long-been harvested in most areas and new supplies are increasingly difficult and expense to bring on line. It has become widely understood and accepted that improving the efficiency of water use and reducing waste and losses are the most affordable and practical solution elements to water scarcity (Wiener 1977; Postel 1992; Lazarova et al. 2000). As Postel (1992, p. 23) expressed, "Doing more with less is the first and easiest step along the path toward water security. By using water more efficiently, we in effect create a new source of supply."

Demand can be reduced through increased efficiency in water use. Efficiency is most simply defined as the ratio of output to input. Efficiency can also be defined as achieving the maximum productivity with minimum wasted effort or expense (McKean 2004). Water use efficiency (WUE) is broadly defined as the output per unit volume of water. Output can be defined in economic terms (i.e., the economic value of goods produced), yield terms (e.g., the mass or volume of crops produced), and for agriculture in terms of transpired water (i.e., water actually used by plants). WUE can be considered for an individual activity or for a community, groundwater basin, or country as a whole. The latter context of efficiency involves consideration of the allocation of water between sectors of society with the goal that the allocation of available water should result in the greatest net benefit to society.

Oweis et al. (1999) recommended use of the term "productivity," where the numerator is a measure of either the quantity of a product or its economic value. Input is normally expressed in terms of a standard volume of water (e.g., m^3, km^3, gallons). WUE or productivity is commonly expressed in terms of the value of goods generated in U.S. dollars or Euros per m^3, acre-ft, or 1000 gallons of water. An essential aim of demand management is to ensure that a given supply of water is distributed more closely to its optimal use pattern, which would maximize productivity (Winpenny 1994).

Demand management can also involve not allowing activities that would increase demands. Water scarcity, defined in terms of per capita availability, is inherently linked to population growth. Local population growth includes internal population growth (i.e., difference between local births and deaths) and net migration into an area. Population growth combined with a drying climate is a recipe for severe water shortages (Gober and Kirkwood 2010).

Population control is an emotion-laden, politically charged issue that is often avoided as it touches upon what are perceived to be basic freedoms. In areas that are already facing water scarcity, limiting inward migration might be a logical step but there is often no legal mechanism to do so. Local economies in some areas (e.g., parts of the southern United States) have a high dependency on population growth and very strong opposition to measures to curb the inflow of new residents would be expected from people and businesses involved in the home building and related industries. In areas of developing countries experiencing severe drought, migration may be a matter of survival.

Water allocation extends beyond economic efficiency into the issue of equity or fairness, which relates to the distribution of water within and across generations of users. Current groundwater users can impact other aquifer users and the ability of future generations to use an aquifer. Few would argue that the poor should not have access to water because it is not an economically efficient use. Gleick (1996, 1998) proposed that a basic water requirement (BWR) of 50 L (13.2 gal) per person per day of clean water is needed to meet the four basic needs, which are water for drinking, human hygiene, sanitary services, and food preparation, and should be considered a fundamental human right. On July 28, 2010, the United Nations General Assembly voted to expand the Universal Declaration of Human Rights to include the right to water and sanitation as a basic human right. However, it has been observed that declaring water a human right does not put pipes in the ground. A number of questions exist concerning the implementation of such a right to water including (Segerfeldt 2005)

- how much water do people have a right to and of what quality
- how is it to be delivered and who pays for it
- who is responsible for implementation of the right and is liable for its non-achievement
- what are the obligations of water users and the timetable for implementation.

Water-policy decisions are inherently value laden and men and women of goodwill often have different priorities, perspectives, and self-interests with respect to the allocation and manner of use water (Maliva and Missimer 2012). Local culture, customs, and societal norms impact water use. The use of water for irrigation of public and private lawns, landscaping, and golf courses may seem trivial, frivolous, and wasteful, particularly when compared to potable use, but people long accustomed to these uses would strongly oppose their prohibition.

It is also becoming widely understood and accepted that improving the efficiency of water use on a national level in countries and regions facing severe water scarcity must involve a comprehensive analysis and that implementation of programs should consider all aspects of water management (Wiener 1977). Demand management should involve exploring the opportunities for water savings in all the major water-using sectors of society. The greatest potential benefits of demand management lie in the sectors of the economy where there is the greatest use of water. Pie diagrams of total water use and consumptive freshwater use in the United States (Fig. 10.1) show that thermoelectic cooling is the largest total water use but is a very small consumptive use of freshwater as saline water is commonly used in coastal areas and most (~97%) of the water is returned to its source (Dieter et al. 2018). Agriculture is far by the largest consumptive use of freshwater water followed by public water supply. Dieter et al. (2018) assumed that 62% of irrigation water is consumptively used (i.e., removed from availability owing to evaporation, transpiration, or incorporation into crops) with most of the remainder (e.g., irrigation return flows) recharging shallow aquifers or returning to surface water bodies.

Control of water use can be a particularly daunting problem in developing countries where water governance is poor to non-existent. Shah (2009, p. 8) observed, for example, that "India has witnessed growing discussions on how best

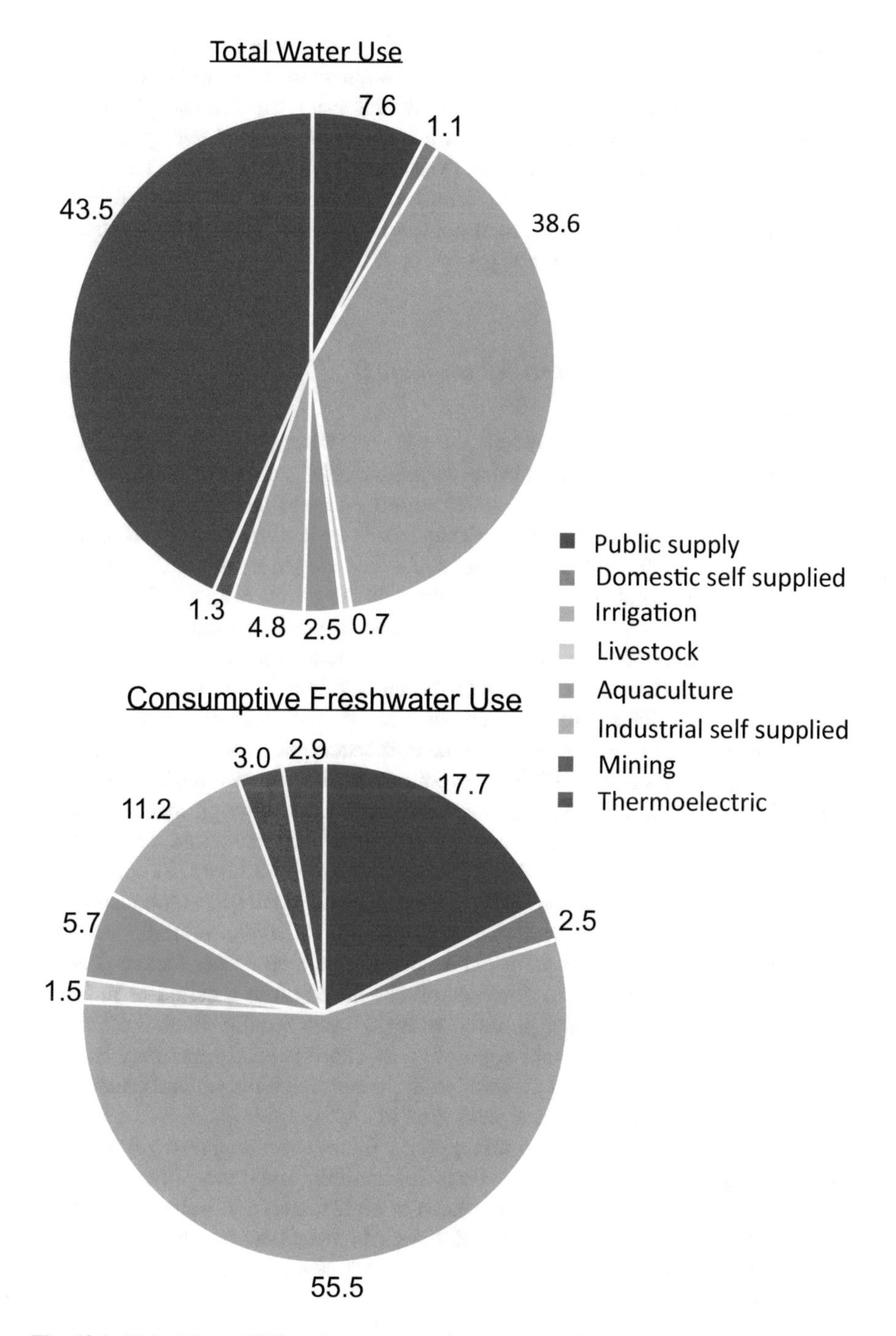

Fig. 10.1 United States 2015 total water use and consumptive freshwater use (Data from Dieter et al. 2018)

to manage the runaway expansion in demand for agricultural groundwater and for the energy needed to pump it. Laws and administrative regulations—such as licensing—have been extensively discussed and even tried; however, the key challenge is in enforcing these on several tens of millions of widely dispersed pumpers in a vast countryside" and that "Groundwater irrigation is the mainstay of India's small farmers and rural poor; therefore, governments and political leaders are reluctant to adopt a heavy-handed approach to curtail groundwater demand." Other countries face similar challenges.

10.2.2 Irrigation Demand Management

Irrigation is defined as the artificial provision of water to foster plant growth. Irrigation can result in a great increase in productivity compared to rainfed (dry-land) agriculture and can allow a greater variety of crops to be grown in a given area. Irrigation is essential for economically viable agriculture in arid and semiarid regions (Pescod 1992) and there is no doubt that irrigation is vital for meeting current and future food demands. Supplemental irrigation (i.e., application of a limited amount of water when rainfall is insufficient) can be highly efficient as a relatively small amount of water can substantially increase production compared to non-irrigated areas (Oweis et al. 1999; Oweis and Hachum 2006). Irrigation entered its heyday in the 1950s through a combination of cheap energy and improved pumping technologies and has become the cornerstone of global food security by increasing crop yields (Postel 1992, 1993). For countries with economies dependent largely on agriculture, increasing agricultural productivity through irrigation is the greatest opportunity for reducing poverty and increasing economic development (Comprehensive Assessment of Water Management in Agriculture 2007).

The amount of water required for irrigation depends on crop evapotranspiration (ET) requirements, which must be adjusted for effective rainfall, leaching requirements, application losses (irrigation inefficiency), and other factors (Pescod 1992). The timing and amount of irrigation should be those required to maintain soil moisture at levels sufficient to fully meet the water requirements of plants for optimal growth. Some additional application of water beyond plant requirements may be required for some deep percolation to prevent salt build-up in the plant root zone, which can adversely impact plant growth.

Fedoroff et al. (2010) observed that meeting future water supplies will require a radical rethinking of agriculture in terms substantially improving both land productivity (i.e., crop production per unit area of arable land) and water productivity (i.e., crop production per unit volume of water). In areas now facing water scarcity or are projected to face greater scarcity in the future, several options are available to reduce irrigation water demands:

- reduce local crop area (e.g., conversion of agricultural land to urban development) and relocate agriculture to areas with more abundant water resources

- switch to crops with lesser irrigation water requirements or greater drought resistance
- irrigate with reclaimed water rather than fresh groundwater and surface water
- increase the efficiency of irrigation.

Wallace and Batchelor (1997) identified four main improvement categories for increasing irrigation efficiency at a field level: agronomic, technical, managerial, and institutional. Agronomic options are measures to improve crop yields, such as improved crop husbandry, introduction of higher-yielding varieties, and planting crops to coincide with times of low potential ET and high rainfall. Technical options involve the adoption of more efficient irrigation technologies. Managerial options include adoption of demand-based irrigation schedules using soil-moisture sensors and improved maintenance of equipment. Institutional options include creating incentives and disincentives that encourage efficient water use and discourage waste, and improving training and education.

Numerous tools are available to increase irrigation efficiency and water productivity. The optimal choice of techniques for a given area will depend upon local conditions. Technical and managerial modifications of conventional irrigation methods and alternative irrigation methods that can increase the efficiency of irrigation include (Postel 1992, 1993):

- surge irrigation in which water is applied to alternative furrows at pre-established time intervals
- low-energy precision application (LEPA) sprinkler systems, which reduce the energy required and water losses of standard overhead sprinklers
- laser leveling of farm fields, which increases the efficiency of furrow irrigation
- recycling of water
- soil moisture monitoring—irrigation times and amounts are scheduled based on actual crop water needs
- microirrigation such as drip irrigation—water is applied using porous or perforated piping or hoses at or just below land surface (Fig. 10.2)
- improved overall system management to discourage excess irrigation applications by some users so that all users receive adequate water
- conjunctive use of groundwater and surface water
- development of salt-tolerant, drought-resistant, and water-efficient plant breeds.

The "paradox of irrigation efficiency" is the observation that greater irrigation efficiency does not necessarily mean that saved water will become available for reallocation to other sectors (e.g., cities and the environment; Grafton et al. 2018). It has been observed by a number of workers that water saved by increasing irrigation efficiency can induce increases in irrigated area and consumptive use of water (Huffaker and Whittlesey 1995; Whittlesey 2003; Contor and Taylor 2013; Grafton et al. 2018). Farmers with access to a given amount of water will rationally tend to try to maximize their total incomes by growing as much crops as possible with that resource. In terms of food security, increases in irrigation efficiency (IE) may allow more food to be grown with a given volume of water, so the expansion of irrigated area with increased IE could be beneficial.

Fig. 10.2 Drip irrigation at a commercial aloe vera plantation in Curacao

Excess irrigation water may recharge underlying shallow aquifers or be collected in drainage systems and thus is not necessarily lost from future beneficial uses. Irrigation return flows are often captured for use by downgradient users. In some areas where surface water is used for irrigation, irrigation return flows are the primary source of local recharge. Increased IE will decrease the amount of local recharge and potentially also impact the quality of the recharged water as salts and chemicals become more concentrated. Grafton et al. (2018, p. 749) noted the monitoring is a key requirement for improved irrigation water management and more specifically that a "key constraint to better decision-making is inadequate estimates of water inflows and outflows at watershed and basin scales. This analysis of water accounts is essential to demonstrate when IE policies are or are not in the public interest." Water use that is inefficient on the field scale may be efficient on the basin scale if return flows are being used by downgradient users.

10.2.3 Residential Water Demand Management

The World Health Organization (WHO) derived a minimum water requirement of 100 L per day (lpd) per person (26.4 gallons per day per person) and above as the optimum domestic water supply to promote good health (Howard and Bartram

2003). Per capita water use in many areas is several times the WHO minimum requirement. Total domestic per capita water use in the United States in 2010 averaged 88 gallons per day (gpd) or 333 L per day (lpd) and ranged between states from highs of 167 and 168 gpd (632 and 638 lpd), respectively, in the states of Idaho and Utah, to lows of 51 and 55 gpd (193 and 208 lpd), respectively, in the states of Wisconsin and Maine (Maupin et al. 2014). Arizona, the driest state in the nation, had a relatively high per capita use rate of 147 gpd (556 lpd).

Some states have shown impressive decreases in domestic per capita water use. Domestic per capita water use in Florida decreased from about 122 gpd (462 lpd) in 1985 to 85 gpd (322 lpd) in 2010 (Marella 2014), a decrease of about 30% (Fig. 10.3). However, Florida's population over the same period increased from 11.35 million people to 20.21 million people, an increase of 78% (Macrotrends 2020). Per capita domestic water use in the United States peaked in the 1980s through early 2000s (Donnelly and Cooley 2015).

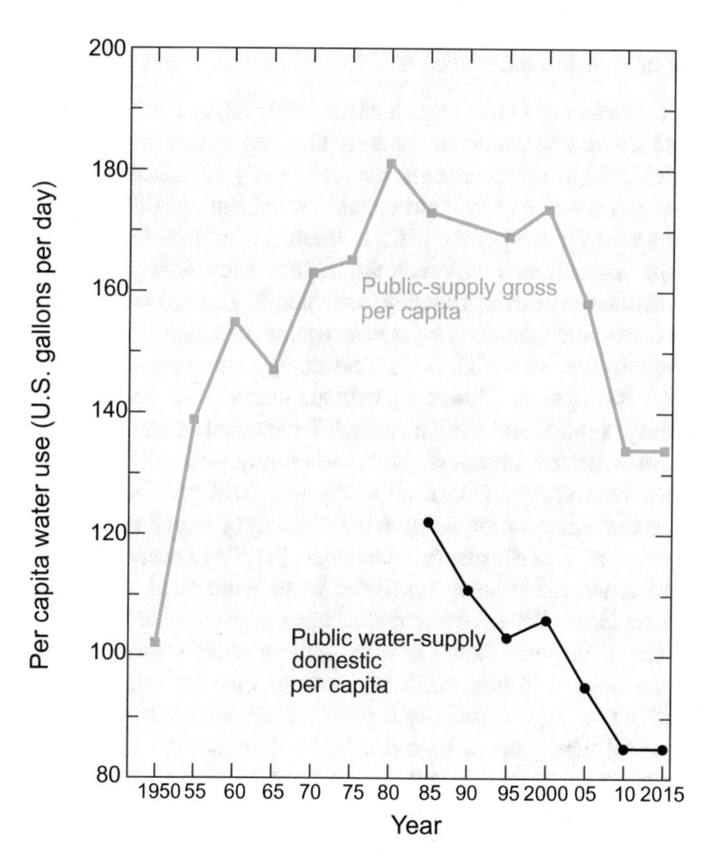

Fig. 10.3 Historical public-supply gross and domestic per capita water use in Florida, 1950–2015. Gross per capita water use is the total public supply water use (including commercial and industrial uses) divided by population (Modified from Marella 2020)

Large variations in domestic water use occur between countries. The average per capita use of tap water for various household purposes in the European Union was reported to be 31.7 gpd (120 lpd), ranging from only 13.2 gpd (50 lpd) in Malta to 64.2 gpd (243 lpd) in Italy (European Parliament 2020). Some of the most water scarce regions of the world have very high household use rates. Kuwait was reported in 2011 to have the highest water consumption rate in the world at about 132 gpd (500 lpd) (Toumi 2011). The average household use rate in Saudi Arabia was reported to be 70 gpd (265 lpd), twice the global average (Tago 2014).

Per capita consumption includes uses beyond the minimum required to promote good health. The main domestic water uses in developed countries are:

- outdoors lawn and landscape irrigation
- clothes washing
- sanitary—toilet flushing
- sanitary—showering, baths, washing, shaving, and brushing teeth
- potable use, food preparation, and dish washing
- leaks
- washing of vehicles, sidewalks, etc.

There is a plethora of books (e.g., Vickers 2001; Allen 2015), papers, brochures, and websites discussing numerous actions that can reduce residential water use. Water utilities and local governments very commonly provide educational information on water conservation to their customers and residents. Multiple opportunities for water savings exist for each of the listed domestic water uses. Outdoors water use is commonly the largest household use in the United States. Water savings are possible through the avoidance of over-irrigation, reduction in irrigated areas, and xeriscaping (i.e., use of native and adaptive plants that require very little water to thrive).

Use of appliances with high water and energy efficiencies can result in substantial water savings and lower operational costs. It is recognized that high water-efficiency fixtures and appliances will be adopted in the marketplace only if they result in equal or improved homeowner satisfaction (Selover 2010). The United States Environmental Protection Agency (USEPA) WaterSense Program encourages water conservation through the labeling of plumbing fixtures that meet both efficiency and performance standards (http://www.epa.gov/WaterSense/). Products and services that have earned the WaterSense label are certified to be at least 20 percent more efficient without sacrificing performance. There are numerous small changes in personal behavior (e.g., taking shorter showers, not allowing faucets to run while brushing teeth or shaving) that can cumulatively result in substantial water savings, especially if they become widely adopted.

From a water management perspective, the fundamental challenge is how to motivate water consumers to adopt conservation measures. Some people have a strong conservation ethic and will be inclined to be as water and energy efficient as possible as a moral principle. However, widespread adoption of water conservation requires other motivations and education. As a general principle, development of a water conservation ethic is more effective than coercion through legal mandates.

10.2.3.1 Economic Incentives

Professionals in the water field appreciate that water is grossly underpriced when one considers what is involved in producing and distributing high-quality water and the true value of having safe and readily available water reliably delivered to one's home. For most water users in developed countries, water is a small part of their monthly budgets. Market forces can act to reduce wasteful water use when consumers bear the full costs of water, particularly developing new supplies. Low costs send the economic signal that water has a low value.

Water demand is elastic beyond the minimum amount required to meet basic human needs. Increases in the price of water will tend to affect overall consumer behavior by reducing non-essential (discretionary) water use. Conservation rate structures provide water users with pricing information that conveys (at least in part) the marginal cost of developing new water resources. Conservation rate structure include charging higher rates during peak demand periods, inverted block rates, and excess-use rates (Cuthbert and Lemoine 1996; AWWA 2000). Inverted-block rate billing structures incorporate increasing rates with higher use. The first block of water, which is deemed sufficient for non-discretionary uses, is billed at the lowest rate, and subsequent blocks of water, which consist mostly of discretionary use, are billed at higher rates.

Municipal water bills in the United States, which may also include sewer service, commonly include a monthly fixed service charge that covers the cost of operating the systems and a volume of water sufficient to meet basic needs. As an example, the author's home in Florida is served by Lee County Utilities, which employs an inverted rate structure starting at US\$ 3.27 per 1000 gallons (\$0.86/m^3) for a monthly usage of 1000 to 6000 gallons (3.78–22.71 m^3), increasing to \$6.54 per 1000 gallons (\$1.72/m^3) for monthly usages of 18,001 gallons (68.14 m^3) and over. The Phoenix, Arizona, monthly service charge includes six (6) units of water (4488 gallons; 1 unit = 748 gallons, 10 ft^3) with usage within the city charged at \$3.20 per unit (equivalent to \$4.28/1000 gal, \$1.13/1000 m^3) in the winter low season and \$4.09/unit (\$5.47/1000 gal, \$1.44/m^3) in the summer high season. Although it is difficult to directly compare water rates because of differences in how rates are structured, taxes, and subsidies, water rates in Europe tend to be higher than those in the United States. National average water rates in Europe range from about 1.2 to over 8 €/m^3 (US\$1.4 to > \$9.4/m^3) with the highest rate reported for Denmark (EurEau 2017).

The response to conservation rates (elasticity) tends to be greater over the long run (Carver and Boland 1980). Most consumers are probably unaware of the quantities of water used for various tasks and it may take some time for them to obtain an understanding of the effects of rate changes on their water bills and the benefits of conservation. Homeowners also tend to gradually invest in more water efficient fixtures and landscaping (Carver and Boland 1980; Chesnutt and Beecher 1998). Appliances have a relatively long life cycle and consumers will usually not prematurely replace a functional major appliance with a more water or energy efficient model unless the replacement cost will be recouped by operational cost savings. Conversely, after the initial impact of a rate increase, consumers may come to accept the increase as normal and revert to previous behavior. Some users will use excessive amounts of water regardless of its cost.

Conservation rates can benefit utilities in two manners (Mann and Clark 1993). If future water use is insensitive to the imposition of surcharges, then customers will provide funding for projects that expand the water supply. Alternatively, if future water demands are sensitive to the surcharges, then the utility and its customers will benefit by avoiding the high marginal costs associated with constructing new water treatment and distribution infrastructure.

10.2.3.2 Legal Mandates

Water savings can be forced upon consumers through laws or ordinances that impose penalties for the improper or excessive use of water and through building codes that mandate the use of water efficient fixtures, appliances, and plumbing designs. In regions suffering from drought and dwindling water supplies, restrictions on outdoor water use are often implemented, which may include prohibitions against vehicle washing (in non-recycling systems) and limitations of lawn and landscape irrigation to a few hours on selected days each week or even an outright ban if the shortage is very severe.

Conservation ordinances and statutes may fail to be effective because of often little or no enforcement (Pape 2010). Water restrictions may be more effective as an educational tool in convincing people that a water shortage is indeed a serious problem. It has been the author's observations that most people will obey water restrictions because they believe it is the right thing to do and they are by nature law-abiding rather than because they fear sanctions for non-compliance. However, a minority of people are intransigent. During the 2014 drought in California, some municipalities initiated or increased "water-wasting patrols" (referred to as "water cops"), whose job was mostly education in that warnings were given for first offenses rather the fines (Becerra and Stevens 2014). In some communities, residents were active in the enforcement of water use restrictions by reporting violations, such as lawn watering outside of permitted times. During the 2014 drought, a smart phone app was developed to allow people to "drought-shame" their neighbors, i.e., report on their violations of water conservation rules (Wells 2014).

Green building codes have the advantage of being able to take advantage of existing code enforcement structures (e.g., local building code and enforcement departments; Pape 2010). Building codes apply to new construction and remodeling that requires a permit, but typically do not mandate the retrofitting of existing buildings.

10.2.3.3 Consumer Education

Consumer education is a key element in prompting a water conservation ethic. Often water consumers are not aware of wasteful practices or conditions, such as leaks. Water conservation education can include a wide variety of tools such as (Maliva and Missimer 2012):

- flyers included in utility bills
- mass market advertisements (television, radio, newspapers ads, billboards)
- presentations to clubs and organizations
- presentations to schools
- tours of water treatment facilities
- presentations (booths) at public gatherings
- xeriscape demonstration gardens
- in-home visits and water audits.

Consumers have a wide range of responsiveness to passive educational materials. Keen et al. (2010) documented the results of an in-home visit program tested in the City of Phoenix, Arizona, that provided personalized water conservation tips and technologies. High-volume users were targeted and offered tips and guides for conservation. The overall reduction was about 4% above the reductions that occurred during similar climate conditions in non-visited households. The education program was most effective in reducing the watering of lawns and landscaping that were previously overwatered and in prompting the repair of indoors and outdoors leaks (Keen et al. 2010).

10.2.4 Water Utilities Leakage and Non-revenue Water

Non-revenue water (NRW) is the difference between water put into a distribution system and the amount billed to customers, and includes the following elements (Kingdom et al. 2006; Thornton et al. 2008; AWWA 2016a, b):

- physical losses, mainly from leakage in the distribution network and theft
- commercial losses (also referred to as apparent losses)—unbilled or underbilled customer use
- unbilled authorized uses (e.g., firefighting flows and utility uses).

Leak detection and repair can increase the supply of water available to customers and reduce the need to seek new supplies to meet increases in demand. Where water is unbilled or under-billed, there is little, if any, economic incentive to use water frugally. NRW can seriously affect water utilities through lost revenues and increased operational costs. Leakage of water may cause rising groundwater levels that adversely affect below-ground infrastructure by causing flooding or corrosion.

A World Bank study (Kingdom et al. 2006) estimated that NRW in the developing world is probably more than 40–50%, and around 35% in developed countries and has a worldwide cost of US$ 14 billion with one third borne in developing countries. The mean NRW for urban utilities in the United States was estimated to be about 16% (Thornton et al. 2008; Chini and Stillwell 2018). The mean NRW in Europe was estimated to be 23% (EurEau 2017). Such large losses have a particularly great impact on developing countries that are starving for additional revenues to finance maintenance of existing services and expansion of services (Kingdom et al. 2006; Mutikanga et al. 2009).

Although leak detection and repair are an obvious water saving strategy, it is complicated by the water losses usually occurring through many small leaks throughout the distribution system. Reducing physical losses from leaks can be expensive, requires significant technical knowledge to locate the leaks, and must be carried out extensively to bring results (Kingdom et al. 2006). In cities in developing countries where utilities are often greatly underfunded, high NRW is the result of a lack of technical capacity, missing management focus and lack of individual (management and staff) incentives, opposition from those benefiting from commercial losses by receiving free or discounted water, and a lack of flexibility in implementing key elements of NRW reduction (Kingdom et al. 2006).

Books, guidance documents, and software are available for utility water audits and water loss control (e.g., Thornton et al. 2008; Mathis et al. 2008; USEPA 2010; AWWA 2016b). Leak quantification is usually performed using water budget methods (e.g., night flow analyses) and measurements of flow and pressure drops along mains. Acoustic detection remains the primary means of pinpointing the location of individual water main leaks.

10.3 Supply Augmentation

In areas using groundwater for water supply, the historical response to increased demands from population and economic growth has been to pump more groundwater. When groundwater use is constrained by specific local hydrological or environmental impacts, some additional groundwater might be obtained by installing new wells in less sensitive areas or otherwise shifting production away from the sensitive areas.

Declining groundwater levels in numerous aquifers worldwide indicate that current groundwater use rates in many areas are not sustainable. Unsustainable water use can be maintained or increased in the short-term, but eventually a day of reckoning will be reached at which water levels and/or quality has declined to the point where groundwater resources can no longer be economically used. Groundwater withdrawals in the United States and other countries may be locally limited due to environmental concerns, such as requirements to maintain environmental spring flows and wetland hydration. Hence, on the supply-side, a key part of meeting increasing future demands from population growth, economic development, and drier future climate conditions will be developing new alternative sources of water and optimizing the use of all available existing water resources.

Optimization is generally defined as making the most effective use of a resource. Discussions of optimization of water resources usually leads to integrated water resources management (IWRM), which has become a widely espoused paradigm for the management of water resources. IWRM has been defined as (Global Water Partnership 2000, p. 22):

> a process which promotes the coordinated development and management of water, land, and related resources, in order to maximize the resultant economic and social welfare in an equitable manner without compromising the sustainability of vital ecosystems.

IWRM and related concepts such as "total water management" and "holistic water management" share the overriding goals of improving overall water managements, balancing of interests and interdependencies in water decisions, and increasing shareholder participation in the decision-making process (Griggs 1999, 2008; Thomas and Durham 2003). IWRM is a response to the historically fragmented management of water in which the various components of water management have been regulated (to the extent that they had been regulated at all) in isolation from the other aspects. Groundwater, surface water, wastewater, and land use are still commonly managed separately even though they are closely interrelated. Bauer (2004, p. 9) cautioned that IWRM is an ideal concept rather than a set of specific guidelines and practices, and that like "sustainable development" it is "a phrase that can mean all things to all people; because it seems to mean everything, it may end up meaning nothing" Medema et al. (2008, p. 3) observed that "IWRM is not an end state to be achieved, it is a continuous process of balancing and making trade-offs between different goals and views in an informed way."

From a practical implementation perspective, potential water sources are identified and the single or combination of options that can least expensively and most reliably provide the target volume of water are selected with consideration of environmental, sociopolitical, and other pertinent constraints. Conjunctive use of surface water and groundwater (Chap. 11) is an integral element of IWRM that involves consideration of all available water resources and developing strategies to obtain the maximum value from the combined resources.

A fundamental distinction in water management is between the physical scarcity of water and the scarcity of inexpensive water. Any country with access to the coast can have an essentially unlimited supply of water through desalination if it has the economic resources and willingness to pay for it. Many coastal and inland regions are underlain by brackish and saline groundwater resources that can be desalted to provide freshwater. The main water supply challenge is obtaining water supplies that are affordable considering the low economic productivity of water for some uses, particularly agriculture. Water supply alternatives to fresh groundwater that are potential adaptation options to climate change (as well as to current water scarcity) are summarized below with discussions of key constraints.

10.3.1 Desalination

Desalination includes a broad range of technologies used to produce freshwater from saline and brackish waters. Because of the enormous global economic importance of desalination, a great amount of research is on-going on the optimization of current desalination technologies and the development of new and modifications of existing technologies. Conventional desalination is energy intensive and thus has a large carbon footprint when fossil fuels are used to provide the thermal and electrical energy needed for the process. An important research focus is on how to reduce the energy consumption of desalination and make the process more efficient. The National Research Council (2008, p. 12) concluded that the "potential for desalination to meet water demands in the United States is constrained not by the source water resources or the capabilities of current technology, but by a variety of financial, social, and environmental factors." Similar constraints are operative elsewhere in the world.

Detailed reviews of desalination have been provided by Buros et al. (1980), Buros (2000), Watson et al. (2003), National Research Council (2008), World Health Organization (2007), Gude (2018), and Kucera (2019). An enormous number of papers have been published on all aspects of desalination, and several dedicated journals are now published. Five elements of desalination projects affect their technical viability, economics, and environmental impacts:

- raw water intake
- pretreatment of the raw water
- desalination process
- post-treatment of desalted water
- concentrate management (residuals).

Intakes for seawater desalination systems were reviewed by Missimer et al. (2015). Brackish groundwater desalination systems typically are sourced using conventional vertical wells constructed with corrosion resistant materials. Seawater intakes include both open, surface-water intakes (offshore pipes with some type of screen system) and subsurface intakes (e.g., beach, slant, and horizontal wells, and offshore galleries). Intakes are designed to most cost effectively obtain the required volume of water at the highest quality practicable and to minimize environmental impacts. Entrainment and entrapment of marine organisms is a major environmental concern for seawater intakes. Subsurface intakes may be preferred for some facilities because they have lesser environmental impacts and produce higher water quality as beach and seabed sediments and aquifer materials provide natural filtration. Pretreatment of the raw water is required in order to protect the desalination system (particularly membranes) from fouling and failure. The degree and type of pretreatment depend on the quality of the raw water.

The desalination processes that are widely used for water supply can be divided into distillation and membrane processes. Distillation is in essence a vaporization-condensation process in which relatively pure water is vaporized by either heating the raw water or reducing its pressure (or both). Upon cooling, the water vapor condenses and is collected. The main distillation technologies used for large-scale water production are multi-stage flash distillation (MSF), multiple effect distillation (MED), and less commonly vapor compression distillation. Distillation plants may be collocated with power plants to take advantage of excess heat from the power plant to heat water.

Membrane processes are widely used in water and wastewater treatment and in many industrial processes. Semipermeable membranes can selectively remove particles and dissolved constituents of various sizes and chemistries. The four primary membrane processes (microfiltration, nanofiltration, ultrafiltration, reverse osmosis) remove particles, molecules, and ions of progressively smaller sizes. Only reverse osmosis (RO) is capable of desalting seawater. A fifth membrane technology type, electrodialysis and electrodialysis reversal, use membranes in combination with electrical separation to treat brackish water.

Osmotic membranes allow water to pass through but retain most dissolved constituents. Water will tend to naturally flow from the side of a membrane with low salt concentrations to the side with higher salt concentrations so as to equalize the salinity on both sides of the membrane. Membrane desalination reverses the natural osmotic flow by pressurizing the high salinity side of the membrane. Sufficiently high pressure is applied to overcome the natural osmotic pressure and to force freshwater to pass from the seawater (feed) side of the membrane to the freshwater (discharge) side. The pressure required depends on the salinity of the feed solution and the properties of the membranes. RO is the least energy intensive of the main commercial desalination technologies currently used and its use for seawater desalination has been increasing relative to thermal desalination over the past several decades. RO is now the standard technique for desalting brackish groundwater.

Desalted water tends to be too pure for human consumption. Consuming water with very low dissolved solids (particularly calcium) could lead to health problems as the result of the flushing of minerals and salts from the body with reduced replacement (Cotruvo 2006; World Health Organization 2005, 2006a). Very pure water also corrodes steel pipe. Post-treatment of desalted water involves the addition of some dissolved solids to make the water safe for potable use and less corrosive.

The salt-rich residuals rejected during the desalination process are referred to as "concentrate." Concentrate wastestreams at times also include much smaller amounts of cleaning and treatment chemicals. Concentrate from desalination facilities must be disposed of in an environmentally acceptable manner and identification of an economical and environmentally sound disposal method can be a critical feasibility issue for large desalination plants.

Concentrate disposal options and their environmental impacts were reviewed by Mickley (2001, 2009, 2010), Maliva et al. (2011), Ladewig and Asquith (2011), Maliva and Manahan (2014), and Missimer and Maliva (2017). Concentrate from large seawater desalination plants is usually disposed of by some type of ocean outfall. A principal environmental concern is impacts to salinity sensitive organisms. Studies of operational outfalls indicate that properly located and designed open outfalls with diffuser systems tend to have minimal local environmental impacts (Missimer and Maliva 2017). Systems collocated with power plants take advantage of the opportunity to decrease concentrate salinity by blending with the much greater cooling water flows.

Multiple desalination plants operating over very long time periods could cause progressive increases in salinity throughout the receiving surface water body. Surface water bodies that are most vulnerable to salinity increases are those with limited circulation (i.e., hydraulic connection to the oceans) and large present and anticipated future seawater desalination capacities. The Mediterranean Sea, Arabian-Persian Gulf, and Red Sea are particularly vulnerable to salinity increases caused or exacerbated by large scale implementation of desalination combined with decreasing freshwater inflows and warmer temperatures due to climate change (Missimer and Maliva 2017).

Disposal of concentrate at inland brackish groundwater desalination plants can be a much more difficult technological challenge than in coastal areas. Discharge of saline water to surface freshwater bodies is now usually not permitted. Blending with wastewater is also usually not practical for all but the smallest plants because associated increases in the salinity of the untreated wastewater could upset biological treatment processes. Increases in the salinity of treated wastewater may impair its suitability for reuse. A variety or zero liquid discharge (ZLD) technologies have been developed but suffer from the disadvantages of very high costs and the need to dispose of large quantities of salts unless the salts can be processed into marketable products (Bond and Veerapaneni 2007, 2008).

A remaining option is deep well injection, which can be an environmentally sound disposal mechanism provided that local hydrogeological conditions are suitable for the practice (Maliva et al. 2011; Maliva and Manahan 2014). Deep well disposal of desalination concentrate is widely practiced in South Florida and the Tampa Bay region of the state where injection zones capable of accepting large flows are available. Large brackish groundwater desalination plants in El Paso and San Antonio, Texas, also use deep well injection for concentrate disposal.

Desalination is not sensitive to climate and thus can significantly contribute to increased water supply resiliency. It has the disadvantage of being energy intensive and thus contrary to climate change mitigation goals unless systems are powered by non-fossil fuel sources. Desalted water is considerably more expensive that fresh surface water and groundwater. Total water costs for desalination systems depend on numerous factors including system design, water quality (salinity and pretreatment requirements), local energy costs, and system size and thus economy of scale. Ghaffour et al. (2013) reported that the total water costs for some large seawater desalination facilities ranged from US$ 0.63/m^3 to 1.20/m^3 and that the cost has fallen below US$ 0.50/m^3 for a large-scale seawater reverse osmosis plants at a specific location. The total water costs for two brackish water desalination plants were reported to be US$ 0.31/m^3 and 0.41/m^3.

10.3.2 Managed Aquifer Recharge

Managed aquifer recharge (MAR) has been described as the "intentional banking and treatment of waters in aquifers" (Dillon 2005) and as the "the purposeful recharge of water to aquifers for subsequent recovery or environmental benefits" (Dillon 2009). MAR includes methods intended to increase the volume of water in storage, such as recharge by infiltration and using wells, and techniques used to treat water (e.g., bank filtration). MAR technologies were reviewed by Huisman and Olsthoorn (1983), National Research Council (1994), Pyne (2005), Maliva and Missimer (2010), and Maliva (2019).

MAR has great value as an adaptation to climate change impacts on groundwater in that it can increase the supply of groundwater. The importance of groundwater as a buffer for managing increased variability of surface-water supplies combined with MAR as a means to supplement groundwater supplies for use during droughts is being increasingly recognized (Dragoni and Sukhija 2008; Kundzewicz and Döll 2009; Green et al. 2011; Van der Gun 2012; Scanlon et al. 2012; Taylor et al. 2013). MAR is a key element of some conjunctive use schemes (Chap. 11). Excess water that might otherwise not be beneficially used is captured and stored underground when available during wet or low-demand periods and later recovered for use during dry or high-demand periods.

MAR includes a broad suite of technologies of varying scales. Water storage-type MAR techniques vary primary based on whether recharge is performed by either applying water onto a land surface (surface spreading), subsurface discharge into the vadose zone using wells, galleries, or trenches, or by injection using wells directly into either confined or unconfined aquifers (Table 10.1). The decentralized nature and low technical requirements of some MAR techniques can be important advantages. Small-scale MAR systems implemented at the local level in developing countries can have marked water supply and public health benefits. MAR can be performed using treated or untreated surface water, reclaimed water, and stormwater. Small-scale stormwater MAR techniques include elements of stormwater "best management practices" (BMPs), "green infrastructure," "low-impact development" (LID), and rainwater harvesting techniques that are intended to infiltrate water closer to the site of rainfall and reduce off-site runoff.

Surface spreading tends to be the most cost-effective means for recharging shallow aquifers, especially if suitable land is economically available. Surface spreading may be performed using constructed facilities (infiltration basin complexes; Fig. 10.4), modification of channels by the construction of levees and dams, modification of land surfaces to slow runoff and increase infiltration, or by controlled discharges to ephemeral stream (i.e., wadi or arroyo) channels. Land surfaces may be modified to locally increase runoff that is to be captured for MAR.

Wells are used for recharge where the receiving aquifer is confined, low permeability strata are present between land surface and the water table, or insufficient land is available or affordable for surface spreading. The main disadvantage of using wells for recharge is that they are prone to clogging and associated loss of

Table 10.1 Water storage-type MAR techniques (Maliva 2019)

Technique	Description
Aquifer storage and recovery	Injection of freshwater into an aquifer and its later recovery using either the same well or, less commonly, a nearby well
Aquifer recharge using wells	Injection of water into an aquifer with the goal of increasing overall aquifer water levels; water may be recovered elsewhere in the aquifer
Infiltration basins	Constructed basins into which water is diverted to recharge an underlying water table aquifer
Dry wells, Infiltration galleries, pits, soakaways, and trenches	Shallowly excavated structures used for subsurface infiltration into the vadose zone
In-channel infiltration systems	Dams, check dams, and levees constructed in channels to back-up, spread, slow, and retain water to increase wetted area and the duration of inundation
Percolation tanks	Basins created in ephemeral streams to capture part of monsoon flows for direct use and to recharge the underlying aquifer (term is commonly used in India)
Recharge releases	Slow controlled releases of water from surface reservoirs, used as sedimentation basins, into ephemeral streams to enhance recharge of a downstream shallow aquifer
Sand dams	Low dams constructed on ephemeral streams to capture sand and create artificial aquifers used to store storm flows
Surface flooding and ditch-and-furrow systems	Systems that recharge through sheet flow on land surfaces or flow in off-channel shallow ditch and furrow systems
Land cover changes	Land cover modifications that increase net infiltration, such as phreatophyte vegetation removal
Infiltration-based LID techniques	Land development practices designed to increase on-site infiltration and decrease runoff

capacity. The injected water volume must pass through a much smaller area (the borehole wall) compared, for example, to the bottom and sides of infiltration basins. Subsurface injection is generally regulated more strictly than land application, which can necessitate expensive additional pretreatment with associated costs.

Recharge may be performed either near the water source or at an alternative location that has more favorable hydrogeological characteristics or operational advantages. Recovery is performed using either existing wells, which may be either existing production wells located throughout an aquifer area, or newly constructed wells located near the point of recharge or points of use. Aquifer storage and recovery (ASR) involves the injection and recovery of water usually using the same wells (Pyne 2005).

A fundamental constraint of storage-type MAR systems is that they require a supply of excess (i.e., not immediately needed) water. MAR has limited

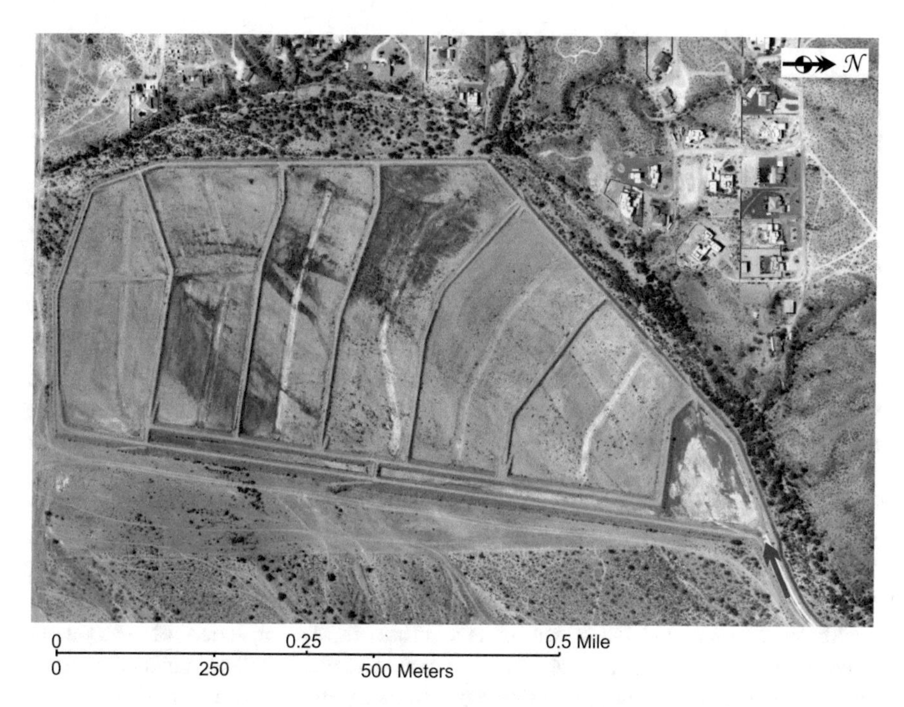

Fig. 10.4 Aerial photograph of the Agua Fria Recharge Project near Peoria, Arizona. The infiltration basin is divided into cells which allow for wetting and drying cycling (*Source* U.S. Geological Survey)

applicability in areas in which water is always scarce. A transition to a more variable climate with more intense precipitation events could locally result in more capturable runoff for MAR.

10.3.3 Wastewater Reuse

Wastewater reuse (recycling) has a long history throughout the world. Wastewater is reused most often for irrigation, both of crops and of landscaping, parks, and golf courses. Irrigation is a preferred use of reclaimed water (treated sewage effluent) because it does not require water of the highest quality and wastewater contains nutrients needed for plants. Wastewater is also recycled for industrial uses (e.g., cooling water) and is increasingly being used to augment potable water supplies. In areas facing water scarcity, wastewater reuse may be the least expensive alternative water source for agricultural, industrial, and urban non-potable purposes (Lazarova et al. 2000; Miller 2006). In many parts of the world, using water once simply is no longer an option (Levine and Asano 2004). Wastewater reuse has six main benefits (Maliva and Missimer 2012; Maliva 2019 and references therein):

1. it can provide additional needed water in water scarce regions where other options are either unavailable or too expensive
2. it can provide needed plant nutrients and thus reduce fertilization costs
3. reuse of wastewater may prevent or reduce the adverse environmental impacts associated with wastewater disposal
4. it is a reliable source of "new" water as its supply from urban areas is year round
5. it is the only source of additional water that increases as population increases
6. reclaimed water supplies have a very low sensitivity to climate change and thus increase the resiliency of water supplies.

The terms "reclaimed water" and "reuse water" refer to treated sewage effluent that is of a quality suitable for reuse. An important characteristic of reclaimed water is that it is a very reliable source of water whose production is relatively constant throughout the year and is almost constant between years (Dillon 2000; Friedler 2001).

Numerous studies have reviewed the health risks associated with wastewater reuse (e.g., Asano and Cotruvo 2004; World Health Organization 2006b; NRMMC-EPHC-AHMC 2006; NRMMC-EPHC-NHMRC 2009; Maliva and Missimer 2012; National Research Council 2012; USEPA 2012). The primary health risk associated with wastewater reuse is the presence of pathogenic microorganisms and, to a much lesser degree, chemical contaminants of variable toxicity. Pathogens present in wastewater can cause illness from a single exposure, whereas chemicals in wastewater are typically present at very low concentrations and would require long-term, chronic expose to cause health impacts. Proven wastewater treatment and purification technologies currently exist to produce water of virtually any quality desired (Asano and Levine 1996; Mujeriego and Asano 1999; USEPA 2012). Wastewater can be treated to such a degree that it is of superior quality to currently used water supplies and poses essentially no public health threat even if the water were to be directly consumed. Asano and Cotruvo (2004, p. 1944) observed that

> The irony is that water derived from 'natural' but obviously imperfect sources, often receives only basic treatment (filtration and disinfection). The final product might not be as high quality as the reclaimed wastewater that has been subject to much more rigorous treatment, water quality control, and management.

The USEPA (2012) noted that the key objective is to achieve a quality of reclaimed water that is appropriate for its intended use and is protective of human health and the environment. Large-scale use of reclaimed water for irrigation often requires that it be treated for unrestricted irrigation so that farmers can grow the crops they want, and the water can be used for other purposes that may involve public contact, such as landscape irrigation (Bouwer 1991). However, the quality of some reclaimed waters may not be ideal for agricultural irrigation and some other uses. Reclaimed water in some arid and semiarid lands and coastal areas has elevated salinities caused by the poor quality of the freshwater supply, evaporative concentration, and infiltration of saline waters into collection systems. Over treatment of reclaimed water for its intended uses is an unnecessary waste of resources that could otherwise be used to provide other societal benefits and can result in water recycling projects becoming economically unviable.

Numerous studies have been performed and papers published on the attitudes of the public toward potable reuse and wastewater reuse in general, which were reviewed in part by Maliva and Missimer (2012) and Maliva (2019). Public acceptance is vital for wastewater reuse as users must be willing to accept the water. Reuse projects can be ranked in terms of their perceived safety and public acceptance (from greatest to least):

1. restricted public access irrigation
2. public access landscape irrigation (e.g., parks, playing fields, and golf courses)
3. agricultural irrigation for non-consumable crops (e.g., cotton), fruits not in contact with the water, and crops that are cooked or peeled
4. agricultural irrigation of fruits and vegetables that may be eaten raw
5. indirect potable reuse
6. direct potable reuse.

Health risks associated with wastewater reuse are reduced through a combination of treatment and exposure control. Countries differ in their technical and economic resources available for wastewater treatment. The World Health Organization (2006b) recognizes the basic reality that more affluent nations can afford higher levels of wastewater treatment than developing nations and that less wealthy nations should focus on the implementation of wastewater treatment and reuse practices as resources become available, focusing first on the steps that could provide the greatest public health benefits. The World Health Organization (2006b, p. 35) advocated that

> the adverse impacts of wastewater use in agriculture should be carefully weighed against the benefits to health and the environment associated with these practices. Yet this is not a matter of simple trade-offs. Whenever wastewater use in agriculture contributes significantly to food security and nutritional status, the point is to identify associated hazards, define the risks they represent to vulnerable groups and design measures aimed at reducing these risks.

Potable reuse is the introduction of wastewater into the drinking water supply. Potable reuse may be either direct or indirect and may occur in either a planned or unplanned manner. Indirect potable reuse (IPR) was defined by the National Research Council (1998, p. 20) as "the abstraction, treatment, and distribution of water for drinking from a natural source-water that is fed in part by the discharge of wastewater effluent." Indirect potable reuse employs an environmental buffer (e.g., aquifer or surface body) between the wastewater treatment process and water treatment system. Environmental buffers allow for the dilution and degradation of any remaining contaminants by physical and chemical attenuation processes, provide a time buffer in the event of a treatment system failure, and help break the psychological connection of the water as being wastewater. Unplanned indirect reuse has long been a common reality. River water used for potable supply by downstream communities often contains wastewater (albeit diluted to varying degrees) discharged into the river by upstream communities.

Direct potable reuse is the introduction of highly treated wastewater either directly into a potable water distribution system downstream of a water treatment plant or into the raw water supply immediately upstream of a water treatment plant without an intervening environmental buffer (Asano et al. 2007). The current state-of-the-art advanced treatment for potable reuse systems is referred to as full advanced treatment (FAT), which is ultrafiltration (UF) or microfiltration (MF) followed by reverse osmosis (RO) and then an advanced oxidation process (AOP). FAT produces water of a quality far beyond that produced by many potable water systems (Markus 2009). AOPs commonly include a UV-based advanced oxidation process for removal of some organic compounds that passed through the RO membranes (e.g., NDMA) and the inactivation of pathogens. Ozone treatment may also be used either before the MF/UF units to prevent membrane fouling or downstream of the RO units for additional organics removal or degradation.

It is increasing being understood that the use of environmental buffers in IPR systems can be quite illogical from a technical perspective considering the high quality to which wastewater can be treated. The USEPA (2012, pp. 3–27) observed that

> Implementation of technologies for increasingly higher levels of treatment for many of these IPR projects has led to questions about why reclaimed water would be treated to produce water with higher quality than drinking water standards, and then discharged to an aquifer or lake.

A related issue is whether to apply advanced treatment technologies (if necessary) to the wastewater or potable water treatment processes (Asano et al. 2007). For example, should FAT be applied to the secondary treated wastewater or to the water-wastewater blend that will be used potable supply?

The fundamental argument for wastewater reuse is that where water is scarce, then the luxury of using water only once is not acceptable. The economic justifications for wastewater reuse include (Khouri et al. 1994; Winpenny et al. 2010; Maliva and Missimer 2012):

- If irrigation water demands exceed supply and no other new sources of water are available, then wastewater reuse may be economical even if treatment costs are borne by irrigators.
- Irrigation with treated wastewater can result in fresh groundwater or surface water otherwise used for irrigation becoming available for higher value urban (potable) and industrial uses.
- Reuse may provide cost savings relative to upgrading wastewater treatment systems to meet environmental standards for wastewater disposal.
- Nutrients in wastewater can result in increased crop yields and reduce fertilization costs.
- Reclaimed water can have a higher economical value to users because of its greater reliability.

A combination of the above factors may be sufficient to justify investment in wastewater reuse systems, even though each benefit alone is not sufficient. A major benefit, in most cases, is the value of freshwater made available for higher-value urban and industrial uses (Winpenny et al. 2010).

A significant constraint on wastewater reuse in some areas is a lack of sufficient potential users that could economically accept reclaimed water flows. Retrofitting urban and suburban areas for residential reuse systems is usually not cost-effective and there may not be sufficient potential large customers to accept all the produced reclaimed water. Reclaimed water users may not be able to accept water flows during wet periods when there is no irrigation need. Aquifer storage and recovery of reclaimed water is a means for managing seasonal variations in reclaimed water supplies and demands.

10.3.4 Rainwater Harvesting

Rainwater harvesting is defined as the collection, storage, and use of rainwater that falls on individual parcels of land. Rainwater harvesting includes small-scale MAR techniques where the primary goal is local aquifer recharge. The related term "water harvesting" refers to the process of concentrating rainfall or runoff from a larger area for use in a smaller target area, typically for agricultural purposes (Oweis et al. 1999). In arid and semiarid lands, the limited annual rainfall often occurs in short-duration, high-intensity events. A solution to the water scarcity is to divide available land areas into catchment and smaller crop areas. Runoff collected from the larger catchment areas combined with the rainfall that directly falls on the crop areas becomes sufficient to meet crop water requirements (Stern 1979). Runoff can be stored in a tank or pond or be used to directly increase the soil moisture of a crop area.

Roof-top rainwater harvesting is performed on some islands and arid regions for potable water supply. For example, roof-top rain water harvesting is mandated by law in Bermuda for all buildings and is the primary source of water for domestic supply (Rowe 2011). Even in areas with abundant rainfall, it is now common to see roof rain gutters connected to rain barrels for non-potable uses, commonly motivated by a conservation ethic (Fig. 10.5). As another example, the city of Tucson, Arizona, adopted an ordinance in October 2008 that requires that 50% of the water used for landscaping on new commercial properties is to come from harvested rainwater (Gaston 2010).

There is a now an enormous amount of popular and technical literature on rainwater harvesting including dedicated books (National Academy of Sciences 1974; Boers and Ben-Asher 1982; Bruins et al. 1986; Pacey and Cullis 1986; Waller 1989; Prinz and Singh 2000; Lancaster 2006, 2008; Waterfall 2006;, Kinkade-Levario 2007; Downey 2010); Daily and Wilkins 2012; Bickelmann 2013; Avis and Avis 2018), brochures, and websites operated by government agencies, non-governmental organizations (NGOs), and a wide variety of contractors and

Fig. 10.5 Rooftop rainwater harvesting system, North Carolina Arboretum, Asheville, North Carolina

commercial suppliers of rainwater collection supplies. Water harvesting for agriculture was reviewed by the UNEP (1984), Critchley and Siegert (1991), and Oweis et al. (1999, 2012).

Rainwater harvesting systems are divided between earthwork (passive) and storage (active) systems. The objective of earthwork systems is to convert "convex" impervious landscapes where rainwater runs off into "concave" pervious landscapes that infiltrate water (Lancaster 2006). Systems are designed to slow the flow of water and spread the harvested water over as much pervious area as possible to increase infiltration into the soil. Storage systems harvest and store rainwater in tanks, cisterns, or reservoirs. Rainwater harvesting ranges in scale from a simple barrel used to store roof runoff collected from gutters for garden irrigation or contouring a property to reduce runoff and increase infiltration, to sophisticated systems involving water treatment facilities.

Rainwater harvesting can be an adaptation to climate change by reducing dependence on vulnerable groundwater resources and increasing groundwater recharge. Rainwater harvesting is appropriate for developing countries as it has the advantages of being small-scale simple operations with high adaptability and low costs, and it can empower local communities to manage their own water resources.

Reductions in runoff from upgradient properties can be beneficial or harmful to downgradient properties. Rainwater harvesting can be beneficial to downgradient users if it reduces flooding and increases groundwater levels. Widespread rainwater harvesting could adversely impact downgradient users of surface waters. One household or farmer in a watershed harvesting rainfall will have minimal impacts (either beneficial and adverse), but widespread implementation could have significant impacts on the amount of water available for downgradient properties and water users.

10.3.5 Transferring Water

Climate change is expected to have both winners and losers as far as changes in precipitation. Irrespective of climate change, a wide variety of schemes have bene proposed to transfer water from water-rich to water-scarce regions, including the towing of icebergs from the Arctic and Antarctic to water-stressed areas (Weeks and Campbell 1973; Spandonide 2009) and the use of enormous fabric bags filled with freshwater (Medusa, n.d.).

Large aqueduct and canal projects have been constructed, or are under development, to transfer fresh water from wetter to drier areas within states and countries. Southern California receives most of its water from the wetter northern part of the state via two aqueduct systems (Central Valley Project and State Water Project-California Aqueduct), the western part of the state (Los Angeles Aqueduct system), and the Colorado River via the Colorado Aqueduct. Similarly, the Phoenix and Tucson areas of Arizona receive Colorado River water via the 336-mile (540 km) long Central Arizona Project. The South–North Water Transfer Project will convey water from the Yangtze River in southern China to the more arid and industrialized northern part of the country.

More recently proposed water transfer schemes have foundered on the rocks of public opposition in the donor region. A pipeline system proposed in the early 2000s that would convey water from the less populated northern part of the state of Florida to the more developed southern cities faced vigorous public opposition in the northern counties who insisted that “Our Water is Not for Sale” (Klein 2006). On a larger scale, rapidly growing states in the drought prone American Southwest and agricultural interests in the High Plains area, which are facing declining groundwater levels, have been eyeing the transfer of water to the region from the Great Lakes, a proposal strongly opposed by the midwestern states and Canada (Reinumagi 2011; Way 2018). A proposed pipeline that would convey groundwater approximately 250 miles (400 km) from northern Nevada to the Las Vegas area faced strong local public opposition because of its impacts to the local rural populations in the wellfields areas. The required permits to pump water out of four basins in northern Nevada were denied but the Southern Nevada Water Authority is still pursuing the project in court (Rothberg 2018; Solis 2020).

Viewed solely from an overall water management perspective, water transfers are an attractive means of optimizing the use of water resources, provided that they can be economically justified (i.e., their benefits exceed costs). However, from a practical perspective, implementing new large-scale water transfers will be extremely challenging because of expected very strong opposition from donor areas.

10.4 Adaptations Options for Rural Areas of Developing Countries

Urban areas of developed countries have the adaptive capacity to address climate change impacts to groundwater supplies. The options may be expensive, and some conservation elements may be unpopular, but the integrity of water supplies will likely be maintained. On the contrary, rural areas of developing countries are especially vulnerable to climate change because of low adaptative capacities. As summarized in a World Bank study (Clifton 2010, p. 25):

> Shallow wells often provide an important source of drinking water for rural populations in developing nations. Increased demand and potentially increased severity of droughts may cause these shallow wells to dry up. With limited alternatives for safe drinking water supplies (surface water may be absent or contaminated and deeper wells may not be economically feasible), loss of groundwater would force people to use unsafe water resources or walk long distances for water.
>
> The livelihoods of rural populations are largely dependent on land, water and the environment with limited alternatives compared to their urban counterparts. Reduced water availability can cause severe hardships. Drying up of pasture and drinking water to livestock can wipe out herds of livestock that are sources of income, family security and food. Small scale irrigation enterprises, usually reliant on shallow groundwater, may also fail.

It must be stressed again that climate change impacts will likely be less than those from existing climate conditions and natural variability and increasing water demands from population growth. Shah (2009, p. 1) wrote with respect to adaptation to climate change impacts on groundwater in India that

> For millennia, India used surface storage and gravity flow to water crops. During the last 40 years, however, India has witnessed a decline in gravity-flow irrigation and the rise of a booming 'water-scavenging' irrigation economy through millions of small, private tube wells. For India, groundwater has become at once critical and threatened. Climate change will act as a force multiplier; it will enhance groundwater's criticality for drought-proofing agriculture and simultaneously multiply the threat to the resource.

India has experienced explosive growth in groundwater pumping over recent decades because aquifers act as a more resilient buffer during dry spells, and groundwater use is expected to expand further in the wake of climate change (Shah 2009). For millennia, wells have been the principal weapon Indian farmers used to cope with droughts (Shah 2009). Shah (2009) concluded that India's adaptation strategy needs to incorporate effective means to manage agricultural water demand as well as to enhance natural groundwater recharge through large MAR investments in both large-scale government systems and small-scale systems constructed and operated by individual farmers or local communities.

Poor areas where water is scarce and self-supplied by individual wells or community wells are particularly vulnerable to the impacts of climate change on groundwater because of the absence of alternative water sources. In some arid and semiarid regions, falling groundwater levels from drought and overdraft have resulted in wells going dry with devastating impacts to poor farmers. Wells running dry and associated growing debt from crop failures in parts of India during a recent drought had triggered a spate of farmer suicides (BBC 2016). A warmer and drier climate would be expected to exacerbate the situation.

Adaptation options to climate change are essentially the same available and utilized to address current water scarcity. In rural areas of developing countries with limited technical and financial resources, adaptation options necessarily must be small-scale and within the means of local populations to at least operate and maintain them, although governmental and NGO support might be needed for their design and construction. The key issue is taking best advantage of all available water resources. Rainwater harvesting includes technologies used throughout history and are well suited to help address current and future rural water scarcity.

Rainwater harvesting has been continuously practiced in India since at least the third millennium BC, as reviewed in great detail in the book "Dying Wisdom" (edited by Agarwal and Narain 1997). Rainwater harvesting was rediscovered in the 1960s in response to declining groundwater availability caused by the rapid expansion of irrigation pumping. The increase in aquifer recharge and rainwater harvesting in India over the past three to four decades is of such an extent that it has been called a "groundwater recharge movement" (Sakthivadivel 2007). A remarkable feature of the groundwater recharge movement in India is that it is largely a "bottom up" movement initiated by local communities, rather than being a "top-down" movement initiated and mandated by the central government (Sakthivadivel 2007). Western India areas hardest hit by groundwater depletion spontaneously created, with the support of NGOs, a massive well-recharge movement based on the principle "water on your roof stays on your roof; water on your fields stays on your fields; and water on your village stays in your village" (Shah et al. 2000, p. 10). The recharge program included modification of some 300,000 wells to divert rainwater into them, and the construction of thousands of ponds, check dams and other rainwater harvesting and recharge structures on the self-help principle to keep rainwater from flowing into the Arabian Sea (Shah et al. 2000). The primary recharge method is surface spreading using percolation tanks (ponds) and check dams constructed across or near streams and drainage channels to impound runoff and retain it for a longer time for recharge (Sakthivadivel 2007). In India there are more than 1.5 million tanks, ponds, and earthen embankments in 660,000 villages (Pandey et al. 2003).

10.5 Adaptations to Saline-Water Intrusion

The impacts of sea level rise on groundwater are discussed in Chap. 6. SLR can contribute to the salinization of coastal aquifers through horizontal saline water intrusion, inundation of low-lying areas, and the infiltration of saline waters during temporary inundation events (e.g., storm surges). Horizontal saline water intrusion is usually caused by excessive groundwater pumping landward of the saline-fresh groundwater interface. Options to manage saline water intrusion were reviewed by Oude Essink (2001), Maliva and Missimer (2012), Maliva (2019), among many, and include:

- reducing fresh groundwater pumping to restore the water balance of an aquifer toward its predevelopment conditions, which may entail developing alternative water sources
- relocating pumping inland, away from the fresh-saline groundwater interface
- creation of a positive salinity barrier (hydraulic mound) by managed aquifer recharge (MAR) on the freshwater side of the fresh-saline groundwater interface (between the interface and production wells)
- MAR inland to increase the flow of freshwater toward the coast
- extraction barriers, which involves pumping on the saline-water side of the interface to pull the interface seaward
- subsurface physical barriers
- scavenger wells, which involves managing vertical saline water intrusion (up-coning) by pumping brackish or saline groundwater below freshwater production zones
- land reclamation—creating a new foreland where a freshwater body may develop.

References

Agarwal A, Narain S (1997) Dying wisdom. Rise, fall and potential of India's traditional water harvesting systems. India Centre for Science and Environment, New Delhi

Allen J (2015) The water-wise home: How to conserve, capture, and reuse water in your home and landscape. Storey Publishing, North Adams, MA

Asano T, Burton FL, Leverenz HL, Tsuchihashi R, Tchobanoglous G (2007) Wastewater reuse. Issues, technologies and applications. McGraw-Hill, New York

Asano T, Cotruvo JA (2004) Groundwater recharge with reclaimed wastewater: Health and regulatory considerations. Water Res 38:1941–1951

Asano T, Levine AD (1996) Wastewater reclamation, recycling and reuse: Past, present, and future. Water Sci Technol 30(11–11):1–14

Avis R, Avis PM (2018) Essential rainwater harvesting: A guide to home-scale system design. New Society Publishers, Gabriola Island, BC

AWWA (2000) Principles of water rates, fees, and charges (AWWA Manual M1). American Water Works Association, Denver

AWWA (2016a) The state of water loss control in drinking water utilities. American Water Works Association. https://www.awwa.org/Portals/0/AWWA/ETS/Resources/WLCWhitePaper.pdf?ver=2017-09-11-153507-487. Accessed 4 June 2020

AWWA (2016b) Water audits and loss control programs (4th edn) (Manual M36). American Water Works Association, Denver

Bauer CJ (2004) Siren Song: Chilean water law as a model for international reform. Resources of the Future Press, Washington, DC

BBC (2016) How drought is changing rural India, 31 May. https://www.bbc.com/news/world-asia-india-36410866. Accessed 4 June 2020

Becerra H, Stevens M (2014) L.A. dramatically increases 'water cops' staffing as drought worsens. *Los Angeles Times*, August 18. https://www.latimes.com/local/lanow/la-me-ln-la-dramatically-increases-water-cops-as-drought-worsens-20140818-story.html. Accessed 4 June 2020

Bickelmann C (2013) Harvesting rainwater. Guide to water-efficient landscaping, 2nd edn. City of Tucson Water Department, Tucson

Boers TM, Ben-Asher J (1982) A review of rainwater harvesting. Agric Water Manag 5(2):145–158

Bond R, Veerapaneni S (2007) Zero liquid discharge for inland desalination. AWWA Research Foundation, Denver

Bond P, Veerapaneni S (2008) Zeroing in on ZLD technologies for inland desalination. J Am Water Work Assoc 100(9):76–89

Bouwer H (1991) Groundwater recharge with sewage effluent. Water Sci Technol 23:2099–2108

Bruins HJ, Evenari M, Nessler U (1986) Rainwater-harvesting agriculture for food production in arid zones: The challenge of the African famine. Appl Geogr 6(1):13–32

Buros OK (2000) The ABCs of desalting, 2nd edn. International Desalination Association, Topsfield, MA

Buros OK, Cox RB, Nusbaum I, El-Nashar AM, Bakish R (1980) The USAID desalination manual. IDEA Publications, Englewood Cliffs

Carver PH, Boland JJ (1980) Short- and long-run effects of price on municipal water use. Water Resour Res 16(4):609–616

Chesnutt TW, Beecher JA (1998) Conservation rates in the real world. J Am Water Works Assoc 90(2):60–70

Chini CM, Stillwell AS (2018) The state of US urban water: Data and the energy-water nexus. Water Resour Res 54(3):1796–1811

Clifton C, Evans R, Hayes S, Hirji R, Puz G, Pizarro C (2010) Water and climate change: Impacts on groundwater resources and adaptation options. Water Working Notes 5507. World Bank Group, Water Sector Board of the Sustainable Development Network

Comprehensive Assessment of Water Management in Agriculture (2007) Water for food, water for life: A comprehensive assessment of water management in agriculture. Earthscan and International Water Management Institute, London and Colombo

Contor BA, Taylor RG (2013) Why improving irrigation efficiency increases total volume of consumptive use. Irrig Drain 62(3):273–280

Cotruvo J (2006) Health aspects of calcium and magnesium in drinking water. In: Proceedings of the International Symposium on Health Aspects of Calcium and Magnesium in Drinking Water, Baltimore, MD, USA, April 2006

Critchley W, Siegert K (1991) Manual for the design and construction of water harvesting schemes for plant production. Food and Agriculture Organization of the United Nations, Rome

Cuthbert RW, Lemoine PR (1996) Conservation-oriented water rates. J Am Water Works Assoc 88(11):68–78

Daily C, Wilkins C (2012) Passive water harvesting. University of Arizona, College of Agriculture and Life Sciences Cooperative Extension, Tucson

Dieter CA, Maupin MA, Caldwell RR, Harris MA, Ivahnenko TI, Lovelace JK, Barber NL, Linsey KS (2018) Estimated use of water in the United States in 2015. Circular 1441. U.S. Geological Survey, Reston

Dillon P (2000) Water reuse in Australia: Current status, projections and research. In: Dillon PJ (ed) AWA Water Recycling Forum, Proceedings of the First Symposium. Adelaide, pp 99–104
Dillon P (2005) Future management of aquifer recharge. Hydrogeol J 13:313–316
Dillon P (2009) Water recycling via managed aquifer recharge in Australia. Boletín Geológico y Minero 120(2):121–130
Donnelly K, Cooley H (2015) Water use trends in the United States. Pacific Institute, Oakland
Downey N (2010) Harvest the rain, how to enrich your life by seeing every storm as a resource. Sunstone Press, Santa Fe
Dragoni W, Sukhija BS (2008) Climate change and groundwater: A short review. In: Dragin W, Sukhija BS (eds), Climate change and groundwater. Special Publications, 288(1). Geological Society, London, pp 1–12
Dronkers J, Gilbert JTE, Butler LW, Carey JJ, Campbell J, James E, McKenzie C, Misdorp R, Quin N, Ries KL, Schroder PC, Spradley JR, Titus JG, Vallianos L, von Dadelszen J (1990) Coastal zone management. In: Climate change: The IPCC response strategies. Intergovernmental Panel on Climate Change, Geneva
EurEau (2017) Europe's water in figures. An overview of the European drinking water and waste water sectors (2017 edition). The European Federation of National Associations of Water Services. http://www.eureau.org/resources/publications/1460-eureau-data-report-2017-1/file. Accessed 4 June 2020
European Parliament (2020) Drinking water in the EU: Better quality and access. https://www.europarl.europa.eu/news/en/headlines/society/20181011STO15887/drinking-water-in-the-eu-better-quality-and-access. Accessed 4 June 2020
Fedoroff NV, Battisti DS, Beachy RN, Cooper PJM, Fischhoff DA, Hodges CN, Knauf VC, Lobell D, Mazur BJ, Molden D, Reynolds MP, Ronald PC, Rosegrant MW, Sanchez PA, Vonshak A, Zhu J-K (2010) Radically rethinking agriculture for the 21st century. Science 327:833–834
Friedler E (2001) Water reuse—an integral part of water resources management: Israel as a case study. Water Policy 3:29–39
Gaston TL (2010) Rainwater harvesting in the southwestern United States. A policy review of the Four Corners states. Unpublished research report
Ghaffour N, Missimer TM, Amy GL (2013) Technical review and evaluation of the economics of water desalination: Current and future challenges for better water supply sustainability. Desalination 309:197–207
Gleick PH (1996) Basic water requirements for human activities: Meeting basic needs. Water Int 21:83–92
Gleick PH (1998) The human right to water. Water Policy 1:487–503
Global Water Partnership (2000) Integrated water resources management, TAC Background Paper No. 4. Global Water Partnership, Stockholm
Gober P, Kirkwood CW (2010) Vulnerability assessment of climate-induced water shortage in Phoenix. Proc Natl Acad Sci 107(50):21295–21299
Grafton RQ, Williams J, Perry CJ, Molle F, Ringler C, Steduto P, Udall B, Wheeler SA, Wang Y, Garrick D, Allen RG (2018) The paradox of irrigation efficiency. Science 361(6404):748–750
Green TR, Taniguchi M, Kooi H, Gurdak JJ, Allen DM, Hiscock KM, Treidel H, Aureli A (2011) Beneath the surface of global change: Impacts of climate change on groundwater. J Hydrol 405:532–560
Griggs NS (1999) Integrated water resources management: Who should lead, who should pay? J Am Water Resour Assoc 35(3):527–534
Griggs NS (2008) Total water management: Practices for a sustainable future. American Water Works Association, Denver
Gude G (ed) (2018) Sustainable desalination handbook: Plant selection, design and implementation. Butterworth-Heinemann, Oxford
Howard G, Bartram J (2003) Domestic water quality, service level and health (WHO/SDE/WSH/03.02). World Health Organization, Geneva

Huffaker RG, Whittlesey NK (1995) Agricultural water conservation legislation: Will it save water? Choices 10:24–26
Huisman L, Olsthoorn TN (1983) Artificial groundwater recharge. Pitman Advanced Publishing, Boston
Keen AH, Keen D, Francis GE, Wolf A (2010) High-contact, hands-on outreach program changes customers' water use behavior. J Am Water Works Assoc 102(2):38–45
Khouri N, Kalbermatten JM, Bartone CR (1994) Reuse of wastewater in agriculture: A guide for planners. UNDEP—World Bank Water and Sanitation Program, Washington, DC
Kingdom B, Liemberger R, Marin P (2006) The challenge of reducing non-renewable water (NRW) in developing countries. How the private sector can help: a look at performance-based service contracting. Water Supply and Sanitation Board Discussion Papers Series, Paper No. 8. The World Bank Group, Washington, DC
Kinkade-Levario H (2007) Design for water: Rainwater harvesting, stormwater catchment, and alternate water reuse. New Society Publishers, Gabriola Island, BC
Klein CA (2006) Water transfers: The case against transbasin diversions in the Eastern States. UCLA J Environ Law & Policy 25:249–282
Kucera J (ed) (2019) Desalination: Water from water, 2nd edn. Wiley, Hoboken
Kundzewicz ZW, Döll P (2009) Will groundwater ease freshwater stress under climate change? Hydrol Sci 54(4):665–675
Kundzewicz ZW, Mata LJ, Arnell NW, Döll P, Jimenez B, Miller K, Oki T, Şen Z, Shiklomanov I (2008) The implications of projected climate change for freshwater resources and their management. Hydrol Sci J 53(1):3–10
Ladewig B, Asquith B (2011) Desalination concentrate management. Springer, Berlin
Lancaster B (2006) Rainwater harvesting for drylands and beyond. Volume 1. Guiding principles to welcome rain into your life and landscape. Rainsource Press, Tucson
Lancaster B (2008) Rainwater harvesting for drylands and beyond. Volume 2. Water-harvesting earthworks Rainsource Press, Tucson
Lazarova V, Cirelli G, Jeffrey P, Salgot M, Icekson N, Brissaud F (2000) Enhancement of integrated water management and reuse in Europe and the Middle East. Water Sci Technol 42 (1–2):193–202
Levine AD, Asano T (2004) Recovering sustainable water from wastewater. Environ Sci Technol 38:201A–208A
Macrotrends (2020) Florida population 1900–2019. https://www.macrotrends.net/states/florida/population. Accessed 4 June 2020
Maliva RG (2019) Anthropogenic aquifer recharge. Springer, Cham
Maliva RG, Manahan WS (2014) Deep injection well disposal of desalination concentrate at inland sites: Opportunities and challenges. AMTA Solut 2014:1–7)
Maliva RG, Missimer TM (2010) Aquifer storage and recovery and managed aquifer recharge using wells: Planning, hydrogeology, design, and operation. Schlumberger Corporation, Houston
Maliva RG, Missimer TM (2012) Arid lands water evaluation and management. Springer, Berlin
Maliva RG, Missimer TM, Fontaine R (2011) Injection well options for sustainable disposal of desalination concentrate. IDA J 3(3):17–23
Mann PC, Clark DM (1993) Marginal-cost pricing: Its role in conservation. J Am Water Works Assoc 85(8):71–78
Marella RL (2014) Water withdrawals, use, and trends in Florida, 2010. U.S. Geological Survey Scientific Investigations Report 2014-5088
Marella RL (2020) Water withdrawals, uses, and trends in Florida, 2015. U.S. Geological Survey Scientific Investigations Report 2019–5147
Markus MR (2009) Groundwater replenishment & water reuse. The Water Report, Issue 59, January 15, 1–9
Mathis M, Kunkel G, Chastain Howley A (2008) Water loss audit manual for Texas utilities (Report 367). Texas Water Development Board, Austin

Maupin MA, Kenny JF, Hutson SS, Lovelace JK, Barber NL, Linsey KS (2014) Estimated use of water in the United States in 2010. Circular 1405. U.S. Geological Survey, Reston

McKean E (2004) The new Oxford American dictionary, 2nd edn. Oxford University Press, New York

Medema W, McIntosh BS, Jeffery PJ (2008) From premise to practice: A critical assessment of integrated water resources management and adaptive management approaches in the water sector. Ecol Soc 13(2): article 29

Medusa (n.d.) Medusa Enterprise. http://medusaenterprise.com/about-us/. Accessed 1 April 2020

Mickley M (2001) Membrane concentrate disposal: Practices and regulation, Report 69. U.S. Bureau of Reclamation, Denver

Mickley M (2009) Desalination and water purification research and development and program no. 155: Treatment of concentrate, Final report. U.S. Bureau of Reclamation, Denver

Mickley M (2010) Overview of global inland desalination concentrate management: Solutions, challenges, and technologies. IDA J Desalination Water Reuse 2(3):48–52

Miller GW (2006) Integrated concepts in water reuse: Managing global water needs. Desalination 187:66–75

Missimer TM, Jones B, Maliva RG (eds) (2015) Intakes and outfalls for seawater reverse-osmosis desalination facilities. Springer, Berlin

Missimer TM, Maliva RG (2017) Environmental issues in seawater reverse osmosis desalination: Intakes and outfalls. Desalination 343:198–215

Mujeriego R, Asano T (1999) The role of advanced treatment in wastewater reclamation and reuse. Water Sci Technol 40(4–5):1–9

Mutikanga HE, Sharma S, Vairamoorthy K (2009) Water loss management in developing countries: Challenges and prospects. J Am Water Works Assoc 101(12):57–68

National Academy of Sciences (1974) More water for arid lands. Promising technologies and research opportunities. National Academy of Sciences, Washington, DC

National Research Council (1994) Ground water recharge using waters of impaired quality. National Academy Press, Washington, DC

National Research Council (1998) Issues in potable reuse: The viability of augmenting drinking water supplies with reclaimed water. National Academy Press, Washington, DC

National Research Council (2008) Desalination—A national perspective. National Academies Press, Washington, DC

National Research Council (2012) Water reuse. Potential for expanding the nation's water supply through the reuse of municipal wastewater. National Academy Press, Washington, DC

NRMMC-EPHC-AHMC (2006) National guideline for water recycling: Managing health and environmental risks (Phase 1). Natural Resource Management Ministerial Council, Environment Protection and Heritage Council, and Australian Health Ministers Conference

NRMMC-EPHC-NHMRC (2009) Australian guidelines for water recycling: Managing health and environmental risks (Phase 2) Managed aquifer recharge. Natural Resource Management Ministerial Council, Environment Protection and Heritage Council, National Health and Medical Research Council

Oude Essink GHP (2001) Improving fresh groundwater supply—problems and solutions. Ocean Coast Manag 44(5–6):429–449

Oweis T, Hachum A, Kijne J (1999) Water harvesting and supplementary irrigation for improved water use efficiency in dry areas. SWIM Paper 7. International Water Management Institute, Colombo, Sri Lanka

Oweis T, Hachum A (2006) Water harvesting and supplemental irrigation for improved water productivity of dry farming systems in West Africa and North Africa. Agric Water Manag 80:57–73

Oweis TY, Prinz D, Hachum AY (2012) Rainwater harvesting for agriculture in the dry areas. CRC Press, Boca Raton, FL

Pacey A, Cullis A (1986) Rainwater harvesting: The collection of rainfall and runoff in rural areas. Intermediate Technology Publications, London

Pandey DN, Guota AK, Anderson DM (2003) Rainwater harvesting as an adaptation to climate change. Curr Sci 85(1):46–59

Pape TE (2010) New codes are making it easier to be green. J Am Water Works Assoc 102(2):57–63

Pescod ME (1992) Wastewater treatment and use in agriculture, FAO Irrigation and Drainage Paper 47. Food and Agriculture Organization of the United Nations, Rome

Postel S (1992) Last oasis: Facing water scarcity. WW Norton, New York

Postel S (1993) Water and agriculture. In: Gleick PH (ed) Water in crises. A guide to the world's fresh water resources. Oxford University Press, Oxford, pp 56–66

Prinz D, Singh A (2000) Technological potential for improvements of water harvesting. Contributing Paper for World Commission on Dams, Cape Town, Africa

Pyne RDG (2005) Aquifer storage recovery: A guide to groundwater recharge through wells. ASR Systems, Gainesville

Reinumagi IU (2011) Diverting water from the Great Lakes: Pulling the plug on Canada. Valpso Univ Law Rev 20(2):299–347

Rothberg D (2018) SNWA board, including Sisolak, approves litigation to pursue 250-mile groundwater project. *The Nevada Independent*, September 13. https://thenevadaindependent.com/article/snwa-board-including-sisolak-approves-litigation-to-pursue-250-mile-groundwater-project. Accessed 4 June 2020

Rowe MP (2011) Rain water harvesting in Bermuda. J Am Water Resour Assoc (JAWRA) 47 (6):1219–1227

Sakthivadivel R (2007) The groundwater recharge movement in India. In: Giordano M, Villholth KG (eds) The agricultural groundwater revolution: Opportunities and threats to development. CAB International, Wallingford, pp 195–210

Scanlon BR, Faunt CC, Longuevergne L, Reedy RC, Alley WM, McGuire VL, McMahon PB (2012) Groundwater depletion and sustainability of irrigation in the US High Plains and Central Valley. Proc Natl Acad Sci 109(24):9320–9325

Segerfeldt F (2005) Water for sale. How business and the market can reduce the world's water crisis. Cato Institute, Washington, DC

Selover G (2010) Ensuring a more energy-efficient future by changing the way homes are built. J Am Water Works Assoc 102(2):46–51

Shah T (2009) Climate change and groundwater: India's opportunities for mitigation and adaptation. Environmental Research Letters 4(3):035005

Shah T, Molden D, Sakthivadivel R, Seckler D (2000) The global groundwater situation: Overview of opportunities and challenges. International Water Management Institute, Colombo, Sri Lanka

Solis J (2020) SNWA loses another round in court. *Nevada Current*, March 10. https://www.nevadacurrent.com/blog/snwa-loses-yet-another-round-in-court. Accessed 4 June 2020

Spandonide B (2009) Iceberg freshwater sustainable transportation, a new approach. In: The Nineteenth International Offshore and Polar Engineering Conference. International Society of Offshore and Polar Engineers

Stern PH (1979) Small scale irrigation. A manual of lost cost water technology. Intermediate Technology Publications, London

Tago AH (2014) KSA water consumption rate twice the world average. *Arab News*, April 11. https://www.arabnews.com/news/532571. Accessed 4 June 2020

Taylor RG, Scanlon B, Döll P, Rodell M, van Beek R, Wada Y, Longuevergne L, Leblanc M, Famiglietti JS, Edmunds M, Konikow L, Green TR, Chen J, Taniguchi M, Bierkens MFP, MacDonald A, Fan Y, Maxwell RM, Yechieli Y, Gurdak JJ, Allen DM, Shamsudduha M, Hiscock K, Yeh PJ-F, Holman I, Treidel H (2013) Groundwater and climate change. Nat Clim Chang 3:322–329

Thomas J-S, Durham B (2003) Integrated water resources management: Looking at the whole picture. Desalination 156:21–28

Thornton J, Sturm R, Kunkel G (2008) Water loss control manual, 2nd edn. McGraw-Hill, New York

Toumi H (2011) Kuwait has world's highest water consumption. *Gulf News*, April 25. https://gulfnews.com/world/gulf/kuwait/kuwait-has-worlds-highest-water-consumption-1.798870#. Accessed 4 June 2020

UNEP (1984) Rain and stormwater harvesting in rural areas. United Nations Environment Programme, Tycooly International Publishing, Dublin

USEPA (2010) Control and mitigation of drinking water losses in distribution systems, EPA 816-R-10-019, November 2010. U.S. Environmental Protection Agency, Washington, DC

USEPA (2012) Guidelines for water reuse (EPA/600/R-12/618). U.S. Environmental Protection Agency, Washington, DC

Van der Gun J (2012) Groundwater and global change: Trends, opportunities and challenges. United Nations World Water Assessment Programme, Perugia, Italy

Vickers A (2001) Handbook of water use and conservation. WaterPlow Press, Amherst

Wallace JS, Batchelor CH (1997) Managing water resources for crop production. Philos Trans R Soc London Ser B: Biol Sci 352(1356):937–947

Waller DH (1989) Rain water—An alternative source in developing and developed countries. Water Int 14:27–36

Waterfall PH (2006) Harvesting rainwater for landscape use. University of Arizona Cooperative Extension, Tucson. https://extension.arizona.edu/sites/extension.arizona.edu/files/pubs/az1344.pdf. Accessed 4 June 2020

Watson IE, Morin OJ, Jr, Henthorne L (2003) Desalting handbook for planner's (3rd ed), Desalination and Water Purification Research and Development Program Report No. 72. U.S. Bureau of Reclamation.Denver

Way R (2018) The great siphoning: Drought-stricken areas eye the Great Lakes. *Star Tribune*, May 25. https://www.startribune.com/the-great-siphoning-drought-stricken-areas-eye-the-great-lakes/483743681/. Accessed June 4, 2020

Weeks WF, Campbell WJ (1973) Icebergs as a fresh-water source: An appraisal. J Glaciol 12 (65):207–233

Wells F (2014) Drought-shaming your neighbor: There's an app for that. *CNBC*. https://www.cnbc.com/2014/08/04/drought-shaming-your-neighbor-theres-an-app-for-that.html. Accessed 4 June 2020

Whittlesey N (2003) Improving irrigation efficiency through technology adoption: When will it conserve water? Dev Water Sci 50:53–62

Wiener A (1977) Coping with water deficiency in arid and semiarid countries through high-efficiency water management. Ambio 6(1):77–82

Winpenny J (1994) Managing water as an economic resource. Routledge, London

Winpenny J, Heinz I, Koo-Oshima S (2010) The wealth of water: The economics of wastewater use in agriculture. Food and Agriculture Organization of the United Nations, Rome

World Health Organization (2006a) Expert committee meeting on health effects of calcium and magnesium in drinking water. WHO Document Production Services, Geneva

World Health Organization (2006b) WHO guidelines for the safe use of wastewater, excreta, and greywater Volume 1, Policy and regulatory aspects. World Health Organization, Geneva

World Health Organization (2005) Nutrients in drinking water. Water, Sanitation and Health, Protection and the Human Environment, World Health Organization, Geneva

World Health Organization (2007) Desalination for safe water supply, guidance for the health and environmental aspects applicable to desalination. Public Health and the Environment, World Health Organization, Geneva

Chapter 11
Conjunctive Use

11.1 Introduction

Conjunctive use is usually defined as the coordinated use of surface water and groundwater but can be more broadly defined as the coordinated use of all available water resources with the objective of optimizing overall water management. Conjunctive use of surface water and groundwater is an essential element of integrated water resources management. The primary objectives of conjunctive use are maximizing water availability (both total supply and reliability) and minimizing capital and operational costs. Conjunctive use schemes are a response to water scarcity and can contribute to resilience to both current climate variability and future climate change. Conjunctive use is an adaptation to climate change in that it enhances the capacity to use groundwater to bridge periods of water scarcity.

Conjunctive use of surface water and groundwater typically involves using surface water during wet periods when supplies are plentiful and reserving groundwater resources for dry or high demand periods. Groundwater has a number of compelling advantages for agricultural and other uses (Shah et al. 2000; Maliva and Missimer 2012):

- it is produced at the point of use and requires little transport
- groundwater is available "on demand"
- it is less capital intensive to develop than surface water systems
- its use is usually self-financed by users
- it is typically a more reliable supply than surface water because of the often vast storage volume of aquifers—it has less temporal and spatial variation in availability
- groundwater often requires minimal treatment prior to potable use—usually just chlorination and perhaps softening for municipal use
- groundwater is less vulnerable than surface water to contamination by agricultural, industrial, and human activities

R. Maliva, *Climate Change and Groundwater: Planning and Adaptations for a Changing and Uncertain Future*, Springer Hydrogeology,
https://doi.org/10.1007/978-3-030-66813-6_11

- its use may entail a significant incremental cost of lift, so farmers tend to be more frugal in its use as opposed to the use of gravity-driven surface water schemes.

The main problems associated with groundwater use are related to excessive pumping:

- aquifer overdraft and associated declining water levels and increased pumping costs
- deteriorating groundwater quality due to horizontal and vertical intrusion of poorer quality (more saline) water
- adverse impacts to groundwater dependent ecosystems, springs, and surface water bodies
- induced land subsidence.

The advantages of groundwater have led to rapid rates of depletion of major aquifers in the arid and semi-arid regions of the world, including the North China Plain, the Canning Basin of Australia, the Northwest Sahara Aquifer System, the Guarani Aquifer in South America, the High Plains and Central Valley aquifers of the United States, and the aquifers beneath northwestern India and the Middle East (Famiglietti 2014). Nearly all of these aquifers that are experiencing rapid depletion (i.e., pumping rates well in excess of replenishment rates) underlie the world's great agricultural regions and are primarily responsible for their high productivity (Famiglietti 2014). Climate change may exacerbate the depletion of these critical aquifers if it reduces their recharge or dwindling rainfall and surface water supplies prompt more groundwater pumping.

The greatest value of groundwater lies not in its continuous use but rather in the stabilization of water supplies (Tsur 1990; Purkey et al. 1998; Reichard et al. 2010). Groundwater can be used to mitigate undesirable fluctuations in surface water supplies and to improve the timing of water application within a growing season. Purkey et al. (1998, p. 52) observed that "In any situation where surface water supplies are variable, the presence of groundwater resources that can be tapped 'as needed' for municipal, agricultural or other uses, carries a stabilization value well beyond that associated with increases in water supply alone." Groundwater can act as a bridge to allow agricultural and other uses to continue during droughts when surface water supplies are disrupted. Groundwater can also be used as an emergency supply in the event of other disruptions in supply, either natural or man-made. For example, Reichard et al. (2010) discussed how groundwater could mitigate the impacts of a major earthquake that disrupts the surface-water importation system into Southern California. Supplemental irrigation, which is the conjunctive use of rain and irrigation, can result in a substantial improvement of yields over rainfed agriculture alone (Oweis et al. 1999; Oweis and Hachum 2006). Supplemental irrigation can be highly water efficient since a relatively small amount of water applied during periods of low rainfall can result in a substantial increase in productivity.

Fresh surface water flows, if not used or stored, are lost to tide, evaporation, or other sinks and are thus not available for future beneficial use. A basic principal of conjunctive use is, therefore, that surface water should be the primary source of water. When surface water is available, groundwater use should be curtailed to allow aquifers to recharge and groundwater levels to recover (or at least decline no further).

Conjunctive use, as broadly defined, extends well beyond alternation between surface water and groundwater use. Managed aquifer recharge (MAR) plays an important role in many conjunctive use schemes by increasing the supply of groundwater available to stabilize water supplies. MAR takes advantage of the enormous storage capacity of depleted aquifers and the variability of surface water flows. In many regions, surface water flows in excess of human and environmental requirements occur at times, which if captured and stored underground would increase the supply of groundwater available in dry and high demand periods. Surface water is conjunctively used during wet periods to meet current demands in lieu of groundwater and to artificially recharge local aquifers. Similarly, reclaimed water should also be used as much as practicable for non-potable purposes to reduce stress on groundwater resources and for managed aquifer recharge.

MAR does not eliminate the need for surface storage. Dam and reservoirs may still be locally needed to impound surface water flows to protect downstream areas from flooding. Reservoirs can play an important role in conjunctive use through enhanced recharge by controlled release to downstream channels and off-stream infiltration basins and other surface spreading areas. Reservoirs act as giant stilling basins to improve the quality (i.e., reduce the suspended solids and thus clogging potential) of water used for MAR. In areas where surface water is flashy, water storage provided by reservoirs can provide more time for MAR.

Conjunctive use ties into groundwater banking, which broadly refers to the large-scale storage of water in aquifers for later use. Excess water (typically surface water) is stored to accommodate seasonal variations in supply and demand and to provide a buffer against droughts and other interruptions in water supplies. Groundwater banking takes advantages of the general MAR benefits of underground storage including (Maliva 2019):

- often lesser costs than surface storage options
- lesser environmental footprint than surface reservoirs and tanks
- avoidance of evaporative losses
- lesser susceptibility of stored water to contamination
- water may be recovered using wells at the point of use, using the aquifer to, in essence, convey water from the recharge location.

Groundwater banking is not synonymous with water banking, which has been defined as "an institutionalized mechanism specifically designed to facilitate the transfer of water use entitlements" (MacDonnell et al. 1994, p. 1–4). Water banking serves to facilitate transfers of water from low-value to high-value uses by bringing buyers and sellers together (Frederick 1995; Clifford et al. 2004).

Conjunctive use schemes have been developed mainly toward alleviating current aquifer overdraft. However, there are risks associated with developing new supplemental groundwater supplies. In water scarce regions, the temptation will exist to continue to use drought relief wells after a drought has ended, which would effectively increase demands on the source aquifer and water use in general. The end result is to make the system more vulnerable to the next drought (Loucks and Gladwell 1999). Similarly, MAR could spur additional groundwater use, including by free riders who did not contribute to the additional recharge.

11.2 Water Governance

Successful implementation of conjunctive use requires effective water governance. Conjunctive management requires a great amount of joint and coordinated effort among individuals and organizations (Blomquist et al. 2004). The governance may be provided by governmental regulatory agencies or through cooperation amongst aquifer users. Conjunctive management is made more complicated in some countries and states where groundwater and surface water are regulated separately and, in some instances, under very different doctrines (Blomquist et al. 2004). In the state of Texas, for example, surface water use is tightly controlled under the prior appropriation doctrine, which has well-defined water rights, whereas groundwater use had been essentially unregulated under the rule of capture doctrine. A major change in groundwater governance occurred in 1985 and 1997 when the Texas Legislature passed legislation encouraging the establishment of groundwater conservation districts that have the authority to promulgate rules for conserving, protecting, recharging, and preventing waste of groundwater.

A key hydrological and regulatory issue is the degree of connection between groundwater and surface water, specifically whether groundwater pumping impacts surface water supplies. A distinction is made in the western United States between tributary and non-tributary groundwater. In the state of Colorado, for example, use of groundwater that is determined to be hydraulically connected to surface water requires a water right within the surface water prior appropriation system, which would now essentially be unobtainable as surface waters are fully allocated. A surface water right is not required for deep groundwater although its use is still regulated. The state of Texas makes the distinction between whether or not groundwater is underflow and thus regulated in the same manner as overlying surface water. Underflow is defined in the Texas Administrative Code (§ 297.1(56)) as "Water in sand, soil, and gravel below the bed of the watercourse, together with the water in the lateral extensions of the water-bearing material on each side of the surface channel, such that the surface flows are in contact with the subsurface flows, the latter flows being confined within a space reasonably defined and having a direction corresponding to that of the surface flow".

Conjunctive management of water has been practiced in some areas of the southwestern United States for about a century. Over the past several decades, the principal driver of conjunctive use has been over exploitation of aquifers with associated large declines in aquifer water levels and accompanying adverse impacts, such as land subsidence. Conjunctive use is less expensive and has lesser environmental damage than the construction and operation of additional surface water infrastructure, which has been the traditional solution to water scarcity in the region (Blomquist et al. 2001, 2004). The question that arises is "why would the implementation of such a straightforward and sensible management approach continue to lag behind its promise, even after decades of support in the water resource management literature" (Blomquist et al. 2004, p. 14).

Blomquist et al. (2004) observed that participants in a 1998 workshop on conjunctive use, convened by the National Water Research Institute, listed and ranked the impediments to conjunctive use. "Every one of the 10 most significant barriers they identified concerned the assignment of rights, risks, and responsibilities; the distribution of costs and benefits; and the opportunities and disincentives for interorganizational cooperation and coordination of activities—in short institutional issues" (Blomquist et al. 2004, p. 15). Key issues noted include protection of those who invest in facilities and store water, and a fair method to distribute costs among those who benefit from the systems (Blomquist et al. 2004).

Mainstream economic thought assumes that people are driven by the rational pursuit of self-interest (Smith 1776). Water users will invest and support conjunctive use schemes if they perceive that they will receive commensurate benefits. Irrigators will generally be receptive to using surface water or high quality reclaimed (particularly for non-food crops) in lieu of groundwater if they see benefits in preserving local groundwater for their future use during dry periods and the associated costs are not unacceptably burdensome. Coordinated management of water tends to be less complex where there are few water users in an area and they share common interests.

Governance of conjugate use systems that include groundwater banking are more complex because additionally recharged water is added to a common pool resource (i.e., aquifers). Banked surface water is comingled with native groundwater. Some groundwater banking schemes recognize "in lieu recharge" in which credits are given for groundwater not used, which is considered hydrologically equivalent to actively recharging water. Reconciling groundwater and surface water rights is a critical institutional issue for groundwater banking systems. Preserving landowners' (and other users') existing rights to access groundwater while preserving stored surface-water for later extraction can be a very complex undertaking (Pinhey 2003).

Purkey et al. (1998) discussed the technical and regulatory issues associated with groundwater banking in the state of California, which are broadly applicable elsewhere. A basic requirement for any groundwater banking scheme, and other conjugate use systems intended to conserve or augment groundwater supplies, is that some mechanism must be in place to control extractions and, in particular, to prevent the additional stored water from being extracted by other aquifer users who are not participating in the scheme. Use of the storage aquifer needs to be tightly regulated and not open to new allocations. There would be little point to a groundwater banking system if new users or non-participating existing users could freely withdraw more water. In economics terms, successful groundwater banking requires a solution to the "free rider problem," which is defined as "the burden on a shared resource that is created by its use or overuse by people who aren't paying their fair share for it or aren't paying anything at all" (Chappelow 2019). Vaux (n. d.) observed with respect to groundwater banking and conjunctive use in California that

> An important consequence of the permissiveness of California's ground water law is that the full promise of ground water banking and conjunctive use cannot be realized in areas that may be subject to competitive exploitation. The reason is that waters banked or stored as part of formal programs are subject, at least in part, to the law of capture and there is thus no assurance that the water banked by one individual or group will not be captured by a competitor who did not participate in the financing of the recharge works.

Groundwater banking and similar conjugate use systems should not contribute to further aquifer overdraft nor cause significant adverse impacts during recovery. In over drafted aquifers, MAR may not actually result in increased water storage, as manifested by increases in aquifer water levels or heads, if even with the additional recharge aquifers water budgets are still out of balance. Recharged water may be extracted by other users, leak out, be discharged from the aquifer, or result in decreased (rejected) natural recharge (Maliva 2014, 2019). Nevertheless, MAR may still be beneficial in reducing the overdraft and preventing aquifer water levels from declining even further.

Groundwater banks may become insolvent if the accumulated credits to recover water (obtained from managed recharge) exceed the quantity of water that can be physically and safely produced from an aquifer when required by users. The insolvency of groundwater banks may become evident during dry periods when both participants in banks want to recover their stored water (i.e., redeem credits for recharged water) and non-participating users are also pumping at high rates (Maliva 2014, 2019). Hence, successful groundwater banking requires careful consideration (and modeling and monitoring) of the hydrological impacts of both recharge and recovery, and aquifer water budgets.

A potentially highly contentious issue for some groundwater banking systems is reconciling the rights of existing groundwater users and system participants (Ward and Dillon 2012). Groundwater withdrawal rates within a given basin may be physically limited or have regulatory restrictions driven by resource and environmental protection concerns. The difficult question is who has priority over the limited extractable groundwater during dry periods when water is in great demand. The question becomes more complex in the common situation where native groundwater extractors are senior to the MAR system operators (i.e., they were extracting groundwater long before the MAR system was constructed).

Ward and Dillon (2009, 2012) observed that to improve the security of water entitlements and encourage MAR, recovery entitlements for MAR systems are likely required to be institutionally differentiated from those governing entitlements to extract native groundwater. Ward and Dillon (2012) proposed that entitlements to recharged water (recovery credits) should be endowed with a higher level of security than entitlements to native groundwater. Allowing groundwater bank participants to, in essence, jump to the head of the queue and have first priority to withdraw groundwater during periods of scarcity would be an important incentive for encouraging groundwater banking. However, giving a higher priority to water bank participants would be extraordinarily contentious where it has the net impact of reducing the ability of entrenched long-existing groundwater users to withdraw water when they need it as they have been historically allowed. Hence, it is desirable to have all the major groundwater users in a basin jointly participate in conjugate use and groundwater banking systems to facilitate the development of equitable management arrangements.

11.3 Implementation of Conjunctive Use

Conjunctive use has been implemented primarily as a response to current water scarcity, and as such, is also an adaptation to decreased availability of water or increasing demands under future climate conditions. The ability to implement conjunctive use and its effectiveness depends upon local hydrological and hydrogeological conditions and the extent to which local water management institutional frameworks and practices are congruent with the practice. A fundamental requirement of conjunctive use is the availability at times of surface water (or reclaimed water) that can substitute for groundwater and thus prolong (ideally indefinitely) the useful life of local groundwater resources. Institutionally, "conjunctive use requires that surface and underground water supplies be treated—in fact if not in law—as more or less interchangeable" (Blomquist et al. 2004, p. 58). Follows are summaries of experiences with conjunctive use in some geographic areas.

11.3.1 Southern California

Southern California is the leader in conjunctive use in the United States due to a combination of semiarid and arid climates, extensive irrigated agriculture, over-drafted aquifers, seasonally variable surface water flows, and decentralized water management. Southern California imports most of its surface water from the wetter northern and eastern parts of the state through the State Water Project and Central Valley Project, the Colorado River via the Colorado River Aqueduct and All-American Canal, and the southern Sierra Nevada via the Los Angeles Aqueducts (Fig. 11.1).

Surface water in California is governed under a complex combination of riparian rights and prior appropriation. Under a riparian right, water can be diverted from a water course if the water is used on adjoining lands and drains back to the lake, river, stream, or creek from which it was taken (California Water Board 2019). An appropriative permit issued by the State Water Board is required for diversion of

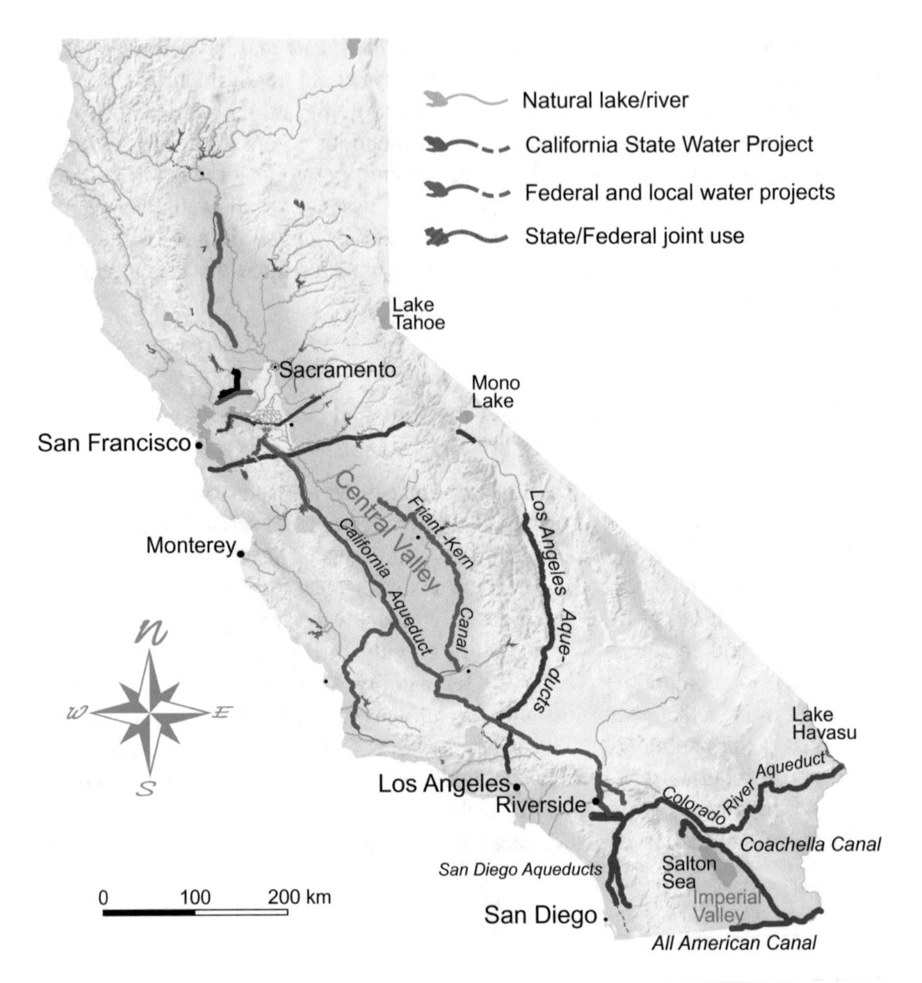

Fig. 11.1 Map showing the locations of major California water supply canals. *Source* California Department of Water Resources

water for use on non-riparian land and for the use of water that would not be present on riparian land under natural conditions (California Water Board 2019).

Groundwater is governed under the correlative rights doctrine in which groundwater rights are proportional to a landowner's share of the land overlying an aquifer. Groundwater is considered to be owned by the landowner, but its use is controlled by the state. The use of the groundwater must be reasonable and beneficial and is correlated with the rights of other groundwater users. In the event of shortages, caused by droughts or excessive aquifer drawdowns, all water users share in a pro-rated reduction in water use related to the amount of land they own; no distinction is made between senior and junior water users or between the uses of the water. Correlative shares of groundwater are not quantified in California unless the groundwater basin has been adjudicated. In an adjudicated groundwater basin, the court decides on the groundwater rights of all landowners, how much groundwater well owners can extract, and who will be the Watermaster. The Watermaster ensures that a basin is managed according to the decrees of the court (California Department of Water Resources 2000, 2004).

California's political system, which embraces local home rule and the creation of special districts and other local governments for the governance of water, has promoted the institutional entrepreneurship that is largely responsible for the conjunctive management and many other water projects developed in California over the past 75 years (Blomquist et al. 2004). Most water districts in California were created under general-purpose enabling acts, in which each act created a class of water districts with a different mix of authority and responsibilities from districts created under other acts (Blomquist et al. 2004).

As observed by (Vaux n.d. p. 4):

> One striking characteristic of California's agricultural lands is the fact that most fall within the boundaries of a special water district formed to acquire and purvey water to local users. These districts, which were characterized early-on as a user cooperatives, have been formed under a variety of provisions in the California Water Code for the general purposes of acquiring, storing, distributing and conserving water. The Water Code permits Districts to tax, to contract with state and federal agencies, issue bonds and receive revenues

The diversity of conjunctive management methods operational in California reflects a tailoring to local circumstances, which is consistent with the state's basin-by-basin decentralized approach to groundwater management (Blomquist et al. 2004). However, the decentralized, home-rule approach that led to local innovation also results in parochial management. For example, "area-of-origin" provisions that ban the extraction of groundwater for use outside of a jurisdiction constrain large-scale conjunctive management (Blomquist et al. 2004).

Kern County is a major agricultural area located at the southern end of the Central Valley of Southern California (Fig. 11.2). Kern County has an arid climate and irrigation is essential for its agricultural operations, which have a gross value of about US $7 billion (WAKC n.d.). Conjugate use and groundwater banking have become so important to Kern County that almost every water district in the County participates in a banking program in some fashion (WAKC n.d.). Kern County groundwater banking programs include (WAKC n.d.):

- Arvin-Edison Water Storage District Water Management Program
- Berrenda Mesa Property Joint Water Banking Project
- Buena Vista Water Storage District Water Management Program
- Buena Vista Water Storage District/West Kern Water District Water Supply Project
- Cawelo Water District/Dudley Ridge Water District Conjunctive Use Program
- Cawelo Water District's Modified Famoso Water Banking Project
- City of Bakersfield 2800 Acre Groundwater Recharge Facility
- Kern Delta Water District's Groundwater Banking Program
- Kern Water Bank

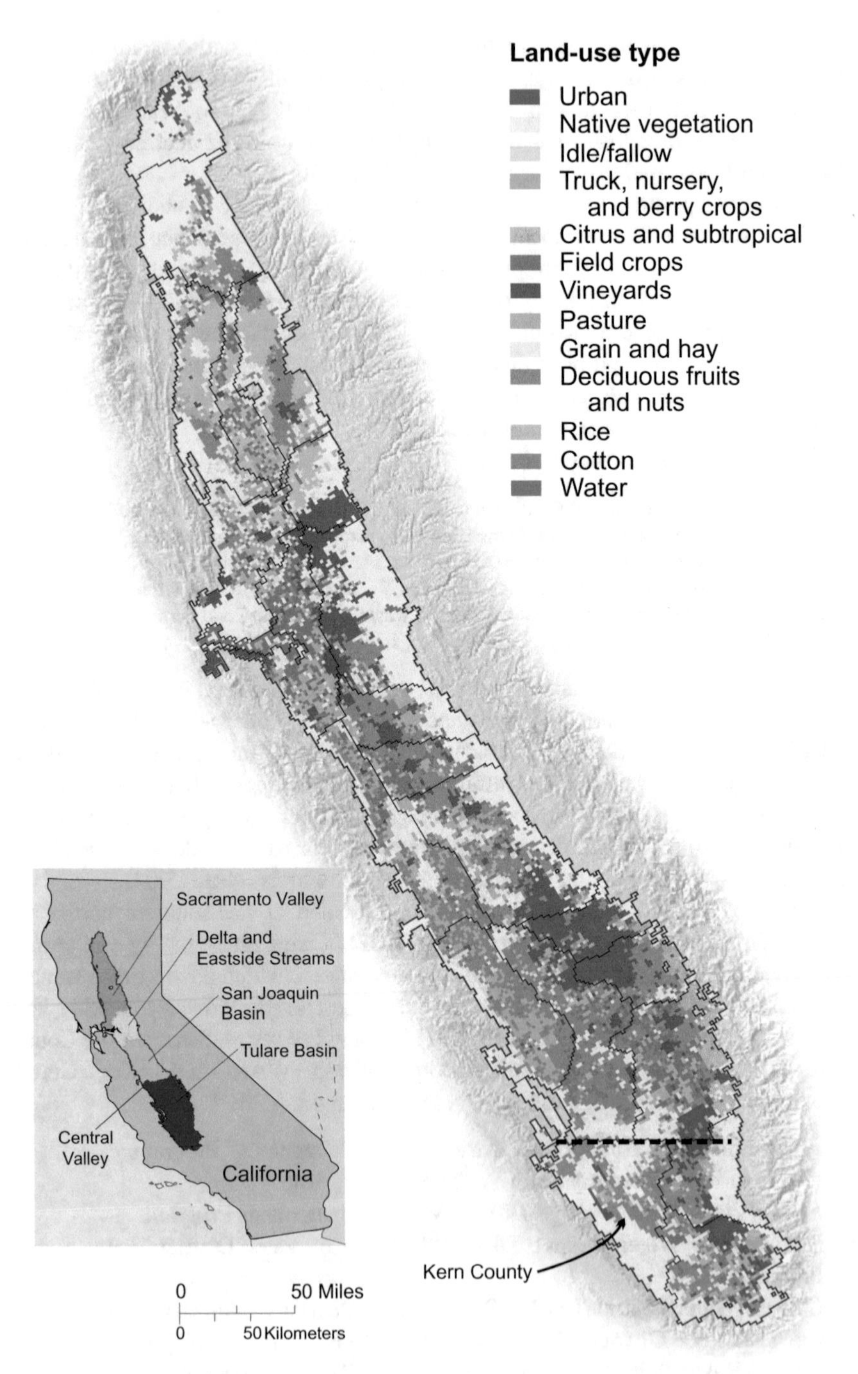

Fig. 11.2 Map showing the location of Kern County in the southern Central Valley of California and land uses. *Source* Faunt (2009)

- North Kern Water District Groundwater Storage Project
- Thomas N. Clark Recharge and Banking Project
- Rosedale-Rio Bravo Water Storage District and Improvement District No. 4 Joint Use Groundwater Recovery Project
- Rosedale-Rio Bravo Water Storage District's Groundwater Banking Program
- Semitropic Groundwater Banking Project
- West Kern Water District's Groundwater Banking Program.

The Arvin-Edison Water Storage District (AEWSD) provides a good example of conjunctive water management in a southern California agricultural district. More than three quarters of the land area of the Arvin-Edison District (40,500 hectares, 100,000 acres) is devoted to irrigated agriculture. Over extension of irrigation had led to rates of ground water depletion that could not be continued for long (Vaux n. d.). Groundwater depletion was occurring throughout the southern Central Valley and continuation of irrigated agriculture became dependent upon the importation of supplemental surface water supplies. Imported surface water was first made available to large areas of the Central Valley through the federal Central Valley Project (CVP), which was constructed by the U.S. Bureau of Reclamation (Vaux n. d.). The AEWSD was organized in 1942 under the California Water Storage District Law (Division 14 of the California Water Code) for the expressed purpose of, among several things, providing an agency to contract with the United States for water service from the CVP and also for contracting for a federal power contract and a federal loan for the construction of new facilities (Provost and Pritchard 2003).

The CVP water service contract provides the AEWSD with a supplemental water supply of which only 11% (the firm supply) would be reasonably reliable, with the remaining 89% being delivered to the District on an "as available" basis that depends primarily on higher than average levels of precipitation. (Vaux. n.d.). As described by Vaux (n.d., p. 6)

> The challenge for the AEWSD was to take a highly variable and uncertain imported surface water allocation and transform it into a more certain (though probably smaller allocation) which would allow irrigation to continue on roughly the same scale as had developed historically while protecting both ground water quantity and quality. This challenge was made more manageable by virtue of the fact that reducing the long-term overdraft would address both the problem of quantitative sustainability and the water quality problem.

The AEWSD conjunctive use program was initiated in 1966. The AEWSD was divided into two service areas, a surface water (40%) and a groundwater (60%) area, with the groundwater area continuing to pump with reduced overall extractions and a formal program of groundwater recharge (Vaux n.d.). Any imported water that is in excess of the immediate demand from the surface water service area is routed to spreading basins (which have a total area of a little more than 200 hectares; 500 acres) to recharge the underlying aquifer. The AEWSD operates and maintains several well fields, which allows it to extract water stored underground and deliver it to the surface water service area at times when surface water deliveries are inadequate to meet prevailing demands (Vaux n.d.). The conjunctive use program, by increasing the reliability of the water supply, has helped to ensure that agriculture in the AEWSD is more productive and profitable than it would have been in the absence of the conjunctive use program (Vaux n.d.).

Groundwater levels in the Central Valley respond to climate-related variations in natural recharge, irrigation return flows, as well as conjunctive use and MAR, making it difficult to isolate the impacts of the latter (Scanlon et al. 2016). Despite widespread implementation of conjunctive use and MAR, the Central Valley had experienced a net groundwater loss between April 2002 and September 2016 (Xiao et al. 2017). Greater groundwater depletion occurred during droughts with some recovery during wetter periods.

California enacted the Sustainable Groundwater Management Act (SGMA) in 2014, which requires by January 31, 2020, that all basins designated as high- or medium-priority (which are subject to critical conditions of overdraft) be managed under a groundwater sustainability plan or coordinated groundwater sustainability plans. The sustainability plans are required to have measurable objectives, as well as interim milestones in increments of five years, to achieve the sustainability goal for the basin within 20 years of the implementation of the plan. Exactly how sustainability will be achieved under the SGMA is still to be determined (Pitzer 2019). The impacts of the sustainability plans are likely to be significant with as much as 185,000 acres of Kern County crop land (roughly 20% of the county's irrigated farm acreage) potentially having to be taken out of production (Pitzer 2019)

A goal of the SGMA is to maintain local governance of water tailored to local circumstances with local institutions supported by the state. The concern has been expressed that the SGMA requirement that the State of California evaluate or assess groundwater sustainability plans will result in a weakening of local control and innovation (Aladjem and Nikkel 2016).

11.3.2 Arizona

Arizona is the most arid state in the United States with an average annual rainfall of 32.3 cm (12.7 in.) and even lower rainfalls in the more populated urban areas in the Sonoran Desert in the south-central part of the state. Groundwater occurs in thick siliciclastic alluvial basins that are surrounded by essentially impervious bedrock mountains. Groundwater has historically been the primary water source in the region. Due to high irrigation use and low recharge rates, the aquifers are prone to overdraft and experienced large historical declines in water levels (Anderson et al. 1992, 2007). Maximum declines in some regions ranged from 30 to 120 m (Anderson et al. 1992). The two most important events in water management in Arizona were the passage by the Arizona Legislature of the Groundwater Management Act (GMA) in 1980 and the completion of the Central Arizona Project (CAP) in 1994.

A key part of the GMA is the designation of the most populous parts of the state that were experiencing serious groundwater overdraft as "Active Management Areas" (AMAs). A primary goal of the GMA is to achieve a safe yield in the AMAs by the year 2025, with safe yield defined as achieving a long-term balance between annual groundwater withdrawals and natural and artificial recharge (Jacobs and Holway 2004; Eden et al. 2007; Pearce 2007; Megdal 2012). The GMA also mandates an Assured Water Supply (AWS) program in which new developments are allowed in AMAs only if the developer can demonstrate an assured water supply for the next 100 years. A 100-year supply of water can be met using groundwater, if groundwater use is offset by the recharge of renewable water, such as reclaimed water or surface water (Megdal 2012). A strong incentive for MAR was created by giving credits for MAR that can be used against groundwater withdrawals.

The Colorado River Compact of 1922 allocated Arizona 2.8 million acre-feet (MAF; 3,450 million cubic meters, MCM) annually of Colorado River water plus half of any surplus additional water in the lower Colorado River basin. At the time of the compact, Arizona had neither the demand for the water or a means to convey the water to the main irrigation areas in the central part of the state. California has the right to any water that Arizona leaves in the river.

The CAP is a 541 km (336 mile) uphill aqueduct that conveys water from the Colorado River to the Phoenix and Tucson metropolitan areas (Fig. 11.3). The CAP began initial deliveries in 1985, but the water supply was substantially under-utilized into the early 1990s (Megdal et al. 2014). The construction of the CAP resulted in a strong impetus for Arizona to vigorously pursue groundwater banking to fully store any excess of its Colorado River allocation, which would otherwise be taken by California. The CAP is administered by the Central Arizona Water Conservation District (CAWCD).

The majority of the CAP water is used for agriculture with large irrigation districts delivering water to farmers (CAP n.d.). The CAP water allows for conjunctive use where it is used in lieu of groundwater. The CAP has been a large supply of water that could be used for MAR. Initially, the demand for CAP water was much less than the available supply leaving an excess that could be used for aquifer recharge. The Arizona Water Banking Authority (AWBA) was created in 1996 to provide long-term underground storage of CAP water that Arizona was allotted but not yet used. The stored water may be used to protect municipal users from droughts or possible CAP disruptions, meet Native American water rights claims, assist in meeting local water management objectives, and facilitate interstate water banking (Jacobs and Holway 2004; August and Gammage 2007; Colby et al. 2007; Eden et al. 2007).

The Central Arizona Groundwater Replenishment District (CAGRD) was established in 1993 to acquire excess Colorado River water and store it underground on behalf of developers without direct access to surface water (August and Gammage 2007; Peace 2007). The CAGRD also stores treated wastewater. Entities that cannot meet the requirements of the AWS program (100-year assured supply) have the option of paying a fee to the CAGRD for the groundwater a subdivision or

Fig. 11.3 CAP canal, near Lake Pleasant, Maricopa County, Arizona

water provider is using or will use in the future. The CAGRD takes responsibility for acquiring and replenishing water to offset the extracted groundwater. The CAGRD is unique in Arizona in that it is allowed to perform replenishment after groundwater withdrawals have occurred (Megdal 2007). The CAGRD must find water and perform replenishment within three years of the excess groundwater use (i.e., use in excess of amounts allowable under rules).

MAR is Arizona is performed mainly using spreading basins (Fig. 11.4) and, less commonly, using vadose wells and river channels. Spreading basins cover an area of $\approx$44 km^2 with 53% of the area located in the Phoenix AMA and 37% in the Tucson AMA (Scanlon et a. 2016). Water delivered for MAR in Arizona totaled 7.3 km^3 from 1994 to 2013, with 75% from the Colorado River via the CAP, 19% reclaimed wastewater, and 6% locally derived surface water (Scanlon et al. 2016).

Characteristics favoring groundwater banking and conjunctive use for water security in Arizona include (Megdal et al. 2014):

- an awareness that augmentation of groundwater resources may be necessary to address aquifer depletion and future imbalances between supply and demand
- availability of a water source that can used for intermittent or continuous recharge
- favorable hydrogeological conditions including suitable storage space and aquifer transmissivity

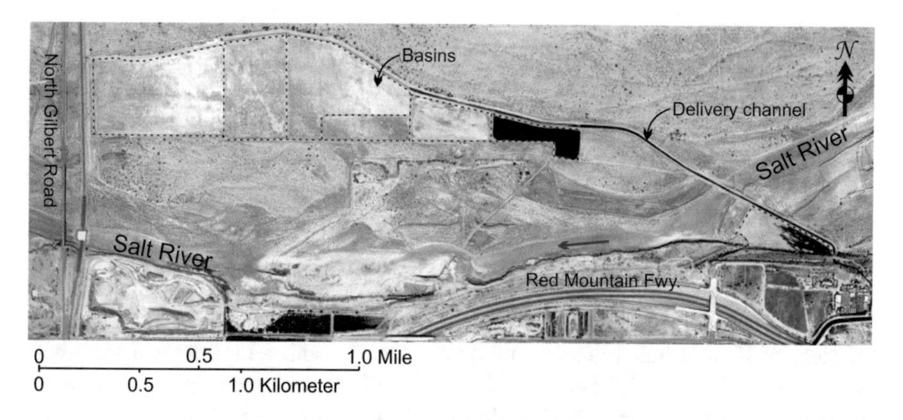

Fig. 11.4 Aerial photograph of the Granite Reef Underground Storage Project near Mesa, Arizona, in which the infiltration basin is divided into cells. *Source* U.S. Geological Survey

- a well-established regulatory framework that is adhered to by water users and ensures that system owners and participants obtain commensurate benefits from the water they recharged
- funding mechanisms that facilitate investments in water banking system planning, construction, and operation
- favorable institutional arrangements that link policy with investments.

The Prescott, Phoenix, and Tucson AMAs are not expected to achieve their safe yield goals by 2025 (Choy 2015). Nevertheless, the GMA has had positive results in stabilizing groundwater levels or reducing the rate of decline of groundwater levels in AMAs compared to nearby basins that do not have access to surface water (Scanlon et al. 2016). The CAP was a lifeline to served areas, but vulnerabilities to their water supplies remain. As CAP contractors' water demands rise, less excess water will be available for the CAGRD and other parties for MAR (Choy 2015). The Arizona Department of Water Resources in the Fourth Management Plan for the Tucson AMA (TAMA) discussed the vulnerabilities of the CAP supply to future drier conditions (ADWR 2016, pp. 8–7):

> The volume of excess CAP water fluctuates depending on the use of CAP subcontracts and allocations and the availability of the overall CAP supply. Based on projections by the US Bureau of Reclamation, there is a probability that CAP shortages may occur in the future. Lower than average precipitation on the Colorado River watershed may increase the likelihood of these shortages occurring. Because CAP delivers mostly lower priority Colorado River water, Colorado River supplies for the CAP (and certain on-river/mainstem users) have a junior priority compared with other on-river/mainstem users. Colorado River supplies for the CAP will be reduced in times of a declared shortage in the Lower Colorado River Basin. As insurance against the impacts of future shortages, unused CAP supplies have been recharged by individual entities within the TAMA holding water storage permits.

The question facing Arizona is where will the "next bucket" of water come from (Jacobs and Megdal 2004). To achieve safe yield, either per capita water use will have to be reduced, water use will need to shift from other sectors (i.e., agriculture) to the municipal sector, or new sources (e.g., desalination and wastewater reuse) be developed (Jacobs and Megdal 2004).

11.3.3 Florida

Unlike Southern California and Arizona, where conjugate use has been embraced, actual implementation of conjugate use has lagged in the state of Florida. Much of the state of Florida has a humid subtropical climate with a high annual state-wide average rainfall of 136.4 cm (53.7 in). Rainfall in peninsular Florida is highly seasonal with most of the precipitation occurring during the summer wet season. The winter and spring dry season coincides with the peak in tourism and seasonal residents, and thus demand for water. Florida is blessed with abundant high-quality groundwater resources. Approximately 63% of the total water supply and 85% of the public water supply in Florida is obtained from groundwater (Marella and Dixon 2018). Exploitation of fresh groundwater resources is believed to be at or close to sustainable limits in much of Florida due to environmental constraints, particularly wetland hydroperiods, lake levels, and spring and stream flows. Fresh surface water is abundant in the wet season but its use for potable supply has been limited because its seasonal availability does not coincide with the peak demand period and its poorer quality and thus greater treatment requirements and costs than for groundwater. Florida's rainfall and demand patterns are ideal for conjunctive use with MAR as there is ample excess water that could be stored in the wet season for later use during the peak demand dry season.

Water supply in Florida tends to be fragmented with individual cities and unincorporated parts of counties operating their own utilities. Local autonomous utilities have worked fine so long as low-cost fresh groundwater is readily locally available. As communities grew and additional local fresh groundwater was no longer obtainable, regional solutions involving alternative water supplies were pursued in some areas.

The Peace River Manasota Regional Water Supply Authority (PRMRWSA) is a regional water supply authority serving parts of west-central Florida. Its mission is "To provide the region with a sufficient, high quality, safe drinking water supply that is reliable, sustainable and protective of our natural resources now and into the future" (PRMRWSA n.d.). The PRMRSWA was established on February 26, 1982, and since 1991, has provided an average of 26 million gallons per day (MGD; 100,000 m^3/d) of drinking water to more than 900,000 people across Charlotte, DeSoto, Manatee, and Sarasota counties in west-central Florida (PRWRWSA n.d.). The Board of Directors (BOD) consist of an elected County Commissioner from each of Charlotte, Manatee, Sarasota, and Desoto Counties.

The PRMRWSA currently provides drinking water to Charlotte, DeSoto and Sarasota Counties, and the City of North Port, and plans are in place to provide water to Manatee County in the future (PRWRWSA n.d.). The customers pay for the operation of the system based on their water allocation at rates determined annually by the BOD. The PRMRWSA also has interconnects with other water systems (Sarasota County Utilities, City of North Port Utilities, City of Punta Gorda Utilities and Englewood Water District) that are available to supply water in the case of a natural disaster, equipment maintenance or failure, resource stress, or unforeseen or unplanned increases in water demand (PRWRWSA n.d.).

The PRMRWSA obtains it water from the Peace River in DeSoto County. The PRMRWSA water use permit allows for the withdrawal of up to 120 MGD (470,000 m^3/d) during high flow periods, but permit conditions prohibit withdrawals during low flow periods. Therefore, large-scale storage is critical for the operation of the water supply system. The PRMRWSA was two off-stream reservoirs that can store 6.5 billion gallons (25.4 million m^3) of untreated water (Fig. 11.5) and an aquifer storage and recovery system with 21 recharge and recovery wells that can store in total more than 6 billion gallons (23.4 million m^3) of treated water (PRWRWSA n.d.). The PRMRWSA obtained an Operational Flexibility Water Use Permit (OFWUP) that authorizes the conjunctive use of interconnected sources, on a short-term basis only, for operational flexibility when regional supplies from the Peace River are insufficient or temporarily unavailable (Atkins 2015). The OFWUP sources include both groundwater and alternative surface water sources.

Fig. 11.5 Peace River Manasota Regional Water Supply Authority reservoir used to store untreated surface water from the Peace River

Tampa Bay Water (TBW) is a wholesale supplier of drinking water to Hillsborough, Pasco, and Pinellas Counties and the cities of New Port Richey, St. Petersburg and Tampa in west-central Florida. TBW was created in 1998 as part of the solution to the "water wars" in the region over access to the fresh groundwater resources of the Upper Floridan aquifer (UFA). Since the early 1970s, densely populated but water-poor Pinellas County had been purchasing land and developing UFA well fields in Pasco and Hillsborough Counties (Dedekorkut 2003). The impact of the wells on local lakes and wetlands and concerns over sinkhole formation prompted Pasco and Hillsborough Counties and their citizens to file numerous court challenges (Dedekorkut 2003).

The creation of the TBW was the culmination of a new alliance between the six governments in west-central Florida to more collectively manage their groundwater resources and to develop new alternative water resources in the rapidly growing region. The region's water is obtained from three different sources: groundwater, river water and desalinated seawater. TBW is now the only water utility in the United States to take advantage of these three sources of water combined (TBW n. d.). The specific main elements of the TBW water supply network are a 120 MGD (470,000 m^3/d) surface water treatment plant, the 25 MGD (98,000 m^3/d) Tampa Bay Seawater Desalination Plant, a 15.5-billion gallon (60.5 million m^3) reservoir (C.W. Bill Young Regional Reservoir), and 120 MGD (470,000 m^3/d) permitted capacity of groundwater from wells (TBW n.d.). The conjunctive use of three water sources provides considerable resilience to the TBW water supply system.

Polk County in central Florida has also relied essentially exclusively on fresh groundwater from the UFA for its water supply. Population growth in the county is projected to outstrip its UFA water supply mainly due to regulatory constraints over minimum flows and levels (MFLs) in surface water bodies. The Polk Regional Water Cooperative (PRWC) was established in 2017 by 16 member governments, consisting of 15 cities and Polk County, to identify sustainable groundwater sources, develop strategies to meet the future water demands of its members, to determine infrastructure needed for treatment and distribution, and to establish consistent rules for fairly meeting all water supply needs across the County (PRWC n.d.). The priority of the PRWC members is to maximize the use of the UFA as it is by far the lowest cost water supply option. However, water demand and supplies projections indicate the alternative water supplies will be needed in the future. Four alternative water supply projects were selected for investigation and preliminary design, which are two brackish groundwater desalination projects (West and Southeast Lower Floridan Aquifer Water Production Facilities) and two surface water projects that would capture seasonally available excess surface water from the Peace Creek and Peace River and recharge the UFA.

The principal impediments to conjunctive use in Florida are the technical challenges and costs of seasonally storing and treating surface water. The very subdued topography of much of peninsular Florida and the often high permeability of shallow strata are not optimal for surface reservoir construction. Surface water also has greater treatment costs than the usually very high-quality UFA water. Treated surface water can be stored underground in ASR systems, such as is successfully performed by the PRMRWSA. Brackish groundwater desalination is currently the preferred alternative water source for most utilities because of the widespread occurrence of brackish groundwater and its usually greater reliability than surface water options. Groundwater desalination, a climate insensitive source, provides greater resilience to water supplies.

References

ADWR (2016) Fourth management plan Tucson Active Management Area. Arizona department of water resources. http://infoshare.azwater.gov/docushare/dsweb/Get/Document-10038/TAMA_4MP_Complete.pdf. Accessed 6 June 2020

Aladjem DRE, Nikkel ME (2016) California groundwater management: laboratories of local implementation or state command and control? Environ Law News 25(2):3–7

Anderson TW, Freethey GW, Tucci P (1992) Geohydrology and water resources in alluvial basins in south-central Arizona and parts of adjoining states. U.S. Geological Survey Professional Paper 1406-B

Anderson MT, Pool DR, Leake SA (2007) The water supply of Arizona: the geographic distribution of availability and patterns of use. In: Jacobs KL, Colby BG (eds) Arizona water policy: management innovations in an urbanizing, arid region. Resources for the Future, Washington, D.C., pp 45–60

Atkins (2015) Integrated regional water supply plan 2015. In: Report prepared for the peace river manasota regional water supply authority. https://regionalwater.org/wp-content/uploads/2018/08/PRMRWSA_IRWSMP_April2015_FINAL-WithTMs.compressed.pdf. Accessed 6 June 2020

August JL, Gammage G (2007) Shaped by water. An Arizona historical perspective. In: Jacobs KL, Colby BG (eds) Arizona water policy: management innovations in an urbanizing, arid region. Resources for the Future, Washington, D.C., pp 10–25

Blomquist W, Heikkila T, Schlager E (2001) Institutions and conjunctive water management among three western states. Nat Res J 41(3):653–683

Blomquist W, Schlager E, Heikkila T (2004) Common waters, diverging streams. Linking institutions and water management in Arizona, California, and Colorado. Resources for the Future, Washington D.C.

California Department of Water Resources (2000) Groundwater management in California—six methods under current law. Water Facts no. 8. California Department of Water Resources, Sacramento

California Department of Water Resources (2004) Adjudicated groundwater basins in California, Water Facts, no. 3. California Department of Water Resources, Sacramento

California Water Boards (2019). Water rights. Frequently asked questions. https://www.waterboards.ca.gov/waterrights/board_info/faqs.html#toc178761091. Accessed 6 June 2020

CAP (n.d.). Background and history. Central arizona project. https://www.cap-az.com/about-us/background. Accessed 6 June 2020

Chappelow J (2019) Free rider problem. Investopedia. https://www.investopedia.com/terms/f/free_rider_problem.asp. Accessed 6 June 2020

Choy J (2015) 7 lessons in groundwater management from the Grand Canyon State. Sanford University, Water in the West. http://waterinthewest.stanford.edu/news-events/news-press-releases/7-lessons-groundwater-management-grand-canyon-state. Accessed 6 June 2020

Clifford P, Landry C, Larsen-Hayden A (2004) Analysis of water banks in the Western States, Publication No. 04-11-011. Washington Department of Ecology, Olympia

Colby BG, Smith DR, Pittenger K (2007) Water transactions. Enhancing water supply reliability during drought. In: Jacobs KL, Colby BG (eds) Arizona water policy: management innovations in an urbanizing, arid region. Resources for the Future, Washington, D.C., pp 79–91

Dedekorkut A (2003) Tampa bay water wars: from conflict to collaboration? In: Dedekorkut A, Scholz J, Stiftel B (eds) Adaptive governance and Florida's water conflicts: the case studies. Florida State University DeVoe L. Moore Center for the Study of Critical Issues in Economic Policy and Government, Tallahassee, pp 23–58

Eden S, Gelt J, Megdal S, Shipman T, Smart A, Escobedo M (2007) Artificial recharge: a multi-purpose water management tool: Arroyo. Water Resources Research Center, College of Life Sciences, University of Arizona, Winter

Famiglietti JS (2014) The global groundwater crisis. Nat Clim Change 4(11):945–948

Faunt CC (2009) California's central valley groundwater study: a powerful new tool to assess water resources in California's Central Valley. U.S. Geological Survey Fact Sheet 2009–3057

Frederick KD (1995) Adapting to climate impacts on the supply and demand of water. Clim Change 37(1):141–156

Jacobs KL, Holway JM (2004) Managing for sustainability in an arid climate: lessons learned from 20 years of groundwater management in Arizona, USA. Hydrogeol J 12:52–65

Jacobs K, Megdal S (2004) Water management in the active management areas. In: Arizona's water future: challenges and opportunities, 85th Arizona Town Hall Background Report. University of Arizona, Tucson, pp 71–93. https://wrrc.arizona.edu/sites/wrrc.arizona.edu/files/pdfs/TownHall.pdf. Accessed 6 June 2020

Loucks DP, Gladwell JS (1999) Sustainability criteria for water resources systems. Cambridge University Press, Cambridge, UK

MacDonnell LJ, Howe CW, Miller KA, Rice TA, Bates SF (1994) Water banks in the West. Natural Resources Law Center, University of Colorado, Boulder

Maliva RG (2014) Economics of managed aquifer recharge. Water 6(5):1251–1279

Maliva RG (2019) Anthropogenic aquifer recharge. Springer, Cham

Maliva RG, Missimer TM (2012) Arid lands water evaluation and management. Springer, Berlin

Marella RL, Dixon JF (2018) Data tables summarizing the source-specific estimated water withdrawals in Florida by water source, category, county, and water management district, 2015. Geological Survey data release, US. https://doi.org/10.5066/F7N29W5M

Megdal SB (2007) Arizona's recharge and recovery programs. In: Jacobs KL, Colby BG (eds) Arizona water policy: management innovations in an urbanizing, arid region. Resources for the Future, Washington D.C., pp 188–203

Megdal SB (2012) Arizona groundwater management. In: The water report, pp 9–15

Megdal SB, Dillon P, Seasholes K (2014) Water banks: using managed aquifer recharge to meet water policy objectives. Water 6:1500–1514

Oweis T, Hachum A (2006) Water harvesting and supplemental irrigation for improved water productivity of dry farming systems in West Africa and North Africa. Agric Water Manag 80:57–73

Oweis T, Hachum A, Kijne J (1999) Water harvesting and supplementary irrigation for improved water use efficiency in dry areas. International Water Management Institute, Colombo, Sri Lanka

Pearce MJ (2007) Balancing of competing interests: the history of state and federal water laws. In: Jacobs KL, Colby BG (eds) Arizona water policy: management innovations in an urbanizing, arid region. Resources for the Future, Washington D.C., pp 26–44

Pinhey NA (2003) Banking on the commons: an institutional analysis of groundwater banking programs in California's Central Valley. Doctoral dissertation, University of Southern California, Los Angeles

Pitzer G (2019) As deadline looms for California's badly over drafted groundwater basins, Kern County seeks a balance to keep farms thriving. Water Education Foundation. https://www.watereducation.org/western-water/deadline-looms-californias-badly-overdrafted-groundwater-basins-kern-county-seeks. Accessed 6 June 2020

PRWRWSA (n.d.) Welcome to the Peace River Manasota Regional Water Supply Authority. https://regionalwater.org/. Accessed 20 June 2020

Provost and Pritchard (2003) Arvin-Edison Water Storage District Groundwater Management Plan. In: Report prepared by Provost and Pritchard for the AEWSD

PRWC (n.d.) Polk Regional Water Cooperative. https://prwcwater.org/. Accessed 6 June 2020

Purkey DR, Thomas GA, Fullerton DK, Moench M, Axelrad L (1998) Feasibility study of maximal program of groundwater banking. Natural Heritage Institute, San Francisco

Reichard EG, Li Z, Hermans C (2010) Emergency use of groundwater as a backup supply: quantifying hydraulic impacts and economic benefits. Water Res Res 46, W 09524

Scanlon BR, Reedy RC, Faunt CC, Pool D, Uhlman K (2016) Enhancing drought resilience with conjunctive use and managed aquifer recharge in California and Arizona. Environ Res Lett 11 (3):035013

Shah T, Molden D, Sakthivadivel R, Seckler D (2000) The global groundwater situation: overview of opportunities and challenges. Sri Lanka International Water Management Institute, Colombo

Smith A (1776) An inquiry into the nature and causes of the wealth of nation. Strahan and Cadell, London, pp 1–11

TBW (n.d.). Tampa Bay Water. https://www.tampabaywater.org/. Accessed 6 June 2020

Tsur Y (1990) The stabilization role of groundwater when surface water supplies are uncertain: the implications for groundwater development. Water Resour Res 26(5):811–818

Vaux HV Jr. (n.d.) Innovations in ground water management: the Arvin-Edison Water Storage District of California. California Institute for Water Resources, University of California Division of Agriculture and Natural Resources. http://ciwr.ucanr.edu/files/187211.pdf. Accessed 6 June 2020

WAKC (n.d.) Water banking. Water Association of Kern County. http://www.wakc.com/water-overview/sources-of-water/water-banking. Accessed 6 June 2020

Ward J, Dillon P (2009) Robust design of managed aquifer recharge policy in Australia. Facilitating recycling of stormwater and reclaimed water via aquifers in Australia. In: Milestone Report 3.1. CSIRO, Canberra

Ward J, Dillon P (2012) Principles to coordinate managed aquifer recharge with natural resources management policies in Australia. Hydrogeol J 20:943–956

Xiao M, Koppa A, Mekonnen Z, Pagán BR, Zhan S, Cao Q, Aierken A, Lee H, Lettenmaier DP (2017) How much groundwater did California's Central Valley lose during the 2012–2016 drought? Geophys Res Lett 44(10):4872–4879

Chapter 12
Groundwater Management and Adaptation Decision Making Process

12.1 Introduction

Increasing global temperatures will result in continued rising sea levels and a likely acceleration of the hydrological cycle through increases in evaporation and precipitation rates. Water resources will be locally impacted to varying degrees by spatial and temporal changes in the amount, seasonality, intensity and form of precipitation, melting of glaciers (and associated changes in surface water flows), changes in water demands, and salinization of coastal aquifers caused by rising sea levels (Vörösmarty et al. 2000; Bates et al. 2008; Kundzewicz et al. 2008). As local climate changes, water users and suppliers will have no choice but to adapt in some manner. Clearly, planned or proactive adaptation is preferred (where possible) as it can allow for a more orderly transition to future climate conditions and thus minimize disruptions in water supplies and impacts to infrastructure. However, it is unlikely that any type of adaptive action will be taken solely in response to anticipated climate change (Smit and Wandel 2006). The impacts of climate change on water resources in many areas will occur more slowly than increases in demands associated with population growth and economic development. Climate change is an insidious problem because the rate of change is so slow on an annual basis as to be essentially imperceptible to the average person, which may belie the urgency of taking adaptive actions. The goal is instead to have climate change adaptation mainstreamed as a consideration in the overall water planning process (Smit and Wandel 2006).

Much has been written on climate change adaptation including on how adaptation decisions *should* be performed. As a broad generalization, a consensus in the climate change literature is that adaptation decisions should be science-based, guided by input from the climate change research community, involve all stakeholders, and consider a broad range of issues including environmental impacts and social equity. A common issue raised is the need for greater communication (i.e., to bridge the gap) between the suppliers of climate information (i.e., the research

R. Maliva, *Climate Change and Groundwater: Planning and Adaptations for a Changing and Uncertain Future*, Springer Hydrogeology,
https://doi.org/10.1007/978-3-030-66813-6_12

community) and users of climate information (i.e., water users and suppliers; Feldman and Ingram 2009; Hewitt et al. 2017). With respect to water utilities, support and buy-in from agency executives and elected officials are often necessary for effective climate change adaptation, otherwise it will not be considered a priority (Ekstrom and Moser 2012; Kay et al 2018).

Climate change is just one of multiple factors involved in water supply and management decisions. Key practical issues that have received relatively little attention concerning adaptation to climate change with respect to groundwater (and water in general) are who actually makes water supply decisions, the role of government in water supply planning, how decisions are made, to what extent and how is climate change information received and acted upon, and the time frame used for planning and capital investments.

The basic water supply and climate change decision-making process is reviewed herein. Several regional examples are presented to illustrate to how groundwater is managed, and how sustainability and reliance are incorporated into the decision-making process in consideration of current and potential future climate conditions.

12.2 Water Supply Decision Makers

As described by Adger et al. (2005, p. 79)

> Adapting to climate change involves cascading decisions across a landscape made up of agents from individuals, firms and civil society, to public bodies and governments at local, regional and national scales, and international agencies.

Adger et al. (2005) made the key broad distinction between actions that involve creating policies or regulations to build adaptive capacity and actions that actually implement operational adaptation decisions. With respect to adaptation to climate change in the water sector, various governmental agencies have regulatory authority over water use, perform or sponsor research, provide educational services, and may build and operate large-scale water infrastructure. Individual water users, from households and farmers with self-supply wells, to industrial users, and on to large municipal utilities and regional wholesale water suppliers are principally responsible for obtaining their own water supplies.

Pahl-Wostl et al. (2007) argued that water resources management practices are currently undergoing a major paradigm shift from one “developed and implemented by experts using technical means based on designing systems that can be predicted and controlled” to a multiparty collaborative process in which stakeholder involvement has gained increasing importance. One argument presented for such an approach is based on democratic legitimacy, which “emphasizes that all those who are influenced by management decisions should be given the opportunity to actively participate in the decision-making process” (Pahl-Wostl et al. 2007).

Many others have similarly argued for increasing stakeholder involvement in the adaptation decision-making process. In practice, water management decisions in the water sector still largely fall within the realm of water users and suppliers, who are responsible for finding means to meet their own water needs (and those of their customers), operating under the purview of governmental agencies that have regulatory authority over water use. Opportunities for broad societal input in the decision-making process tend to be quite limited. The general public can influence water management decisions through the political process. Expenditures for large public capital projects usually require approval by elected officials. In some jurisdictions, there are avenues for stakeholders to object to or challenge the issuance of water use permits. Successful challenges to permits in the United States usually require that specific evidence be provided that the issuance of a permit would be contrary to federal or state water or environmental laws and regulations. General public outreach is important where public action is required with the notable example being the implementation of conservation measures.

Adaptation is typically implemented on the scale of an impacted system, which is regional at best, but mostly local (Klein et al. 2007). Most climate adaptation projects will occur at the local level as neither the federal nor state governments can, or will, solve local climate issues (Bloetscher et al. 2014). Water supply planning is normally performed by individual water users and water suppliers, which include water utilities, irrigation districts, and regional wholesale water suppliers. Water suppliers have varied forms of governance including private for-profit ownership, customer or member ownership, and public ownership as governmental agencies (e.g., public water utilities).

Water supply planning by water suppliers is typically performed by in-house technical staff. Depending upon the size of the supplier and in-house technical expertise, water supply planning is often supported by contracted external technical experts (e.g., hydrogeologists and engineers). There is considerable variation in how the decision-making process progresses upward from the technical level. Water supply plans developed by technical staff normally require approval by higher levels of utility or agency management. Implementation of public water supply plans involving expenditures typically require approval by representatives who are either directly elected (e.g., city or county commissioners), selected by residents or water users (e.g., members of an irrigation district), or otherwise are granted governmental authority to approve spending. In some jurisdictions, large water projects financed through the issuance of bonds require direct voter approval (e.g., via ballot propositions). In rural areas with decentralized water supplies, climate change adaptation decision making, to the limited extent that it is being performed, is the responsibility of individual users, subject to any governmental regulatory oversight.

The degree to which climate change adaptation is a consideration of a water utility or other public or private organization depends upon organizational priorities, which are largely driven by organizational leadership engagement. A core issue appears to be facilitating the translation of existing capacity into action (Burch

2010). If an organization's leadership is engaged in climate change issues, then adaptation issues will likely be given a high priority, at least as far investigating vulnerabilities. Burch (2010, p. 287) observed that

> The necessity of an explicitly articulated high-level directive, leadership that stimulates an organizational culture of innovation and collaboration, and the 'institutionalization' of climate change response measures within standard operating procedures emerged as crucial enablers of action. Addressing a lack of technical, financial, or human resources is less a matter of creating more capacity than of facilitating the effective use of existing resources.

Moser and Ekstrom (2010, p. 22029) similarly observed that

> Leadership can be critical at any stage in the adaptation process but maybe most important in initiating the process and sustaining momentum over time. When there is no mandate, law, job description, or public demand yet for adaptation planning, leaders are required to initiate the process. Importantly, we do not restrict this function to formal leadership and certainly not to just one individual, because some adaptation processes will go on for a long period; rather, we view it as a role that can be taken on by individuals in any position. Leadership can help overcome barriers, but lack of or ineffective leadership can also create some.

The role of government in adaptation to climate change in the water sector is ambiguous. Biesbroek et al. (2013, p. 1125) reported that their "results suggest that the role governments play is key in the governance of adaptation and understanding many of the reported barriers; governments at the local, regional, or (supra)national level are considered to constrain, enable, and stimulate adaptation." However, it has been the author's observation that despite recommendations for a broad participatory approach to climate change adaptation, the initial decision-making process in the water sphere tends to be siloed to a small number of technical professionals and that governmental involvement (except of course in the case of public utilities) is indirect.

Governmental involvement in the water supply planning process tends to be through regulatory constraints that often limit the water supply options available to users and suppliers. Governmental rules and policies may act to direct water supply planning in certain directions. Public policy also plays a role in facilitating adaptation to climate change through the provision of information on risks for private and public investments and decision making, and protecting public goods, such as habitats, species and culturally important resources (Adger et al. 2007). Much of climate change research, including the development of general circulation models (GCMs), is funded by governments. Governmental subsidies (including cost-sharing) and direct construction of surface water storage and conveyance infrastructure has profoundly impacted water supply options and development as a whole in some areas (e.g., the western United States).

Frederick (1997, p. 148) argued for more of a free market approach to adaptation, noting that "With or without the prospect of climate change, allowing supplies to flow to the highest value uses reduces the overall social risks and costs attributable to supply variability and uncertainity." He further noted (Frederick 1997, pp. 141–142) that "some adaptation would occur automatically without specifically designed policy changes. Farmers would adjust cropping patterns, planting and harvesting dates, and other practices as evidence of persistent changes in temperature and precipitation accumulates. And industry would conserve and develop new supplies in response to higher water costs or to interruptions."

Governments have the varied authority to declare water emergencies during droughts and to require utilities to develop drought preparedness plans. Engle (2013) compared the adaptive capacity of states and local community water systems (CWSs) in relation to extreme drought events, focusing on two contrasting cases, Arizona and Georgia, both of which have experienced intense, multi-year droughts over the past decade. Tension was observed between state and local influences on adaptive capacity, which Engle (2013) concluded was not simply a matter of states disengaging and removing themselves as barriers to CWSs. Engle (2013, p. 303) observed that "constraining CWS adaptive capacity might not be universally due to too much regulation, but in some instances too little regulation and direction". In the case of Georgia, Engle (2013, p. 303) concluded that "determining generic drought responses at the state-level, rather than locally relevant responses, fails to capture and mobilize the system specific adaptive capacity at the local level; instead relying on crisis management that is reactive rather than proactive." Not taking advantage of local adaptive capacity leaves CWSs feeling frustrated and overlooked, which might not bode well for future state attempts to motivate CWSs into action during drought emergencies (Engle 2013).

12.3 Water Supply Decision-Making Process

12.3.1 Basic Decision-Making Process

The processes by which adaptation decisions are actually made is a subject that has received relatively little attention in the climate change literature. A central element of the decision-making process for climate change adaptation is the time frame under consideration. For example, water supply plans prepared by utilities and municipalities in the state of Florida usually look forward 20-years, during which period the climate change signal will likely be within the current range of climate variability. Hence, climate change is seldom a primary consideration in water planning. However, elements of water supply plans that incorporate increasing demands from population growth and current climate variability can have the secondary benefit of increasing resilience to climate change. Longer planning horizons are used for expensive infrastructure that will have long operational lives and some water suppliers are planning now on how to meet the demands of its customers 50 years into the future (e.g., Polk Regional Water Cooperative, Florida). Water utilities serving coastal regions consider sea level rise in the location and construction of new infrastructure.

Mainstreaming refers to "the incorporation of climate change considerations into established or on-going development programs, policies or management strategies, rather than developing adaptation and mitigation initiatives separately" (Bockel 2009, slide 7). The objective of mainstreaming is to have risks (and opportunities) associated with climate change (and other environmental changes) actually addressed in decision making at some practical level (Smit and Wandel 2006). Practical climate change adaptation initiatives are invariably integrated with other programs (Smit and Wandel 2006). With respect to water, climate change, to the degree that it is formally considered, is usually integrated into existing water supply planning processes.

Water supply planning on all levels normally starts with a needs and sources evaluation. The amount of water needed over a planning period is first determined and then potential sources of water that can meet the identified current and future needs are evaluated. For municipal water utilities, demand estimation starts with projections of the population or the number of households or homes (equivalent residential units; ERUs) within a service area and historical water use rates per person or ERU. Most states in the United States have academic institutes or governmental agencies that collect and analyze state-wide and local demographic data. For example the mission of the Florida Bureau of Economic and Business Research of the University of Florida is to collect, analyze and generate economic and demographic data on Florida and its local areas, conduct economic and demographic research that will inform public policy and business decision making, and distribute data and research findings throughout the state and the nation (BEBR n.d.). Similar organizations and data sources exist in other states (e.g., Arizona Office of Economic Opportunity, California Department of Finance, Texas Demographic Center). Commercial, industrial, and recreational demands are similarly estimated from anticipated increases in the number and types of facilities and their historical use rates.

Agricultural water demands are estimated from existing and forecasted irrigated area, crop type, climate (rainfall and temperature) data, soil types, and irrigation method. The Blaney-Criddle method (and modifications thereof) is widely used to determine supplemental irrigation requirements (Blaney and Criddle 1950; Allen and Pruitt 1986; Brouwer and Heibloem 1986). Irrigation demands may be estimated for average and drought years. For example, demand calculations and water use permit allocations in South Florida are calculated using a modification of the Blaney-Criddle method for a 1-in-10 drought year with permit allocations including a maximum month or daily amount and an annual allocation.

Once demands are determined, evaluations are made of potential water sources that can reliably provide the required water. Supplies are evaluated based upon physical availability, regulatory availability (permittability), and costs. Additional groundwater and surface water resources may be physically available, but their use is not permitted because of concerns over impacts to the environment, existing users, and the resource (e.g., saline water intrusion). As a generalization, most large water users and suppliers are keenly aware of the water resources available in their area and any physical and regulatory restrictions on the uses of the various sources. Utilities that use groundwater for their supply normally look to install additional wells and pump more groundwater to meet projected increases in demand unless there are regulatory prohibitions against the additional pumping.

Water supply plans tend to be specifically designed to meet permitting criteria (i.e., to not exceed regulatory limits). Numerical groundwater modeling is now the standard tool for evaluating the impacts of proposed groundwater withdrawals and whether regulatory limits will be exceeded. For example, in many parts of Florida additional groundwater withdrawals are now often restricted due to concerns over the dehydration of wetlands and maintaining regulatory minimum flows and levels in springs and surface water bodies. Restrictions on water availability tend to drive the water supply planning process. The author has performed numerous modeling studies in support of water supply plans and permits for private and public clients. The selection of aquifers and locations of wells were often driven by modeling-based investigations to identify logistically preferred locations that meet all regulatory limits (e.g., maximum allowable drawdowns in nearby wetlands). Permits were eventually obtained because water supply plans and permit applications were specifically prepared that met all regulatory criteria for the issuance of permits. A permit application would not be submitted if the proposed use did not meet regulatory criteria and would therefore be denied.

Where supplies of existing water sources are found to be inadequate to meet demands during a planning horizon, then alternative water supply and demand reduction options are evaluated. Alternative water supplies options include wastewater reuse, desalination of seawater and brackish groundwater, and conjugate use of surface water and groundwater including managed aquifer recharge. Individual small-volume users might consider rainwater harvesting techniques.

The decision-making process for more complex water systems typically involves some form of multi-criteria decision making (MCDM). Multiple water supply scenarios are identified that could potentially meet projected demands. Scenarios may be first screened for fatal flaws, which are factors that render an option unviable. A common fatal flaw is permittability; projects that cannot get regulatory approval are nonstarters. Scenarios that are considered viable are then evaluated using multiple criteria which are weighted based on their relative importance. Project cost is usually given a high weighting. Other variables that are often considered include environmental impacts, public acceptance, land availability, and compatibility with existing infrastructure.

Rigorous analyses of project costs are performed using net present value type approaches that incorporate both initial construction costs and annual operation and maintenance costs. If all options provide the same benefits (i.e., the target supply of water), then a net present cost (or life-cycle cost) methodology is appropriate. Risk and uncertainty, such as associated with climate change, can be incorporated into cost-benefit analysis through the expected value analysis approach (Boardman et al. 1996).

12.3.2 Decision Support Systems

Sophisticated integrated water resources management (IWRM) decision support systems (DSSs) are available that can evaluate a wide variety of hydrological factors, managerial options, and climate-driven changes in demand (Harou et al.

2009; Yates and Miller 2011; Pulido-Velazquez et al. 2016). DSSs are computer programs that use analytical methods and data management systems to formulate and evaluate multiple solutions to problems. The basic components of DSSs are a graphic user interface (GUI), databases, and simulation and optimization models.

DSS models are composed of objects and processes (Jordan 2006). Objects are defined by their properties and behaviors and include various water supply components (e.g., reservoirs, conveyances, treatment systems, and wells) and demand components. Objects are linked by processes that quantitatively describe how objects communicate (interact) with other objects. Objects are related to other objects by inputs and outputs, such as water flow into and out of a reservoir (Jordan 2006). Analytical solutions or look-up tables are used to describe the relationship between objects, such as the relationship between well pumping and groundwater levels, or the costs of pumping water and demand components.

In water resources DSSs, simulation models may include groundwater and surface water flow models of varying degrees of sophistication. Socioeconomic management systems address human demands for water, and how water is stored, allocated, and delivered. The great advantage of DSSs is that they allow for the examination of numerous considerations involved in attempting to design and manage sustainable water resources systems (Loucks and Gladwell 1999) under both present and future climate conditions.

For example, Water Evaluation and Planning Version 21 IWRM model (WEAP21) can be used to evaluate water management scenarios with regard to factors such as supply sufficiency, costs, average costs of delivered water, hydropower production, and meeting in-stream flow requirements (Yates et al. 2005; Sieber and Purkey 2011). WEAP was also modified to produce a tool that can address long-term strategic planning for water utilities (Huber-Lee et al. 2006; Purkey and Huber-Lee 2006).The basic modeling procedure is to first develop a reference ("business as usual") scenario and then develop one or more "what if" policy scenarios with alternative assumptions about future developments. The alternative assumptions can be based on policies, costs, technological development, and other factors that may affect demand, pollution, supply, and hydrology (e.g., climate change). Stochastic simulations of various future climate conditions can be used to evaluate the robustness of water management and supply scenarios.

Gorelick and Zheng (2015, p. 3031) flagged the need for even more advanced simulation and optimization methods. Specifically that

> The compound challenges now faced by water planners require a new generation of aquifer management models that address the broad impacts of global change on aquifer storage and depletion trajectory management, land subsidence, groundwater-dependent ecosystems, seawater intrusion, anthropogenic and geogenic contamination, supply vulnerability, and long-term sustainability.

The main challenges associated with DSSs are gathering the correct information, formulating proper questions, and interpreting the output (Purkey and Huber-Lee 2006). The accuracy of the simulations will depend on the underlying assumptions and the quality of the data used in the simulations.

An important question is whether the effort (and cost) to develop DSS models is commensurate with their benefits. Does a DSS provide new insights or is it used to validate what is already known or could be determined using simpler methods (Maliva and Missimer 2012)? For relatively simple water supply systems, the answers (as far as the optimal system design and operation) may already be known or could be readily determined by decision makers without developing DSSs. In some areas where DSSs can potentially bring great value for managing complex water systems (e.g., Southern California), they have already been developed.

Nevertheless, DDSs can bring value by identifying unexpected vulnerabilities. The availability of easy-to-use DSSs that model surface water and groundwater systems will likely facilitate the move of the technology from academia further into practice (Pulido-Velazquez et al. 2016). DSSs can also be used to graphically illustrate results to non-technical audiences. While DSSs can help educate the public, there is the danger with modeling systems with impressive graphics that their results may have a greater acceptance (particularly by lay people) than is warranted by the accuracy of the raw data used and underlying assumptions and thus model outputs.

12.4 General Public Engagement

Water supply decisions by utilities and regional water providers are usually made under the public radar. The apparent lack of visibility in the process stems much more from a lack of public interest than utility obfuscation. Many larger utilities and regional suppliers engage in active public outreach to educate the public as to their operations, which include websites, tours, public meetings, presentations at conferences and meetings, and educational centers. Under sunshine laws, certain proceedings of government agencies are open or available to the public, which include board of director meetings of water conservation districts and regional water suppliers, and city and county commission meetings. Members of the public may have a limited opportunity to express their opinions. However, public meetings are often discussions of already largely decided upon plans, and budgetary and administrative issues, rather than being a forum of potentially decision-influencing debate.

It has been the author's observation, that public utilities and regional water suppliers are proud of the work they do and are open to educating the public about their operations. However, most customers tend to have little interest in where their water comes from so long as the water supply is safe and reliable, and their rates are not increasing. Even small increases in water rates can elicit strong public opposition, which is ironic considering the enormous societal importance of a safe and reliable water supply and amount of money households tend to spend on other arguably less critical services (e.g., mobile phone service).

Strong public interest has been aroused for projects with perceived or real environmental impacts and for projects involving potable reuse. A textbook example of the latter is the "toilet-to-tap" campaign against a potable reuse project in San Diego, California, in the late 1990s. Charges of environmental racism were raised in that poor people were claimed to be required to drink the effluent of affluent people (Penner and Cavanaugh 2010). Growing public opposition, including by a group self-named the "Revolting Grandmas," ultimately lead to the city council voting to halt the recycled water project (Penner and Cavanaugh 2010; Kenney 2019). However, after a successful educational program, public support for potable reuse in San Diego reached 70% and the city council voted unanimously in 2014 to proceed with its "Pure Water" project (Wiseman 2015; Smith 2017).

Public engagement is much more critical on the demand side of water management where changes in public behavior are sought. Conservation is promoted through public outreach and education, economic incentives, and legal and regulatory means (Sect. 10.2) Changes in public behavior and increases in the water efficiency of appliances and plumbing fixtures can be effective in reducing residential (domestic) water use. National average per capita residential water use in the United States, for example, peaked between 1985 and 2005 at about 100 gallons (378 L) per capita per day (Donnelly and Cooley 2015) and has subsequently declined due to efficiency improvements, falling to about 82 gallons (310 L) per day in 2015 (Dieter and Maupin 2017).

12.5 Engagement of Decision-Makers with the Climate Change Research Community

There is a wide range in the degree of engagement between decision makers involved in water resources planning and the global climate change research community. In developing areas facing water scarcity under currently climate conditions and increasing demands driven by population growth, possible future climate changes are far from a primary concern. Providing water for the present is an all-consuming endeavor. One senses from a review of the climate change literature and mass media reporting that there is frustration on part of the climate change research community that decision makers are not adequately heeding their results, while decision makers may complain that climate change research results are too uncertain to be actionable.

Climate change has also become a political issue, which has perhaps occurred to the greatest degree in the United States where even the reality of anthropogenic climate change is denied by many representatives of a major political party. President Donald Trump in 2019 withdrew the United States from the Paris Climate Agreement. Climate change had been politicized in Florida with a former governor even reportedly having banned the use of the words "climate change" and "global warming" in governmental correspondences (Korten 2015). Sea-level rise was to be referred to as "nuisance flooding" (Korten 2015). It must be stressed again that climate change adaptation is performed largely on the local scale (level of impacted parties). Irrespective of state-level and national-level politics, local communities in Florida that are most vulnerable to climate change were and continue to be actively engaged in assessing their vulnerabilities to climate change, particularly sea level rise, and at least investigating options to increase their resilience. In a marked change in attitude, Florida lawmakers created a new statewide Office of Resiliency in 2019 and established a task force to investigate how best to protect the state's 2173 km (1350 miles) of coastline from rising sea levels.

There is an enormous academic research interest in climate change in general and specifically with water-related issues. NOAA's Regional Integrated Sciences and Assessments (RISA) program supports research teams with the goal of expanding and building the nation's capacity to prepare for and adapt to climate variability and change in general (NOAA n.d.). The ten RISA teams focus on creating lasting relationships with decision makers from the public and private sectors, including local, regional, and state governments, federal agencies, tribal governments, utilities, the business community, and national and international non-profit organizations (NOAA n.d.).

Many universities have established internal climate and water research institutes and programs, such as, for a few examples, Water in the West (Stanford University), University of Florida Water Institute, Water School (Florida Gulf Coast University), Water and Climate Impacts Research Centre (University of Victoria), Texas Water Resources Institute (Texas A&M University), Water Resources Institute (California Stated University—San Bernardino) and Pacific Northwest Climate Impacts Research Consortium (Oregon State University). Multiple university and governmental climate change research consortiums have been established (e.g., Research Coalition on Climate, Southwest Climate Adaptation Science Center, Arizona Water Institute; Goyder Institute for Water Research—South Australia). The Florida Water and Climate Alliance (FloridaWCA), which is facilitated by the University of Florida Water Institute, is a "stakeholder-scientist partnership committed to increasing the relevance of climate science data and tools at relevant time and space scales to support decision-making in water resource management, planning and supply operations in Florida" (FloridaWCA 2020). The FloridaWCA collaborators include six major public water supply utilities, three water management districts in Florida, local government representatives, and several academic organizations (FloridaWCA 2020).

Climate change issues are frequently the subject of newspaper articles and television news stories and are addressed in annual local, statewide, and national water and utilities conferences, so general awareness of climate change among decision makers in developed countries is generally high. Water utilities and suppliers tend to be aware of conditions that could impact their water supplies and demands. Surveys and interview studies have confirmed that water utility professionals, in general, are aware of climate change issues but some were unsure as to how to best interact with climate change scientists and incorporate climate change into their planning process (e.g., White et al. 2008; Danilenko et al. 2010; World

Bank 2010; Economist Intelligence Unit 2012; Moser and Ekstrom 2012; Archie et al. 2014; Baker et al. 2018; Raucher et al. 2018; Kay et al. 2018).

In an international survey of utilities by Danilenko et al. (2010), more than 30 percent of the utilities raised concerns about the limited accuracy of climate modeling for their long-range planning, and the vast majority of utilities' responses to climate change have relied on no-regrets short-term strategies to reduce water consumption, improve watershed management, and reduce non-revenue water losses, instead of comprehensive planning for the long-term consequences of climate change (World Bank 2010).

The degree of engagement of water utilities and water suppliers with the climate change research community tends to depend upon their perceived vulnerability to climate change and sea level rise and the size of the utility, with larger utilities and regional water suppliers having greater technical and financial resources. Major water suppliers in the United States are involved in national and regional collaborative efforts to share experiences and expertise and to interact with the climate change research community. The Water Utility Climate Alliance (WUCA) is an organization composed of twelve of the nation's largest water providers, "that is dedicated to enhancing climate change research and improving water management decision-making to ensure that water utilities will be positioned to respond to climate change and protect our water supplies" (WUCA 2020). The WUCA member agencies, which supply drinking water to more than 50 million people are

- Austin Water
- Central Arizona Project
- Denver Water
- Metropolitan Water District of Southern California
- New York City Department of Environmental Protection
- Philadelphia Water Department
- Portland Water Bureau
- San Diego County Water Authority
- San Francisco Public Utilities Commission
- Seattle Public Utilities
- Southern Nevada Water Authority
- Tampa Bay Water.

Major water utilities, suppliers, regulatory and water management districts, and communities in the United States have collaborated with or contracted governmental agencies (e.g., U.S. Geological Survey), university researchers, and consultants to perform climate-change related studies.

A survey of utilities in California indicated that "It is clear that the state takes climate change and adaptation seriously" (Ekstrom et al. 2017, p. 487). However, it was noted that small water systems are largely left to their own choices and devices on whether and how to adapt and that support is especially needed for smaller utilities (Ekstrom et al. 2017). As a generalization, large water utilities, suppliers, and users in developed countries have ample access to technical resources on climate change and adaptation options and their technical staff have at least a general awareness of potential impacts in their areas of operation. Smaller utilities, communities, and users may have much lesser degrees of engagement (if any) with the climate change research community, but do not operate in a vacuum and, based on survey studies, have some awareness of relevant climate change issues.

12.6 Decision-Making Planning Horizon

Climate change is a gradual process and its relevance to human systems depends upon the rate of change and the planning horizon. Planning horizons for water users and suppliers may be regulatorily dictated through permit durations, related to the operational life of infrastructure, or specified for mandated water supply, land use, and development plans. Specific aspects of climate change may be either explicitly considered in the planning and design process or implicitly considered through the design of robust infrastructure. The prime example of the former is coastal construction for which it would be grossly negligent to not specifically consider the impacts of future higher sea levels. In the case of water supply, designing and constructing infrastructure that is robust to non-climate change-driven variability in supplies and demands is an autonomous adaptation to climate change in which the impacts of specific future anthropogenic climate change conditions are not necessarily considered.

A key issue with respect to planning of water supply systems for climate change is whether the magnitude and direction of the anthropogenic climate change signal falls outside of the range of current variability. Where the rate of climate change is slow, and the planning horizon is short, creeping maladaptation may occur where climate change is occurring but is not adequately considered in the planning process. It is also important to appreciate that most professionals involved in planning of water infrastructure are not so myopic as to be unaware as to the potential impacts of climate change beyond any formal planning horizon.

Water supply decision making occurs in several contexts. Water demand projections and proposed water sources and extraction rates are an integral part of the water use planning and permitting process. Identification of future long-term water supplies may be part of land use and development planning processes. Municipal capital improvement plans may include expenditures for water supply, treatment, and distribution infrastructures. Follows is a summary of water planning decision-making horizons in some states in the United States facing water scarcity from aridity and rapid population growth. As a general caveat, even though water utilities and regional suppliers may operate under a prescribed planning horizon, it should not be assumed that longer time periods are not being considered where appropriate.

12.6.1 Florida

Water use is Florida is governed under the reasonable use doctrine and regulated by five regional water management districts (Districts) that have the responsibility of addressing issues such as water supply, drainage/flood protection, water quality, and protection of natural resources (Davis et al. 2018). With the exceptions of self-supplied domestic and home irrigation uses and water used for firefighting, a consumptive (water) use permit (CUP) is required for the use groundwater and surface water. Permit applicants must demonstrate that the proposed use of water is reasonable and beneficial, will not interfere with any presently existing legal use of water, and is consistent with the public interest (Florida Statutes FS 373.223). CUP water allocations are not owned by the applicant and permits have a finite duration. CUPs normally have a duration of 20 years if the applicant demonstrates reasonable assurance that the proposed use meets all the conditions for issuance. Longer duration permits may be obtained for alternative water supply projects (e.g., groundwater desalination systems).

Each District is required to prepare regional water supply plans (RWSPs) that must be based on at least a 20-year planning period and be reviewed every five years. The RWSPs are required to provide estimates of future water demands and a determination of whether existing sources of water are adequate to supply all existing and future reasonable-beneficial uses and to sustain the water resources and related natural systems for the planning period. RWSPs are also required to include "a list of water supply development project options, including traditional and alternative water supply project options that are technically and financially feasible, from which local government, government-owned and privately owned utilities, regional water supply authorities, multijurisdictional water supply entities, self-suppliers, and others may choose for water supply development" (Florida Statutes 373.709). Recent RWSPs address climate change with varying degrees of detail and specificity. The general climate change-related elements included in some recent RWSPs include engagement with the climate change research community, monitoring, and encouraging the development of diversified and redundant water supplies that have greater resilience to climate changes.

Local governments are required to prepare comprehensive plans, which provide the policy foundation for local planning and land use decisions on capital improvements, conservation, intergovernmental coordination, recreation, open space, future land use, housing, transportation, coastal management (where applicable), and public facilities including water supply (Carriker 2006). Water supply elements of comprehensive plans are required to evaluate current and projected industrial, agricultural, and potable water needs and sources for at least a 10-year period. Within 18 months after a District governing board approves an updated RWSP, the water supply element of comprehensive plans must be updated to incorporate an alternative water supply project or projects from those identified in the RWSP or alternative projects proposed by the local government. Comprehensive plans are required to contain a capital improvements element designed to consider the need for and the location of public facilities over at least a 5-year period and include a schedule of when the improvements are needed and their estimated costs and funding options. Capital improvement plans are updated annually.

The formal planning horizon for water supply in Florida is mainly 10–20 years, during which time increases in demands associated with population growth are expected to greatly exceed climate change impacts. It is important to stress that although the formal water supply planning horizon may be 20 years, vulnerable larger utilities and regional water suppliers are looking well beyond that time period as appropriate.

12.6.2 Texas

The Texas Water Development Board (TWDB) was established in 1957 to develop water supplies and prepare plans to meet the state's future water needs (TWDB 2018). The state of Texas was geographically divided into 16 regional water planning areas, each of which has a planning group, made up of about 20 members, that represents a variety of interests, including agriculture, industry, the environment, municipalities, business, water districts, river authorities, water utilities, counties, groundwater management areas, and power generation. Each planning group is tasked with developing regional water plans every five years that, among a number of items, are required to (1) quantify current and projected population and water demand over a 50-year planning horizon, (2) evaluate and quantify current water supplies, surpluses, and needs, (3) evaluate water management strategies and prepare plans to meet identified needs, and (4) evaluate the impacts of water management strategies on water quality and agricultural and natural resources, as well as the water resources of the state (TWDB 2018).

At the end of each five-year regional water planning cycle, TWDB staff compiles information from the approved regional water plans and other sources to develop the State Water Plan (SWP). The planning horizon of the SWP is also 50 years and the most recent SWP (TWDB 2017) addresses demands and sources through year 2070. The 2017 SWP does not specifically mention climate change but notes that

> Texas' water plans are based on benchmark drought of record conditions using historical hydrological data. While we recognize that the full sequence of hydrologic events in our history will never be repeated exactly, the droughts that have occurred have been of such severity that it is reasonable to use them for the purpose of planning. There are currently no forecasting tools capable of providing reliable estimates of changes to future water resources in Texas at the resolution needed for water planning. In order to provide the best available, actionable science, grounded in historical data and patterns, the TWDB continues to collect data and consider potential ways to improve estimates of water supply reliability in the face of drought. (TWDB 2017, p. 75)

Surface water use is Texas is tightly controlled under the prior appropriation doctrine. Groundwater use, on the contrary, has been historically much more loosely governed under the rule of capture (doctrine of absolute ownership), in which a landowner owns the groundwater underlying his or her property and can withdraw as much water as desired without any liability for impacts to adjacent and more distant landowners. Absolute ownership has great disadvantages in that it offers no encouragement for conservation and provides for no long-term security as one's water supply can be compromised by someone installing nearby wells or through cumulative overdraft of an aquifer.

Texas Senate Bill SB-2 (Legislative Session 77R), passed in 2001, holds that groundwater conservation districts (GCDs) are the state's preferred method of groundwater management through rules developed, adopted, and promulgated by each district. The principal power that GCDs have to prevent the waste of groundwater is to require that all wells (with certain exceptions) be registered and permitted. Permitted wells are subject to GCD rules governing spacing, production, drilling, equipping, and completion or alteration (TCEQ n.d.). GCDs are also mandated to develop comprehensive management plans and to adopt the necessary rules to implement the management plans (TCEQ n.d.).

Texas House HB 1763 (Legislative Session 79R), enrolled in 2005, transitioned from each GCD defining their own groundwater availability, which was included in their groundwater management plans, to a more regionalize approach were the GCDs in each groundwater management area jointly determine the "desired future conditions" for their groundwater resources (Mace et al. 2008). Desired future conditions are "the desired, quantified conditions of groundwater resources (such as water levels, water quality, spring flows, or volumes) at a specified time or times in the future or in perpetuity" (Mace et al. 2008, p. 3). The specified time horizon of desired future conditions extends through at least the period that includes the current planning horizon for the development of regional water plans (TWDB n.d.), which is 50 years. Examples of possible desired future conditions given by Mace et al. (2008, p. 3) are "(1) water levels do not decline more than 100 feet in 50 years, (2) water quality is not degraded below 1000 milligrams per liter of total dissolved solids for 50 years, (3) spring flow is not allowed to fall below 10 cubic feet per second in times during the drought of record for perpetuity, and (4) 50% of the water in storage will be available in 50 years."

Once the "desired future conditions" are determined, then the TWDB through groundwater modeling determines the "managed available groundwater" (i.e., allowed groundwater pumping), which is to be used by the GCDs in their groundwater management supply plans and by the regional water planning groups in their regional water plans (Mace et al. 2008).

Water planning is Texas has a 50-year horizon. Historically climate change has been a politically charged issue in the state, which is the headquarters for much of the nation's oil and gas industry. Nevertheless, the potential impacts of future drier climate change are implicitly considered in water planning through drought preparedness.

12.6.3 Arizona

Surface water is Arizona, and most of the western United States, is governed under the doctrine of prior appropriation in which earlier water rights have priority to available water over later granted rights. Groundwater, on the contrary, is governed under the doctrine of reasonable use, which as applied in Arizona allows landowners to withdraw groundwater for reasonable and beneficial use on their property with neighboring landowners having no claim for damages even if the groundwater withdrawals adversely affect water levels under their property (Staudenmaier 2006). The arid climate of Arizona results in high irrigation water requirements and low recharge rates, which combined caused aquifer depletion and large drawdowns.

The Groundwater Management Act of 1980 (GMA) enacted a comprehensive statutory scheme to regulate groundwater rights and uses in Arizona (Staudenmaier 2006). Most of the regulatory provisions of the GMA apply only within the five designated "Active Management Areas" (AMAs), which encompass the areas of the State where the most significant groundwater use was occurring and groundwater overdraft was greatest. Four initial AMAs were established in the GMA: Phoenix AMA (Phoenix metropolitan area, Tucson AMA (Tucson metropolitan area), Prescot AMA, and the Pinal AMA, which encompasses an area of large-scale agricultural production between Phoenix and Tucson. The fifth AMA, the Santa Cruz AMA, was created in 1994 by splitting off the southern portion of the original Tucson AMA. The AMAs include 80% of Arizona's population and 70% of the state's groundwater overdraft (ADWR n.d.a).

Arizona Groundwater Management Code contains the following key provisions (ADWR n.d.a):

- establishment of a program of groundwater rights and permits
- a provision prohibiting irrigation of new agricultural lands within AMAs
- requires preparation of a series of water management plans for each AMA designed to create a comprehensive system of conservation targets and other water management criteria
- development of a program requiring developers to demonstrate a 100-year assured water supply for new growth
- a requirement to meter/measure water pumped from all large wells
- a program for annual water withdrawal and use reporting.

A fundamental requirement of the GMA is that, with the exception of groundwater withdrawn from exempt wells, groundwater use within AMAs is generally subject to water conservation and management standards promulgated by Arizona Department of Water Resources (ADWR; Staudenmaier 2006; ADWR n.d.a). Management goals were established for each AMA, which for the Phoenix, Tucson and Prescott AMAs is achieving by 2025 "safe-yield," defined as achieving and maintaining a long-term balance between annual groundwater withdrawals and natural and artificial recharge in the AMA (Staudenmaier 2006). The management goal for the Pinal AMA is referred to as "planned depletion" because it allows continued access to groundwater for both irrigation and increasing amounts of non-irrigation uses while water tables in parts of the AMA continue to decline (Staudenmaier 2006).

The ADWR is required to publish a series of management plans that impose water conservation measures on groundwater users in each AMA. Each management plan governs a period of ten years, except for the fifth management plan, which will apply to the years 2020 through 2025 when the Phoenix, Tucson and Prescott AMAs are supposed to achieve their safe-yield goals (Staudenmaier 2006). The ADWR performs groundwater modeling to support long-term groundwater management decisions.

The Assured Water Supply (AWS) program requires for real estate developments that involve subdivision of land into six or more lots, a demonstration that necessary water supplies to serve all current and future water demands of the development have been secured for a period of 100 years. The AWS demonstration can be made by obtaining a "commitment to serve" from a designated water provider (city, town or private water company), or by submitting the necessary information to obtain a "certificate of assured water supply" specific to the individual development. Demonstration of assured water supplies are based on five criteria (ADWR n.d.b): physical water availability, continuous water availability, legal water availability, water quality, and financial capability.

The GMA is an admirable effort at implementing long-term planning of groundwater use in Arizona and has been effective is greatly reducing aquifer overdraft rates, but it is open to question whether safe yield can be fully achieved and maintained in the AMAs. Aquifer augmentation (managed aquifer recharge) has been increasingly relied upon to balance groundwater pumping. It is recognized that augmentation efforts depend upon the continued availability of renewable water for replenishment and that the primary source, Colorado River water conveyed by the Central Arizona Project (CAP), is vulnerable to climate change (Choy 2015; ADWR 2016).

12.6.4 California

Groundwater use in California is governed under the correlative rights doctrine in which groundwater rights are proportional to the landowner's share of the land overlying the aquifer. The water is considered to be owned by the landowner, but its use is controlled by the state. In the event of shortages, caused by droughts or excessive aquifer drawdowns, all water users share in a pro-rated reduction in water use related to their land owned.

Groundwater management in California has historically been highly decentralized with local groundwater use managed by local entities (e.g., irrigation districts) established across the state for this purpose. When groundwater users cannot collectively agree on the management of pumping in a basin, parties within the basin can file a lawsuit asking the court to define the rights of the various entities using groundwater resources, which is referred to as adjudication. Adjudications can cover an entire basin, part of a basin, or a group of basins and all non-basin locations in between (CDWR n.d.b). The court will decide who are the extractors (owners), how much groundwater those well owners can extract, and assign a "Watermaster" who will be responsible to ensure that the basin or portion of the basin is managed in accordance with the court's decree (CDWR n.d.b). Watermasters have the responsibility of ensuring that groundwater pumping conforms to the limits defined by the court and must periodically report to the court. Groundwater overdraft has been severe in southern California with resulting large declines in water levels and associated land subsidence.

The Sustainable Groundwater Management Act (SGMA) is a three-bill package passed by the California state legislature and signed into California state law by Governor Jerry Brown in September 2014. The SGMA requires governments and water agencies of high and medium priority basins to halt overdraft and bring groundwater basins into balanced levels of pumping and recharge. Groundwater sustainability agencies for each medium- or high-priority basin are required to develop and implement a groundwater sustainability plan. Under SGMA, the included basins should reach sustainability within 20 years of implementing their sustainability plans. For critically over-drafted basins, that will be 2040. For the remaining high and medium priority basins, 2042 is the deadline (CDWR n.d.a). Sustainable yield is defined in the California Water Code (Division C, Part 2.74, Chap. 2, 10721) as "the maximum quantity of water, calculated over a base period representative of long-term conditions in the basin and including any temporary surplus, that can be withdrawn annually from a groundwater supply without causing an undesirable result," with undesirable results including "Chronic lowering of groundwater levels indicating a significant and unreasonable depletion of supply if continued over the planning and implementation horizon."

The SGMA requires that the State of California "evaluate" or "assess" a groundwater sustainability plan, rather the using the term "approve," which implies maintenance of local control (Aladjem and Nikkel 2016). Aladjem and Nikkel (2016, p. 5) noted that "Some fear that the state's statutory authority to evaluate, and the regulatory power to impose exacting requirements on local agencies, will render SGMA's local control a myth."

The SGMA has a very aggressive 20-year planning and implementation horizon. The SGMA allows for an extension of up to 5 years beyond the 20-year sustainability timeframe upon a showing of good cause and granting of a second extension of up to five years upon a showing of good cause if the groundwater sustainability agency has begun implementation of the work plan.

12.7 Summary

Water planning is largely the responsibility of water users and suppliers with the decision making as to supply options in the hands of a relatively small number of professionals, subject to the approval upper management in the private sector and elected or appointed governmental boards and directors in the public sector. The results of surveys and anecdotal observations is that climate change has become such a high-profile issue that technical staff of water utilities and regional water suppliers have at least a general awareness of potential impacts in their areas. Large utilities and water suppliers often have some engagement with climate researchers either through direct contacts or through participation in national, state, or regional groups. The impacts of sea level rise on coastal groundwater resources tend to receive far greater attention than climate-change induced impacts to recharge rates and water demands because the direction and historical rates of change are known and potential vulnerabilities are more obvious, whereas there is often considerable uncertainty in projections of local changes in precipitation patterns and aquifer recharge.

With respect to groundwater, the primary water management challenge in most areas is addressing current overdraft under existing climate conditions and identifying new supplies where current withdrawals are at, or close, to what are considered sustainable limits. Projected anthropogenic climate change impacts with respect to groundwater for the common 20–50-year planning horizons are likely small relative to natural climate variability and increases in demands due to population growth. Surface water supplies, particularly in snowmelt dominated hydrologic regimes, generally have a greater short- and medium-term vulnerability to climate change and natural variability because of their lesser storage volumes than groundwater systems. Nevertheless, planning for responding to future droughts, including those greater than recent historical experiences, is an implicit adaptation to anthropogenic climate change.

References

Adger WN, Arnell NW, Tompkins EL (2005) Successful adaptation to climate change across scales. Glob Environ Change 15(2):77–86

Adger WN, Agrawala S, Mirza MMQ, Conde C, O'Brien K, Pulhin J, Pulwarty R, Smit B, Takahashi K (2007) Assessment of adaptation practices, options, constraints and capacity. In: Parry ML, Canziani OF, Palutikof JP, van der Linden PJ, Hanson CE (eds) Climate change 2007: impacts, adaptation and vulnerability. Contribution of working group II to the fourth assessment report of the Intergovernmental Panel on Climate Change. Cambridge University Press, Cambridge UK, pp 717–743

ADWR (2016) Fourth management plan tucson active management area, 13 May 2016. Arizona Department of Water Resources. https://new.azwater.gov/ama/management-plan/fourth-management-plan. Accessed 7 June 2020

ADWR (n.d.a) Overview of the Arizona groundwater management code. Arizona Department of Water Resources. http://www.azwater.gov/AzDWR/WaterManagement/documents/Groundwater_Code.pdf. Accessed 7 June 2020

ADWR (n.d.b) Assured and adequate water supply. Arizona Department of Water Resources. https://new.azwater.gov/aaws. Accessed 7 June 2020

Aladjem DRE, Nikkel ME (2016) California groundwater management: laboratories of local implementation or state command and control? Environ Law News 25(2):3–7

Allen RG, Pruitt WO (1986) Rational use of the FAO Blaney-Criddle formula. J Irrig Drainage Eng 112(2):139–155

Archie KM, Dilling L, Milford JB, Pampel FC (2014) Unpacking the 'information barrier': comparing perspectives on information as a barrier to climate change adaptation in the interior mountain West. J Environ Manage 133:397–410

Baker Z, Ekstrom J, Bedsworth L (2018) Climate information? Embedding climate futures within temporalities of California water management. Environ Sociol 4(4):419–433

Bates B, Kundzewicz Z, Wu S, Palutikof JP (eds) (2008) Climate change and water. IPCC technical paper VI. Intergovernmental Panel on Climate Change. IPCC Secretariat, Geneva

BEBR (n.d.) Bureau of economic and business research. https://www.bebr.ufl.edu/. Accessed 7 June 2020

Biesbroek GR, Klostermann JE, Termeer CJ, Kabat P (2013) On the nature of barriers to climate change adaptation. Reg Environ Change 13(5):1119–1129

Blaney HF, Criddle WD (1950) Determining water requirements in irrigated area from climatological irrigation data. US Department of Agriculture, Soil Conservation Service, Washington D.C.

Bloetscher F, Hammer NH, Berry L (2014) How climate change will affect water utilities. J Am Water Works Assoc 106(8):176–192

Boardman A, Greenberg D, Vining A, Weimer D (1996) Cost-benefit analysis: concepts and practice. Prentice Hall, Upper Saddle River

Bockel L (2009) How to mainstream climate change adaptation and mitigation into agriculture policies. In: EASYPol. Food and Agriculture Organization of the United Nations. http://www.fao.org/docs/up/easypol/778/mainstream_clim_change_adaptation_agric_policies_slides_077en.pdf. Accessed 6 June 2020

Brouwer C, Heibloem M (1986) Irrigation water management: irrigation water needs. In: FAO irrigation water management training manual no. 3. Food and Agriculture Organization of the United Nations, Rome

Burch S (2010) Transforming barriers into enablers of action on climate change: insights from three municipal case studies in British Columbia, Canada. Glob Environ Change 20(2):287–297

Carriker R (2006) Comprehensive planning for growth management in Florida (Document FE642). University of Florida IFAS Extension, Gainesville

Choy J (2015) 7 lessons in groundwater management from the Grand Canyon State. Sanford University, Water in the West. http://waterinthewest.stanford.edu/news-events/news-press-releases/7-lessons-groundwater-management-grand-canyon-state. Accessed 7 June 2020

CDWR (n.d.a) SGMA groundwater management. https://water.ca.gov/Programs/Groundwater-Management/SGMA-Groundwater-Management. Accessed 7 June 2020

CDWR (n.d.b) Adjudicated areas. https://water.ca.gov/Programs/Groundwater-Management/SGMA-Groundwater-Management/Adjudicated-Areas. Accessed 7 June 2020

Danilenko A, Dickson E, Jacobsen M (2010) Climate change and urban water utilities: challenges and opportunities (water working notes no. 24). World Bank, Washington, D.C.

Davis J, Borisova T, Olexa MT (2018) An overview of Florida water policy framework and institutions. University of Florida IFAS Extension, Gainesville

Dieter CA, Maupin MA (2017) Public supply and domestic water use in the United States, 2015. U.S. Geological Survey Open-File Report 2017–1131

Donnelly K, Cooley H (2015) Water use trends. Pacific Institute, Oakland, California

Economist Intelligence Unit (2012) Water for all? A study of water utilities' preparedness to meet supply challenges to 2030. The Economist, Economist Intelligence Unit, London
Ekstrom J, Moser S (2012) Climate change impacts, vulnerabilities, and adaptation in the San Francisco Bay area. Publication number CEC-500-2012-071. California Energy Commission, Sacramento
Ekstrom JA, Bedsworth L, Fencl A (2017) Gauging climate preparedness to inform adaptation needs: local level adaptation in drinking water quality in CA, USA. Clim Change 140(3–4):467–481
Engle NL (2013) The role of drought preparedness in building and mobilizing adaptive capacity in states and their community water systems. Clim Change 118(2):291–306
Feldman DL, Ingram HM (2009) Making science useful to decision makers: climate forecasts, water management, and knowledge networks. Weather Clim Soc 1:9–21
FloridaWCA (2020) The Florida Water & Climate Alliance (FloridaWCA). http://www.floridawca.org/. Accessed 12 Jan 2020
Frederick KD (1997) Adapting to climate impacts on the supply and demand for water. Clim Change 37(1):141–156
Gorelick SM, Zheng C (2015) Global change and the groundwater management challenge. Water Resour Res 51:3031–3051
Harou JJ, Pulido-Velazquez M, Rosenberg DE, Medellín-Azuara J, Lund JR, Howitt RE (2009) Hydro-economic models: concepts, design, applications, and future prospects. J Hydrol 375(3–4):627–643
Hewitt CD, Stone RC, Tait AB (2017) Improving the use of climate information in decision-making. Nat Clim Change 7(9):614–616
Huber-Lee A, Schwartz C, Sieber J, Goldstein J, Purkey D, Young C, Soderstrom E, Henderson J, Raucher R (2006) Decision support systems for sustainable water supply planning. AWWA Research Foundation, Denver, Colorado
Jordan D (2006) Understanding decision support systems: a tool for analyzing complex systems. Southwest Hydrol 5(4):16–34
Kay R, Scheuer K, Dix B, Bruguera M, Wong A, Kim J (2018) Overcoming organizational barriers to implementing local government adaptation strategies (CNRA-CCA4-2018-005), California's fourth climate change assessment. California Natural Resources Agency, Sacramento
Kenney S (2019) Purifying water: responding to public opposition to the implementation of direct potable reuse in California. J UCLA J Environ Law and Policy 37(1):85–122
Klein RJT, Huq S, Denton F, Downing TE, Richels RG, Robinson JB, Toth FL (2007) Inter-relationships between adaptation and mitigation. In: Parry ML, Canziani OF, Palutikof JP, van der Linden PJ, Hanson CE (eds) Climate change 2007: impacts, adaptation and vulnerability. Contribution of working group II to the fourth assessment report of the Intergovernmental Panel on Climate Change. Cambridge University Press, Cambridge UK, pp 745–777
Korten T (2015) In Florida, officials ban term 'climate change'. Florida Center for Investigative Reporting. https://fcir.org/2015/03/08/in-florida-officials-ban-term-climate-change/. Accessed 8 Feb 2020
Kundzewicz ZW, Mata LJ, Arnell NW, Döll P, Jimenez B, Miller K, Oki T, Şen Z, Shiklomanov I (2008) The implications of projected climate change for freshwater resources and their management. Hydrol Sci J 53(1):3–10
Loucks DP, Gladwell JS (1999) Sustainability criteria for water resources systems. Cambridge University Press, Cambridge, UK
Mace RE, Petrossian R, Bradley R, Mullican WF, Christian L (2008) A streetcar named desired future conditions: the new groundwater availability for Texas (revised). In: Proceedings state bar of Texas, the changing face of water rights in Texas, 8–9 May 2008, Bastrop, Texas. http://www.twdb.texas.gov/groundwater/docs/Streetcar.pdf. Accessed 7 June 2020

Maliva RG, Missimer TM (2012) Arid land water evaluation and management. Springer, Berlin

Moser SC, Ekstrom JA (2010) A framework to diagnose barriers to climate change adaptation. Proc Natl Acad Sci 107(51):22026–22031

Moser SC, Ekstrom JA (2012) Identifying and overcoming barriers to climate change adaptation in San Francisco Bay. Results from case studies. White paper for the California Energy Commission's California Climate Change Center (July 2012), Sacramento

NOAA (n.d) About the regional integrated sciences and assessments program. National Oceanic and Atmospheric Administration. https://cpo.noaa.gov/Meet-the-Divisions/Climate-and-Societal-Interactions/RISA/About-RISA. Accessed 7 June 2020

Pahl-Wostl C, Craps M, Dewulf A, Mostert E, Tabara D, Taillieu T (2007) Social learning and water resources management. Ecol Soc 12(2):5

Penner G, Cavanaugh M (2010) Political analysis: the legacy of toilet to tap, 4 Aug 2010. KPBS. https://www.kpbs.org/news/2010/aug/04/political-analysis-legacy-toilet-tap/. Accessed 7 June 2020

Pulido-Velazquez M, Marques GF, Harou JJ, Lund JR (2016) Hydroeconomic models as decision support tools for conjunctive management of surface and groundwater. In: Jakeman AJ, Barreteau O, Hunt RJ, Rinaudo JD, Ross A(eds) Integrated groundwater management. Springer Nature, Cham, pp 693–710

Purkey DR, Huber-Lee A (2006) A DSS for long-term water utility planning. Southwest Hydrol 5 (4):18–31

Raucher K, Raucher R, Ozekin K, Wegner K (2018) The opportunities and needs of water utility professionals as community climate–water leaders. Weather, Clim Soc 10(1):51–58

Sieber J, Purkey D (2011) WEAP: Water evaluation and planning system. User guide. Stockholm Environment Institute, US Center, Somerville, MA. http://weap21.org/downloads/WEAP_User_Guide.pdf

Smit B, Wandel J (2006) Adaptation, adaptive capacity and vulnerability. Glob Environ Change 16(3):282–292

Smith JW (2017) Focus: San Diego will recycle sewage into drinking water, mayor declares, 10 May 2017. The San Diego Union Tribune. https://www.sandiegouniontribune.com/news/environment/sd-me-pure-water-recycling-20170510-story.html. Accessed 7 June 2020

Staudenmaier LW (2006) Arizona groundwater law. The Water Rep 33:1–10, Nov 2006

TCEQ (n.d.) What is a groundwater conservation district (GCD)? Texas Commission on Environmental Quality. https://www.tceq.texas.gov/assets/public/permitting/watersupply/groundwater/maps/gcd_text.pdf. Accessed 7 June 2020

TWDB (2017) 2017 State water plan. Water for Texas. TWDB, Austin. http://www.twdb.texas.gov/waterplanning/swp/2017/doc/SWP17-Water-for-Texas.pdf

TWDB (2018) Regional water planning in Texas. Texas Water Development Board, Austin. https://www.twdb.texas.gov/waterplanning/rwp/index.as

TWDB (n.d.) Desired future conditions. http://www.twdb.texas.gov/groundwater/management_areas/DFC.asp. Accessed 6 June 2020

Vörösmarty CJ, Green P, Salisbury J, Lammers RB (2000) Global water resources: vulnerability from climate change and population growth. Science 289(5477):284–288

White DD, Corley EA, White MS (2008) Water managers' perceptions of the science–policy interface in Phoenix, Arizona: Implications for an emerging boundary organization. Soc Nat Resour 21(3):230–243

Wiseman R (2015) San Diego to spearhead direct potable water reuse, 19th Jan 2015. In: Water world. https://www.waterworld.com/international/wastewater/article/16200909/san-diego-to-spearhead-direct-potable-water-reuse. Accessed 7 June 2020

World Bank (2010) Climate change and urban water utilities: challenges and opportunities. P-Notes issue 50 (June 2010). The World Bank, Washington D.C. http://documents.worldbank.org/curated/en/472251468338472294/pdf/558280BRI0PNOT1Box349457B001PUBLIC1.pdf . Accessed 7 June 2020

WUCA (2020) Water Utility Climate Alliance. https://www.wucaonline.org/. Accessed 20 Jan 2020

Yates D, Miller K (2011) Climate change in water utility planning: decision analytic approaches. Water Research Foundation and University Corporation for Atmospheric Research, Denver

Yates D, Sieber J, Purkey D, Huber-Lee A (2005) WEAP21—A demand-, priority-, and preference-driven water planning model. Water Int 30(4):487–500

Chapter 13
Regional Hydrological Impacts of Climate Changes and Adaptation Actions and Options

13.1 Introduction

The impacts of climate change on water resources will vary between regions. Some areas are projected to get drier, whereas others will see little change in precipitation or may benefit from greater rainfall. Whereas climate models consistently predict higher temperatures in the future, there is much greater variation in both the direction and magnitude of projected local changes in precipitation. Temperature and precipitation change will increase with greenhouse gas concentrations and associated radiative forcings. The Coupled Model Intercomparison Project 5 (CMIP5) ensemble modeling results presented in the IPCC Fifth Assessment Report (AR5; IPCC 2013a) most consistently project decreases in mean annual precipitation in the Mediterranean region, the southwestern United States, Central America, and southern Africa (Fig. 13.1). The median of the CMIP5 projections also suggests that much of Australia and northeastern South America may also become drier. As would be expected, projected changes in precipitation increase with time and higher emissions scenarios (Representative Concentration Pathways, RCPs) from RCP2.6 through RCP8.5. Users of climate modeling results need to appreciate that there is considerable scatter in projected changes in precipitation amongst the CMIP5 ensemble models and that median projections should not be construed as most likely future conditions. Coarse-scale projections of future precipitation from general circulation models (GCMs) likely have a lesser accuracy on more local scales. Users need to consider the totality of historical climate data and modeling results to discern the likely range of future projections and the width of the envelope or cone of uncertainty in the projections.

The vulnerability of different areas to climate change will depend on both the direction and magnitude of changes and the capacity of societies to adapt to the changes, which will vary between different sectors of a society. Impacts to groundwater, and water resources in general, will depend upon changes in both the mean annual and seasonal amounts of precipitation, the intensity and duration of

R. Maliva, *Climate Change and Groundwater: Planning and Adaptations for a Changing and Uncertain Future*, Springer Hydrogeology,
https://doi.org/10.1007/978-3-030-66813-6_13

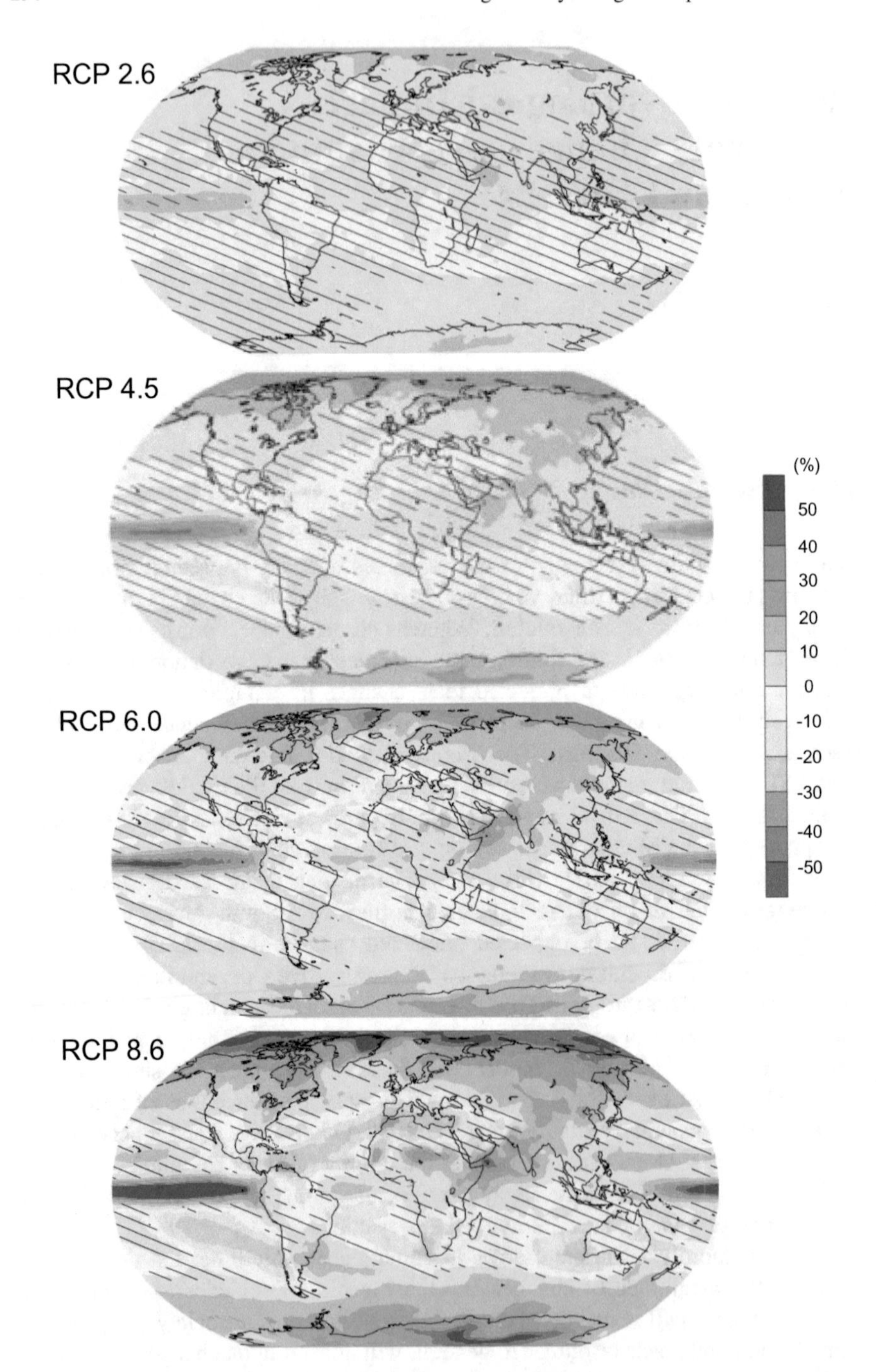

Fig. 13.1 Annual precipitation changes for the median (50%) of the CMIP5 ensemble for RCPs 2.6, 4.5, 6.0, and 8.5 in 2081–2100 relative to 1985–2005. Hatching indicates regions where the magnitude of the change of the 20-year mean is less than one standard deviation of model-estimated present-day natural variability of the 20-year mean differences (i.e., the change is relatively small or there is little agreement between models on the sign of the change. *Source* IPCC (2013b, c, d, e, Figs. AI.SM2.6.9, AI.SM4.5.9, AI.SM6.0.9, and AI.SM8.5.9, respectively).

precipitation events, and the interannual variability in precipitation and temperature, which will affect both aquifer recharge and water demands. Tendencies toward longer and more intense droughts could pose a much greater risk to local water resources than a gradual drift in mean total annual precipitation.

A climate change "hot spot" has been defined as "a region whose climate is especially responsive to climate change" (Giorgi 2006, p. 1). Alternative definitions of hot spots also consider vulnerability to climate change (which depends on adaptive capacity) and impacts to particular systems or sectors, such as agriculture and water (Giorgi 2006; De Sherbinin 2014). Climate change hot spots can also be defined or mapped as areas that have a high vulnerability to anthropogenic sea level rise, which would be areas with large populations (and associated commercial, industrial, and transportation assets) occupying coastal areas with low elevations.

De Sherbinin (2014) addressed the strengths and weaknesses of hotspot mapping. A primary value of hot spot mapping is that by "identifying likely climate change impacts and conveying them in a map format with strong visual elements, hotspots maps can help to communicate issues in a manner that may be easier to interpret than text" (De Sherbinin 2014, p. 23). De Sherbinin (2014, p. 35) also concluded that "Efforts to date can largely be characterized as supply-driven academic exercises rather than responding to demands from the policy community." Nevertheless, climate change hot spot mapping can serve a useful function in risk communication by drawing the attention of policy and decision makers and the general public to high local vulnerabilities.

Physical response-based hot spots are identified from statistical analysis of the results from multi-model ensembles of climate change simulations considering both changes in mean values and interannual variability in temperature and precipitation (Giorgi 2006). Giorgi (2006) in a study utilizing the CMIP3 ensemble results (which supported the IPCC 4th Assessment Report; AR4) identified the Mediterranean and northeastern European regions as the primary hot spots followed by the high latitude regions of the northern hemisphere and Central America. The Mediterranean region is projected to experience decreases in mean precipitation and greater precipitation variability in the summer dry season. The northern hot spot areas are projected to experience a large increase in dry (cold) season precipitation and increased variability.

A study that focused on the continental United States, also using the CMIP3 results, identified the southwestern United States, particularly southern California, southern Arizona, and the southern High Plains as being the most persistent hot spots (Diffenbaugh et al. 2008; Kerr 2008; Fig. 13.2). Northern Mexico is also mapped as a major hot spot. The northern and eastern parts of the United States appear to have a relatively low responsiveness to climate change. The pattern of hot spots was found to be shaped primarily by changes in interannual variability of the contributing parameters rather than by changes in long-term means (Diffenbaugh et al. 2008). The predicted hot spots in the southwestern United States bear a strong resemblance to the drying and warming that have been recently occurring in the region suggesting that we are seeing some emerging hot spot patterns (Kerr 2008).

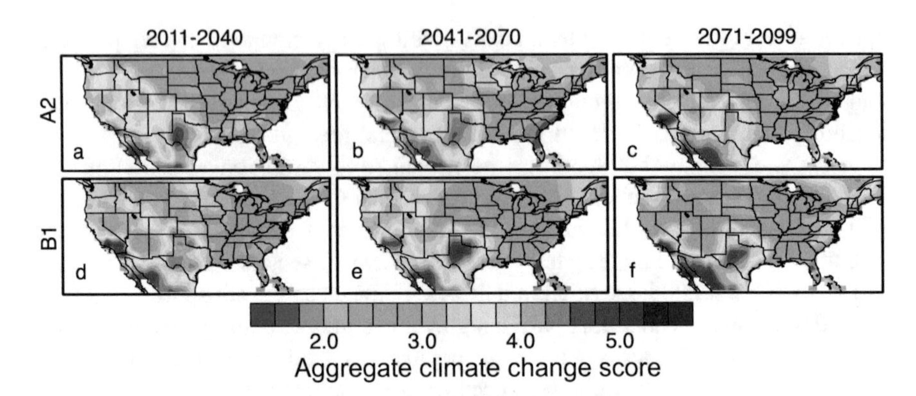

Fig. 13.2 Climate change hot spots map for the United States. Climate change sensitivity for the A2 (top) and B1 (bottom) emissions scenarios were mapped using aggregate standard Euclidean distance, which is based on the distance traveled in multivariate climate space between the future and present periods. *Source* Diffenbaugh et al. (2008)

Diffenbaugh and Giorgi (2012) in an updated study based on the CMIP5 results identified areas of the Amazon, the Sahel and tropical West Africa, Indonesia, and the Tibetan Plateau as persistent regional climate change hotspots throughout the twenty-first century for the RCP8.5 and RCP4.5 forcing pathways. Areas of southern Africa, the Mediterranean, the Arctic, and Central America/western North America were also identified as prominent regional climate change hotspots in response to intermediate and high levels of forcing (Diffenbaugh and Giorgi 2012). This chapter examines the vulnerability of groundwater resources to climate change in some areas that have been flagged as climate change hot spots or otherwise have a high dependence on groundwater. Adaptation options available and implemented to date are considered.

13.2 Southwestern North America

The southwestern North America (SWNA), which includes the southwestern United States and northern Mexico, already faces water scarcity and projected climate change is expected to exacerbate the scarcity. The CMIP5 ensemble results for the high emissions RCP8.5 indicate likely decreasing precipitation throughout southwestern North America, Central America, the Caribbean, and northeastern South America (Fig. 13.3). The Third National (U.S.A.) Climate Assessment observed that

> The Southwest is the hottest and driest region in the United States, where the availability of water has defined its landscapes, history of human settlement, and modern economy. Climate changes pose challenges for an already parched region that is expected to get hotter and, in its southern half, significantly drier. (Garfin et al. 2014, p. 463)

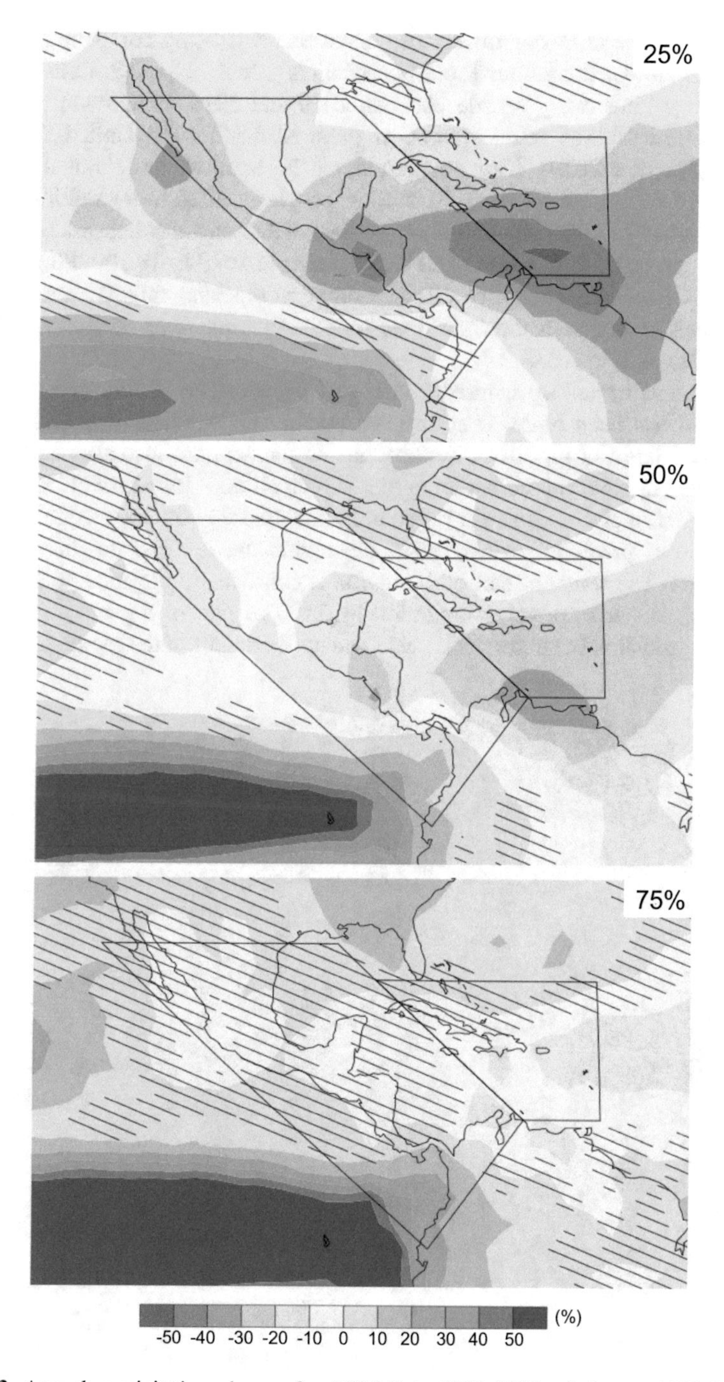

Fig. 13.3 Annual precipitation change for RCP8.5 in 2081–2100 relative to 1985–2005 in southern North America, Central America and the Caribbean. The 25, 50, and 75 percentiles of the CMIP5 multi-model ensemble are shown. *Source* IPCC (2013e, Fig. AI.SM8.5.049)

Water scarcity in the southwestern United States is being compounded by historical and projected continued rapid population growth. The U.S. Census Bureau reported that “For every decade between 1950 and 2010, the growth rate of the Desert Southwest was at least twice as great as that for the United States as a whole” (Mackun 2019). The population of the southwestern United States is expected to increase 68% from 56 million people in 2010 to 94 million in 2050 (Garfin et al. 2014). The population of the state of Arizona, the most arid state in the nation, is projected to increase from 7.29 million in 2020 to 10.10 million in 2050 (Arizona Commerce Authority 2020). Some people who migrate into this arid region seek to maintain their previous more water intensive lifestyles including irrigated lawns (Fig. 13.4).

SWNA has heated up markedly in recent decades (Fig. 13.5), and the period since 1950 has been hotter than any comparably long period in at least the last 600 years (Garfin et al. 2014). Anthropogenic temperature increases and drought are believed to have caused earlier spring snowmelt and shifted runoff to earlier in the year (Garfin et al. 2014). As is generally the case for climate change modeling, projections of precipitation change in the region are less certain than those for temperature. The southern part of the region is consistently projected to experience reduced winter and spring precipitation by 2100 as part of a general global precipitation reduction in subtropical areas. The northern part of the region is projected

Fig. 13.4 Housing development in Phoenix, Arizona, with man-made lakes and grass lawns. View from Thunderbird Conservation Park

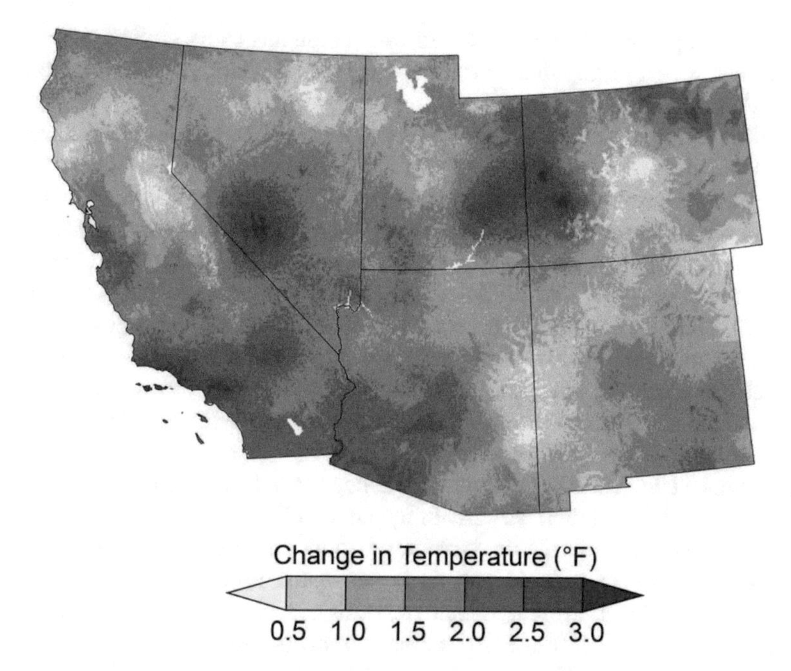

Fig. 13.5 Map showing the difference in average temperature between 1986–2016 and 1901–1960. The greatest increases occurred in southern California, southwestern Nevada, and western Colorado. *Source* Gonzalez et al. (2018) adapted from Vose et al. (2017)

to experience winter and spring precipitation changes that are smaller than the natural variations. Summer and fall changes in precipitation are also projected to be smaller than natural variations throughout the region (Garfin et al. 2014).

The principal hydrological impacts of climate change in the southwestern United States will likely be on surface waters. The region depends on winter snowpack in the Rocky Mountains, Sierra Nevada, and other mountain ranges to provide a major portion of the surface water on which it depends. The spring snowmelt flows into the Colorado, Rio Grande, Sacramento, and other major rivers, where it is captured and stored in reservoirs and then conveyed by systems of canals and pipelines great distances to agricultural and urban demand centers (Gonzalez et al. 2018). Increased temperatures, especially an earlier occurrence of spring warmth, have already significantly altered the water cycle in the southwestern United States by decreasing the snowpack, causing an earlier peak in spring stream flow, and increasing the proportion of rain to snow (Gonzalez et al. 2018).

The greatest vulnerability to water supplies is from droughts. Garfin et al. (2014, p. 463) cautioned that

> Severe and sustained drought will stress water sources, already over-utilized in many areas, forcing increasing competition among farmers, energy producers, urban dwellers, and plant and animal life for the region's most precious resource.

Higher temperatures are projected to cause droughts to become more frequent, intense, and longer lasting than in the historical record (Garfin et al. 2014; Gonzalez et al. 2018).

Starting with the pioneering work of Douglass (1929), tree-ring data have been used to identify past drought periods in the southwestern United States. Harsher growing conditions during droughts are manifested by thinner annual growth rings. The tree-ring data for the Four Corners region of the southwestern United States record a period of great aridity in the late thirteenth century. Frequent droughts were identified between 1247 and 1299 AD, with a period of great drought between 1276 and 1299. Drought conditions, especially in the cool season, have generally dominated over much of southwestern North America since at least 1999 (Woodhouse et al. 2010). However, the recent drought, thus far, pales hydrologically in comparison to the worst-case drought documented in the medieval period in both spatial extent and duration (Woodhouse et al. 2010).

Williams et al. (2020) presented modeling data that indicate that the early twenty-first century drought was driven by natural climate variability superimposed on drying due to atmospheric warming. Anthropogenic atmospheric warming enhanced evaporative demand and early snowpack loss. Williams et al. (2020, p. 314) concluded that "Anthropogenic trends in temperature, relative humidity, and precipitation estimated from 31 climate models account for 47% (model interquartiles of 35–105%) of the 2000–2018 drought severity, pushing an otherwise moderate drought onto a trajectory comparable to the worst SWNA megadroughts since 800 CE." It was cautioned that natural variability may well end the twenty-first century drought and that the transition may be under way after a wet 2019 (Williams et al. 2020).

Reoccurrence of severe drought conditions, such as those in the mid-12th century, which were more extensive and much more persistent than any modern drought experienced to date, would result in cumulative streamflow deficits in the Colorado and Sacramento Rivers that would severely tax the ability of water providers to meet demands throughout the Southwest (Woodhouse et al. 2010). Woodhouse et al. (2010, p. 21283) cautioned that

> The severity, extent, and persistence of the 12th century drought that occurred under natural climate variability, have important implications for water resource management. The causes of past and future drought will not be identical but warm droughts, inferred from paleoclimatic records, demonstrate the plausibility of extensive, severe droughts, provide a long-term perspective on the ongoing drought conditions in the Southwest, and suggest the need for regional sustainability planning for the future.

Scott et al. (2012) investigated the effects of climate changes and population growth on the Santa Cruz River Valley Aquifer, which straddles the border between Arizona and the Mexican state of Sonora. Projected precipitation changes in the aquifer area over the twenty-first century have a high uncertainty, ranging from −21 to +20%, depending on the emissions scenario. The results of this investigation indicate that urban water use will experience a greater percentage change than climate-induced recharge, which will remain the largest single component of the water balance (Scott et al. 2012). Increasing temperatures are expected to increase the demand for groundwater while diminishing recharge (Scott et al. 2012). The Santa Cruz River Valley Aquifer area may also experience both longer, more severe droughts, and due to increased tropical storms, an intensification of precipitation and associated increased future flooding.

Groundwater resources in the Santa Cruz River area are already under pressure and demands will increase with projected population growth. In the absence of new (as yet unidentified) sources of supply, supply options are largely restricted to the reallocation of water currently used in agriculture and wastewater reuse for human purposes, such as for landscaping irrigation and indirect (or direct) potable use (Scott et al. 2012). The water supply situation in Sonora was reported to be even more challenging due to larger increases in absolute population numbers and the fact that close to half of the water demand is already being met through inter-basin transfer from the Los Alisos basin to the south (Scott et al. 2012). The Santa Cruz River Valley Aquifer situation is fairly typical of the challenges and options available in other groundwater basins in the SWNA.

13.3 High Plains (Western United States)

The High Plains Aquifer, also known as the Ogallala Aquifer, underlies an area of 450,000 km^2 (174,000 mi^2) in eight states: Colorado, Kansas, Nebraska, New Mexico, Oklahoma, South Dakota, Texas, and Wyoming (Fig. 13.6). The High Plains aquifer region is referred to as "America's breadbasket" as it is responsible for one-fifth of the wheat, corn, cotton and cattle produced in the United States.

The High Plains aquifer has been historically used primarily for irrigation, which began in the late 1800s when windmills were first utilized as a source of power to pump water from scattered irrigation wells (Miller and Appel 1997). The drought of the 1930s spurred a rapid expansion of groundwater irrigation. Exploitation of the aquifer generally began in Texas, adjacent parts of New Mexico, and in major stream valleys in other states in the 1930s and progressed northward (Miller and Appel 1997). The development of center-pivot irrigation systems during the 1960s was partially responsible for a large increase in the number of wells drilled for irrigation between 1952 and 1978 (Miller and Appel 1997). The expansion of irrigation results in continued aquifer overdraft and associated declining groundwater levels, which are greatest in the southern part of the aquifer where irrigation demands are greater, and recharge is less (Fig. 13.7).

The area-weighted, average overall water-level decline in the High Plains aquifer from predevelopment to 2011 was 4.8 m (15.8 ft) with local declines by well reaching 71.3 m (234 ft; McGuire 2017). Total recoverable water in the aquifer in 2015 was about 2.91 billion acre-feet (3589 km^3), which was a decline of about 273.2 million acre-feet (337 km^3) since predevelopment (McGuire 2017). The decline from 2013 to 2015 was 10.7 million acre-feet (13.2 km^3; McGuire 2017). Groundwater depletion in the irrigated High Plains and California Central Valley

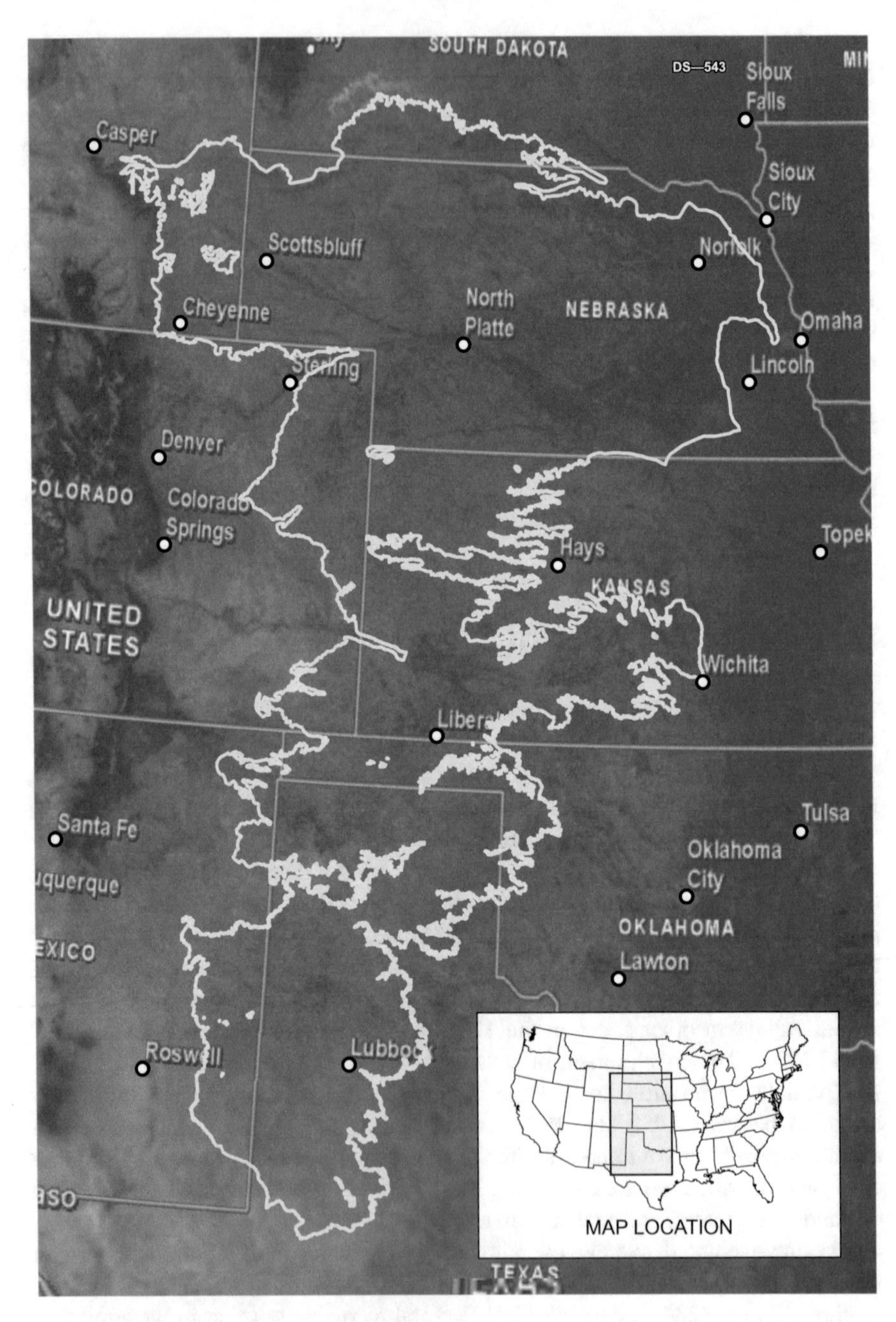

Fig. 13.6 Map showing the boundaries of the High Plains aquifer. Modified from Qi (2010)

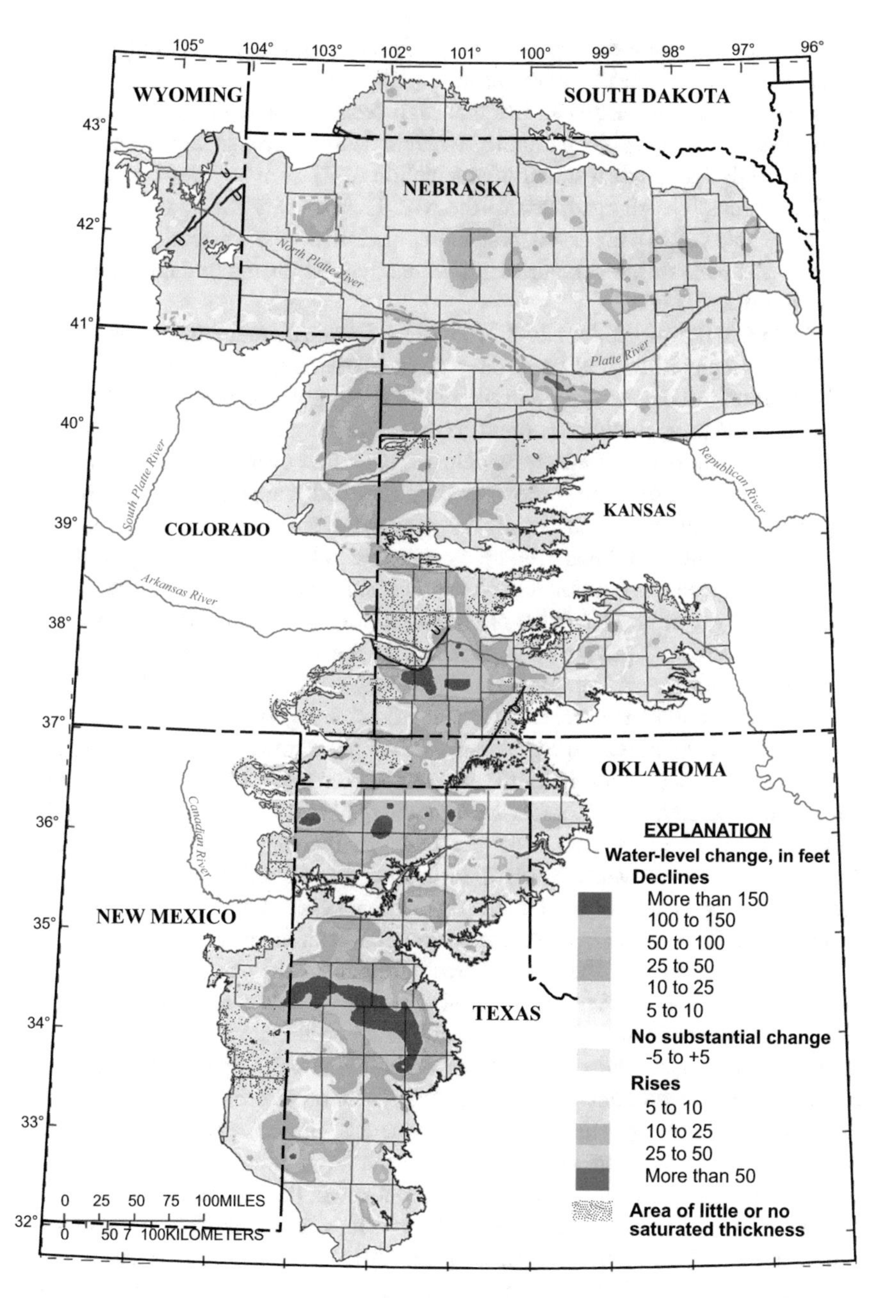

Fig. 13.7 Water-level changes in the High Plains aquifer from predevelopment (about 1950) to 2015. *Source* McGuire (2017)

was reported to account for approximately 50% of the groundwater depletion in the United States since 1900 (Scanlon et al. 2012).

Recharge mapping results shows that high recharge rates in the northern High Plains result in sustainable pumping, whereas lesser recharge in the central and southern High Plains has resulted in a depletion of 330 km^3 of fossil groundwater, which was mostly recharged during the past 13,000 years (Scanlon et al. 2012). Scanlon et al. (2012) observed that extrapolation of the current depletion rate suggests that 35% of the southern High Plains will be unable to support irrigation within the next 30 years.

Median CMIP5 projections in the IPCC AR5 indicate that annual and seasonal changes in precipitation will be in the −10 to +10% range with the southern High Plains tending to become drier and the northern High Plains wetter (IPCC 2013a). The changes in precipitation are small compared to the natural variation. Rosenberg et al. (1999) cautioned based on modeling results that even in scenarios in which precipitation increases, the accompanying elevated temperatures will increase evapotranspiration (ET), consuming the excess water that might otherwise go to recharge the aquifer, and that climate change forced by global warming will make use of the High Plains aquifer water even less sustainable than it is now.

Lauffenburger et al. (2018) investigated the impacts of climate change on irrigated agriculture and groundwater recharge in the northern High Plains aquifer. Their results indicate an important threshold may occur between the low and high warming scenarios that if exceeded could trigger a significant shift in year 2050 hydroclimatology and recharge gradients. Future northern High Plains temperatures will resemble present central High Plains temperatures and future recharge rates in the east will resemble the present lesser recharge rates in the western part of the northern High Plains aquifer. The decreases in recharge rates could accelerate the decline of aquifer water levels if irrigation demand reduction and other management strategies are not implemented (Lauffenburger et al. 2018)

The High Plains area is vulnerable to droughts with the most notorious example in historical times being the "Dust Bowl" drought of the 1930s. The IPCC AR5 (2014, p. 53) has consistently cautioned that there is low confidence in observed global-scale trends in droughts because of limited observational data, dependencies of inferred trends on the choice of the definition for drought, and geographical inconsistencies in drought trends. The term "meteorological drought" is used to describe conditions of precipitation deficit, whereas "agricultural drought" describes conditions of soil moisture deficit. Although the High Plains area may not experience anthropogenic changes in precipitation outside of the normal variation, warmer temperatures will result in increased ET, which would exacerbate soil moisture deficits (Wehner et al. 2017). Wehner et al. (2017) suggested that future temperature increases will likely lead to greater frequencies and magnitudes of agricultural droughts throughout the continental United States as the resulting increases in ET outpace projected precipitation increases. Increases in ET will compound the effects of decreasing precipitation in areas experiencing drying trends.

Much has been written in the media on the potential for the southern High Plains aquifer to "run dry" and the impending severe impacts on local agriculture-dependent economies (e.g., Little 2009; Galbraith 2015; Miller 2018). Current groundwater use is not sustainable and climate change is expected to exacerbate conditions to some degree. In the absence of new large surface water sources for aquifer augmentation, there are no obvious options for increasing water supplies in the High Plain aquifer (Scanlon et al. 2012). Popper and Popper (1987) published the controversial "Buffalo Commons" concept in which agriculture and ranching would be abandoned in large parts of the Great Plains through government purchase (i.e., deprivatization) and the land restored to its pre-white settlers condition. Grandiose schemes were investigated in the 1980s to convey freshwater from the Great Lakes to replenish the High Plains aquifer (Quinn and Edstrom 2000), which, as would be expected, were strongly opposed by the states and provinces that adjoin the lakes.

On the more practical side, the rate of depletion of High Plains aquifer may be reduced through a conversion to dryland farming and increased irrigation efficiency (Scanlon et al. 2012; Stover and Buchanan 2017). However, it was cautioned that increases in irrigation efficiencies may act mostly to reduce agricultural return flows and have often promoted increases in irrigated area (Scanlon et al. 2012). Scanlon et al. (2012) concluded that because of the very low recharge rates (i.e., exploitation of fossil groundwater) of the southern High Plains aquifer, reducing irrigation withdrawals could extend the lifespan of the aquifer but would not result in sustainable long-term management of the aquifer.

13.4 Florida

Florida has been described as "ground zero" in the climate crisis in the United States because of its vulnerability to sea level rise resulting from its more than 1000 miles (1600 km) of tidal coastline and low elevations (Union of Concerned Scientists 2019). The highest land surfaces in the southern part of the state are sanitary landfills. Florida has also been experiencing one of the fastest growth rates in the nation with its population increasing from 12.94 million people in 1990 to an estimated 21.21 million in 2019 (BEBR 2019). Much of Florida's population in concentrated in coastal counties.

Florida has a humid subtropical to tropical climate with an annual state-wide average rainfall of 136.4 cm (53.7 in). There is a strong seasonality to the precipitation with most of the rainfall in the peninsular part of the state occuring during the summer wet season. The winter and spring are relatively dry and coincide with the peak in tourism and seasonal residents from the north (referred to as "snowbirds") and associated water demands. Florida relies heavily on groundwater for its water supply, particularly for potable use, because of its high-quality (and thus low treatment costs), geographic availability across the state, and its being a reliable, year-round source.

Exploitation of fresh groundwater resources is believed to be at or close to sustainable limits in much of the state under current climate conditions due to environmental constraints. Groundwater pumping induced lowering of water levels in wetlands and lakes and decreasing spring and stream flows are the principal regulatory concerns. The Florida Climate Change Task Force concluded that "it is clear that groundwater resources are inadequate now to meet future water demand in many parts of the state, and climate changes adds to the vulnerability of groundwater supplies and uncertainty of supply and demand. Therefore, developing alternative water supplies is a priority" (Berry et al. 2011, p. xii). Yoder (2017) observed with respect to southeastern Florida, that the region will never run out of drinking water, but it could certainly run out of inexpensive drinking water.

Surface water has been a less attractive source for public supply because of is lesser quality, and thus higher treatment costs, and seasonal availability. Storage in reservoirs or underground in aquifer storage and recovery systems is necessary for the use of surface water for large-scale year-round supply. Withdrawals of surface water are limited during dry periods due to physical availability, water quality constraints, and regulatory requirements to maintain specified minimum flows and levels.

Florida's water supply may be impacted by climate change through: (a) changes in rainfall and evapotranspiration patterns that alter the water budget and result in more frequent or prolonged periods of both drought and flooding, (b) sea level rise causing accelerated saline water intrusion in the coastal belt, impacts to coastal flood protection infrastructure, increased storm surge elevation and inland extent, and inundation of coastal ecosystems and real estate, and (c) changes in tropical storm and hurricane frequency and strength (Obeysekera et al. 2015, 2017; Bloetscher 2009; Berry et al. 2011). The greatest impacts of sea level rise will be episodic and permanent direct marine and groundwater inundation of low-lying areas.

Southeastern Florida, the Florida Keys and southwestern Florida are far more vulnerable to sea level rise and storm surges than the rest of the state because these three areas have a combination of dense populations and low topography (Bloetscher et al. 2017). The northwest, northeast, north central ridge and Kissimmee River valley regions are at lesser risk from climate change because of their higher elevations and lesser population densities, but their future water supplies may be impacted by precipitation changes (Bloetscher et al. 2017). Broward, Miami-Dade, Monroe, and Palm Beach Counties joined in January 2010 to form the Southeast Florida Regional Climate Change Compact (SEFRCCC) as a way to coordinate mitigation and adaptation activities across county lines (SEFRCCC n.d.). For planning purposes, the SEFRCCC Sea Level Rise Ad Hoc Work Group projected from historical trends and existing modeling results and projections that relative local sea level will rise (above 2000 levels) between 25 and 43 cm (10–17 in) by 2040, between 53 and 137 cm (21–54 in) by 2070, and between 102 and 345 cm (40–136 in) by 2120 (SEFRCCC Sea Level Rise Work Group 2019). A sea level rise of 1 m or more would have catastrophic human and environmental impacts to the region.

Heimlich et al. (2009) and Bloetscher et al. (2010) investigated the vulnerability and adaptation options of southeastern Florida utilities to climate change using the city of Pompano Beach as a case study. Pompano Beach is located in a low-lying coastal area of Broward County (maximum elevation above sea level of 12 ft; 3.6 m) and is reliant on groundwater from a surficial aquifer (Biscayne Aquifer) for its water supply. Key vulnerabilities of Pompano Beach to sea level rise include (Heimlich et al. 2009; Bloetscher et al. 2010)

- inundation of low-elevation coastal areas with saltwater
- higher water table elevations throughout the service area (resulting in a lesser capacity to store rainfall, and more surface runoff, erosion, and ponding of standing water)
- risk of groundwater contamination from seawater inundation and hurricane storm surges
- saltwater migration toward eastern municipal well fields
- inundation of land on which septic tanks and onsite treatment and disposal systems operate (which could compromise the operability of these systems and may contaminate groundwater as water tables rise)
- increased access hole leakage and inflow to wastewater mains from surface flooding and saltwater intrusion.

Water resources adaptation alternatives identified are (1) water conservation, (2) protection of existing water sources from saltwater intrusion, (3) development of alternative water sources, (4) wastewater reclamation and reuse, and (5) stormwater management.

The U.S. Army Corps of Engineers was tasked to identify solutions to protect 2.8 million people and $311 billion in property value in Miami-Dade County alone from projected more destructive hurricanes and rising seas levels, which in South Florida could reach more than three feet (0.9 m) by 2080 (Harris 2020). The initial unofficial plans envision construction of 10–13-foot (3–4 m) high walls, massive moveable storm surge barriers at the mouths of Miami's rivers and major drainage canals, and the elevation of 10,000 homes and flood proofing of 7000 buildings at a cost of US$ 8 billion dollars, which is subject to change (Harris 2020). Above ground walls would not address rising groundwater levels and a drainage system would presumably be necessary on the landward side of the barrier. The Miami-Dade County sea level rise challenges and adaptation costs are a microcosm of the problems and costs that coastal urban areas will face worldwide.

Saline-water intrusion into coastal wellfields has been long been a problem in Florida coastal communities, which was addressed by moving wellfields inland. Most municipal wellfields are now located far enough inland so that sea level rise-induced saline water intrusion should not be a significant concern over a 20-year planning horizon. As sea level rises, some low-lying areas could become inhabitable, which would impact local freshwater demands. Over a longer term, inland communities could face water supply (and other socioeconomic) challenges through the migration and resettlement of displaced coastal populations.

Climate modeling results do not provide a clear picture of the impacts on climate change on precipitation in Florida. A warmer climate would be expected to increase demands for water. Coarse-scale modeling results presented in the IPCC AR5 show the change in average precipitation for Florida being in the −10 to +10% range for 2081–2100 relative to 1986–2005 under the RCP2.6 and RCP8.5 scenarios (IPCC 2014, Fig. 3.2). Projections in the Third National Climate Assessment indicate a −5 to +3.6% change in water availability in Florida for 2010–2060 with decreases projected in the western panhandle and increases projected in central and upper east coast of the state (Carter et al. 2014). The subsequent Fourth National Climate Assessment emphasized the likely occurrence of more extreme weather events (droughts and floods) in the region (Carter et al. 2018). The wide range of precipitation projections with no clear indication of the sign of change necessitates that the possibility of both negative and positive changes needs to be considered in developing water supply modeling scenarios (Obeysekera et al. 2015).

Climate change is currently not a major issue in water supply planning in Florida, as changes in precipitation and groundwater recharge over the common 20-year planning horizon likely fall within the range in the natural variability and will be dwarfed by expected increases in demand from population growth. Water supply systems are incidentally becoming more resilient to climate change through the increasing utilization of alternative, less climate-sensitive, water sources (particularly brackish groundwater desalination and non-potable wastewater reuse), which is being driven largely by regulatory limitations on additional fresh groundwater withdrawals.

Managed aquifer recharge is also seen as part of the solution to future water shortages in Florida through either local water storage in aquifer storage and recovery systems to meet dry season demands or replenishment of the Upper Floridan aquifer using wells and surface spreading to reduce impacts to sensitive surface environments and water bodies. Recharge projects involving injection of surface water using wells are being constrained by regulatory water quality requirements and associated pretreatment costs. Direct potable reuse is being investigated in some communities, such as Hillsborough County (London 2017) and the city of Daytona Beach (Roque et al. 2017).

Groundwater use in much of Florida is constrained by wetland and surface water body impacts resulting from declining groundwater levels. Groundwater rise associated with SLR could make more groundwater available in shallow aquifers for water supply purposes. Bloetscher et al. (2015) identified infiltration galleries as a tool that has the potential to reduce groundwater levels, which would increase the capacity of soil to absorb water, while creating a constant water supply. The infiltration gallery solution could potentially help many low-lying areas and island communities reduce their flood risks while capturing water to treat for water supply (Bloetscher et al. 2015).

13.5 Mediterranean Region

The Mediterranean region, which includes southern Europe, northern Africa, and the Levant, has been mapped as one of the major global climate change hot spots (Giorgi 2006; Diffenbaugh and Giorgi 2012). The climate of the Mediterranean region is characterized by mild and wet winters and hot and dry summers. The region already faces water scarcity and water demands are increasing due to expansion of irrigated area, urban and industrial development, population growth, social development, and international tourism (Cudennec et al. 2007; Iglesias et al. 2007; Senatore et al. 2011). Agriculture is the main water consumer in the Mediterranean region (Iglesias et al. 2007; Joffe 2017). Irrigation water demand will increase in the region from the combined effects of population growth (and thus demand for crops), increased temperature, and decreased precipitation. Population growth and industrialization have been particularly rapid over the past several decades in the Middle East and North Africa (MENA), which is already the most water stressed region of the world (NIC 2009; Joffe 2017; Cramer et al. 2018). Mediterranean societies thus face the growing challenge of meeting higher water demands from all sectors with decreased available freshwater resources (Cramer et al. 2018). Increasing droughts may make available water resources in the region increasingly unstable and vulnerable (Iglesias et al. 2007). The National Intelligence Council concluded that "Water scarcity, even in the absence of climate change, will be one of the most critical problems facing North African countries in the next few decades" (NIC 2009, p. 42).

Groundwater is the main source of freshwater in many Mediterranean areas and is already under great pressure, and decreased precipitation is expected to decrease aquifer recharge (Moutahir et al. 2017). In North Africa, Libya and to a lesser degree Egypt, non-renewable (fossil) groundwater is being exploited to meet current needs.

A region-specific analysis of climate change in the Mediterranean region utilizing the results of the CMIP3 model ensemble and an ensemble of regional climate models run for the "Prediction of Regional scenarios and Uncertainties for Defining, EuropeaN Climate change and associated risks and Effects" (PRUDENCE) project (Christensen et al. 2002) gave a collective picture of substantial drying and warming, especially in the summer warm season (Giorgi and Lionello 2008). The only exception to the region-wide drying was a projected increase in precipitation over some areas of the northern Mediterranean basin, most noticeably the Alps (Giorgi and Lionello 2008).

The subsequent CMIP5 ensemble results also strongly indicate that the region will likely experience decreased annual precipitation under all the RCPs with the decrease being more profound under higher emissions scenarios (Fig. 13.8). Decreases in precipitation and increases in temperature will compound existing serious water scarcity in much of the region. The Mediterranean region also has great variation in economic resources and thus adaptive capacity. The southern European countries and Israel are much more economically prosperous (in terms of gross domestic product per capita) than the northern African states (Klarić 2017) and thus can more readily afford more expensive water supply options for potable and industrial use, particularly seawater desalination.

A substantial number of investigations have been performed on the potential impacts of climate change in the Mediterranean region, particularly the European Union countries. Follows are summaries of some studies that focused on impacts to groundwater resources.

13.5.1 Alicante, Spain

Moutahir et al. (2017) investigated the likely effects of climate change on groundwater recharge in the province of Alicante in southeastern Spain. Small precipitation events in arid and semiarid regions tend to not result in significant recharge. Heavy precipitation events (HPEs) are rainfalls sufficient to produce appreciable aquifer recharge. Moutahir et al. (2017) determined that the critical HPE threshold in the Alicante study area is ≥ 20 mm/d. Data from nine CMIP5 climate models were downscaled and then used to estimate future changes in HPEs in the 2040–2099 period. The number of HPEs is projected to decrease with climate change with the reduction under the high RCP8.5 scenario (−15%) being close to twice that under the moderate RCP4.5 scenario. The decrease in the frequency of HPEs is projected to be partially offset by an increase in the magnitude of the HPEs. The decrease in the frequency of HPEs will cause an increase in the length of "no aquifer recharge periods" accentuating groundwater droughts (Moutahir et al. 2017).

13.5.2 Southern Italy

Senatore et al. (2011) simulated the impacts of climate change on water availability in the Crati River Basin of southern Italy. The outputs from three regional climate models were applied to a newly developed distributed hydrological model (Intermediate Space Time Resolution Hydrological Model). Projections were made for the 2070–2099 period relative to 1961–1990 for two emissions scenarios (A2 and A1B). The results of the simulations project a decrease in annual precipitation of between 9 and 21% and mean annual reductions in soil moisture of between 12.8 ± 1.9% and 20.7 ± 1.9%, with reductions reaching 37.7 ± 2.4% during the summer. Groundwater storage is estimated to decrease by between 6.5 ± 1.4% and 11.6 ± 1.9%. Senatore et al. (2011) concluded that the hydrological modeling results derived from the different GCMs and SRES emissions scenarios all agree in projecting a general reduction in future water availability.

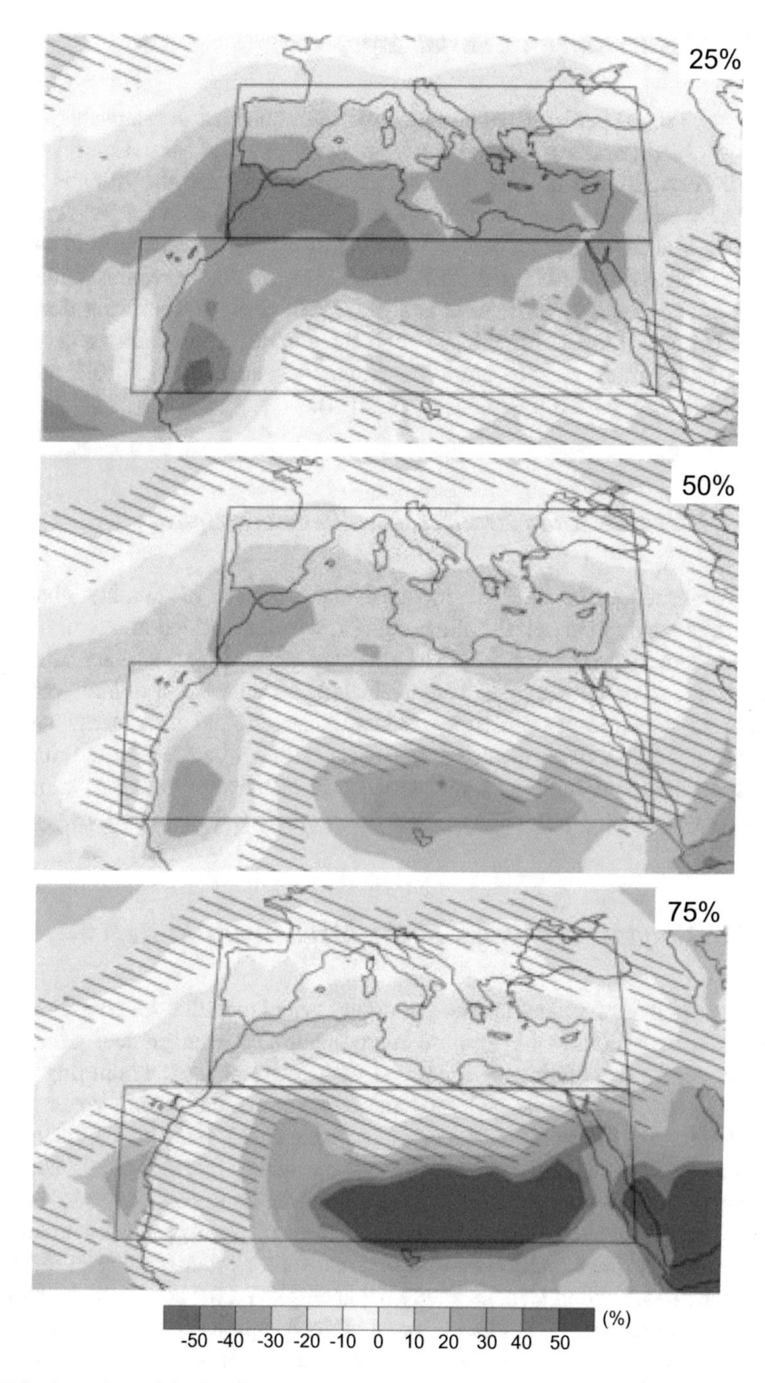

Fig. 13.8 Annual precipitation change for RCP8.5 in 2081–2100 relative to 1985–2005 for the Mediterranean region. The 25, 50, and 75 percentiles of the CMIP5 ensemble are shown. *Source* IPCC (2013e, Fig. AI.SM8.5.049)

13.5.3 Mediterranean Coastal Aquifers

Stigter et al. (2014) performed a comparative assessment of the impacts of climate change on Mediterranean coastal aquifers in three areas: the Central Algarve (southern Portugal), the Ebre Delta (northeastern Spain), and the Atlantic Sahel at the central western coast of Morocco (Sect. 5.8.9). RCM data for the A1B emissions scenario were compared between a reference (present) scenario and two future climate periods (2020–2050 and 2069–2099). The modeling results predict more frequent droughts at the Portuguese and Moroccan sites. A significant decrease in recharge (mean of 25%) is predicted in all three areas. The modeling results also predict steadily increasing in crop water demands, reaching 15–20% additional evapotranspiration by 2100 (Stigter et al. 2014).

13.5.4 Serral-Salinas Aquifer, Southeastern Spain

Pulido-Velazquez et al. (2015) modeled the impacts of climate change on groundwater recharge in Serral-Salinas aquifer, which is located in a semi-arid zone of southeastern Spain (Sect. 5.8.5). The effects of both changes in mean rainfall and the frequency of extreme events were evaluated. An increase in rainfall variability, as expected under future emissions scenarios, was found to increase simulated recharge rates for a given mean rainfall because of an increase in the number of extreme recharge-producing events. Greater decreases in recharge values were obtained by imposing only changes in mean rainfall in the simulations.

13.5.5 Adaptation Options in the Mediterranean Region

Adaptation to more frequent and longer droughts and overall drier conditions in the Mediterranean region would require more integrated management of all water resources and incorporation of improved early warning and monitoring systems (Iglesias et al. 2007). As is generally the case for regions facing water scarcity, water savings (i.e., demand management and increased water use efficiency) is a key strategy for reducing vulnerability to droughts (Iglesias et al. 2007) and climate-changed caused decreases in available water resources. On-going increased use of alternative water sources, particularly wastewater reuse and desalination, are critical elements toward addressing present and future water scarcity in the region.

Seawater desalination is increasingly being employed for potable water supply in the Mediterranean region, but its high cost limits its application to high-value water uses, such as potable supply. Desalination is not economically viable for most irrigation. Desalination is energy intensive and thus has a high carbon footprint unless alternative energy sources are used. Concentrate is typically disposed of using marine outfalls, which when properly designed and located have minimal local environmental impacts (Missimer and Maliva 2018). Surface water bodies with limited circulation (i.e., hydraulic connection to the oceans) and large present and anticipated future seawater desalination capacities, such as the Mediterranean Sea, are vulnerable to long-term increases in salinity to the detriment of salinity-sensitive (stenohaline) marine life (Bashitialshaaer et al. 2011; Missimer and Maliva 2018).

Fossil groundwater is being exploited in the southern Mediterranean states, the most notable example of which is the Great Man-Made River Project (GMRP) in Libya. The GMRP was initiated in 1984 with the goal of transporting non-renewable groundwater from four groundwater basins (Kufra, Sirt, Murzuk, and Hamadah) in the southern part of the country to the northern coastal belt, where most of the population resides and where local shallow aquifers have become increasingly at risk due to saline water intrusion. It is recognized that the GMRP is by no means an everlasting solution to Libya's water supply, but rather a vital transitory stage to a more sustainable water supply, which will involve alternative sources (e.g., desalination and wastewater reuse) and increased efficiency of water use (Bakhbakhi and Salem 2001). The GMRP provided more than 70% of the freshwater supply of the coastal area, but the project is imperiled due to the ongoing civil war. The water flow to Western Libya was reported to have dropped from the normal flow of 1.2 million cubic meters per day to about 800,000 cubic meters per day due to sabotage and lack of funding for maintenance (Cooke 2017; Laessing and Elumami 2019).

13.6 Africa

Africa contains a broad range of climates that will experience greatly different physical impacts from climate change. The impacts of changes in water availability will be superimposed on the more serious effects of rapidly increasing populations, which will place ever-increasing demands on water resources (Carter and Parker 2009; Taylor et al. 2009). Carter and Parker (2009, p. 686) cautioned with respect to Africa that

> Increased urban populations and high population densities will increase domestic and industrial water demands, putting enormous strain on the water resources supplying towns and cities. Urban water demands are likely to increase by at least a factor of 4 by 2050 (compared to 2000). The demands of increasing population are likely to outstrip any problems caused by climate change.

Agriculture in the vast majority of African countries is predominantly rainfed, which results in a vulnerability to existing and future climate variability, such as variations in the starting date of the rains, end date of the rains, dry period duration, and extreme events (Carter and Parker 2009). Hydrological and water use data are sparse in most African nations. It is estimated that probably 50% or more of the continent's population relies on groundwater mostly from shallow aquifers (Carter and Parker 2009). Groundwater quality is also a concern. Inadequate community hygiene in many rapidly urbanizing centers makes potable water supplies derived from shallow aquifers increasingly vulnerable to contamination (Howard et al. 2003; Taylor et al. 2009; Lapworth et al. 2017). In rural areas, increased rainfall intensity could also adversely affect the quality of shallow groundwater (MacDonald et al. 2009).

The findings of the IPCC AR4 indicate that projections of future rainfall change in Africa show much greater variability than temperature, and that the changes will vary by region and season (Carter and Parker 2009). Mean annual rainfall is likely to decrease in Mediterranean Africa, the northern Sahara, and southern Africa, and is likely to increase in east Africa. There is greater uncertainty about changes in the southern Sahara, Sahel, and Guinean coast (Carter and Parker 2009). The subsequent CMIP5 ensemble results summarized in the AR5 also project generally drier conditions in the Mediterranean and northwest Atlantic coastal areas and southern Africa, and a tendency for wetter conditions in the central, eastern, and western parts of the continent (IPCC 2013a, b, c, d, e; Fig. 13.1). A major concern is that climate change may amplify the present-day interannual variability in precipitation and that the frequency of extreme events may increase (Carter and Parker 2009).

The impacts of changes in precipitation on groundwater recharge, and thus groundwater availability, also depend upon the pattern of precipitation events (i.e., frequency, intensity, and duration) and loci of recharge (direct and distributed versus indirect and focused), rather than the mean annual amount. As noted by Carter and Parker (2009, p. 683), for example,

> An extended period of medium-intensity rainfall is more likely to lead to significant recharge than either a period of high-intensity rainfall (which will lead to increased runoff) or a period of low-intensity rainfall (most of which will evaporate). Two years with the same annual rainfall may give very different values of recharge, depending on the daily rainfall distribution.

Indeed Mileham et al. (2009) in a modeling study of a medium-sized (2098 km^2) catchment in the humid tropics of southwestern Uganda showed how transforming the rainfall distribution to account for projected changes in rainfall intensity resulted in simulated increases in recharge and runoff of 53% and 137%, respectively, whereas simulations using the historical (baseline) rainfall distribution grossly underestimated future recharge, predicting a 55% decrease.

Macdonald et al. (2009) examined the impacts of climate change on the sustainability of rural water supplies in Africa, which are commonly obtained from shallow (<50 m) aquifers. MacDonald et al. (2009) concluded that climate change is unlikely to lead to widespread catastrophic failure of improved rural groundwater supplies because they require only an estimated 10 mm of recharge annually to support a hand pump, which should still be achievable for much of the continent. However, it was noted that up to 90 million people may be adversely affected in marginal groundwater recharge areas (200–500 mm annual rainfall).

It was observed during recent droughts that increased demand on dispersed water points, as shallow unimproved sources progressively fail, poses a much greater risk of individual source failure than regional resource depletion (MacDonald et al. 2009). Africa needs to increase food production to meet the needs of a growing population. Groundwater-based irrigation can be element of a sustainable Africa-wide strategy for reducing poverty and increasing food production (MacDonald et al. 2009). Although shallow groundwater resources in most areas will likely be sufficient to meet domestic needs, large increases in abstractions to sustain irrigated agriculture could lead to more widespread over-exploitation of groundwater and threaten the sustainability of domestic water sources (MacDonald et al. 2009).

Adaptation options to address the impacts of climate change on groundwater resources are limited in parts of Africa where increases in demand due to population growth and existing climate variability may exceed climate change impacts. Africa also suffers from a paucity of site-specific hydrogeological data. MacDonald et al. (2009, p. 700) suggested that

> Matching the technology to the groundwater conditions, and siting sources in the most productive parts of the aquifer will improve the security of groundwater supplies. Other options, such as increasing the number of sources in a community, managed aquifer recharge, or relief boreholes may all have benefits.

Taylor et al. (2009) also concluded that there is a need in Africa for more investment in monitoring infrastructure for groundwater resources.

Rainwater harvesting and small-scale managed aquifer recharge are increasingly being recognized as practical and cost-effective elements of both water supply and stormwater management on both the household and community level in rural arid and semiarid lands (Berg 2018). There are considerable opportunities in Africa to apply rainwater harvesting to reverse soil degradation, increase climate-resilient food production, and improve dietary diversity (Berg 2018).

References

Arizona Commerce Authority (2020) Population projections. https://www.azcommerce.com/oeo/population/population-projections/. Accessed 9 June 2020

Bakhbakhi M, Salem O (2001) Why the great man-made river project. In: Proceedings, regional aquifer systems in arid zones—managing non-renewable resources, Tripoli, Libya, 20–24 Nov 1999, IHP-V, Technical documents in hydrology, vol 42. UNESCO, Paris, pp 1–16

Bashitialshaaer RA, Persson KM, Aljaradin M (2011) Estimated future salinity in the Arabian Gulf, the Mediterranean Sea and the Red Sea consequences of brine discharge from desalination. Int J Acad Res 3(1):133–140

BEBR (2019). Florida estimates of population. University of Florida, Bureau of Economic and Business Research

Berg L (2018) Harvesting rain to improve crops across Africa. In: ReThink, Dec 5 2018.https://rethink.earth/harvesting-rain-to-improve-crops-across-africa/. Accessed 21 June 2020

Berry L, Bloetscher F, Hernández-Hammer N, Koch-Rose M, Mitsova-Boneva D, Restrepo J, Root T, Teegavarapu R (2011) Florida water management and adaptation in the face of climate change. Florida Climate Change Task Force, State University System of Florida

Bloetscher F (2009) Climate change impacts on Florida (with a specific look at groundwater impacts). Fla Water Resour J 2009:14–26

Bloetscher F, Meeroff DE, Heimlich BN, Brown AR, Bayler D, Loucraft M (2010) Improving resilience against the effects of climate change. J Am Water Works Assoc 102(11):36–46

Bloetscher F, Hammer NH, Berry L, Locke N, van Allen T (2015) Methodology for predicting local impacts of sea level rise. Br J Appl Sci Technol 7(1):84–96

Bloetscher F, Hoermann S, Berry L (2017) Adaptation to Florida's urban infrastructure to climate change. In: Chassignet EP, Jones JW, Misra V, Obeysekera J (eds) Florida's climate: changes, variations, & impacts. Florida Climate Institute, Gainesville, FL, pp 311–338

Carter RC, Parker A (2009) Climate change, population trends and groundwater in Africa. Hydrol Sci J 54(4):676–689

Carter LM, Jones JW, Berry L, Burkett V, Murley JF, Obeysekera J, Schramm PJ, Wear D (2014) Southeast and the Caribbean. In: Melillo M, Richmond TC, Yohe GW (eds) Climate change impacts in the United States: the third national climate assessment. U.S. Global Change Research Program, Washington D.C., pp 396–417

Carter L, Terando A, Dow K, Hiers K, Kunkel KE, Lascurain A, Marcy D, Osland M, Schramm P (2018) Southeast. In: Reidmiller DR, Avery CW, Easterling DR, Kunkel KE, Lewis KLM, Maycock TK, Stewart BC (eds) Impacts, risks, and adaptation in the United States: fourth national climate assessment, vol II. U.S. Global Change Research Program, Washington, D.C., pp 743–808

Christensen JH, Carter TR, Giorgi F (2002) PRUDENCE employs new methods to assess European climate change. EOS 83:147

Cooke K (2017) Trouble ahead for Gaddafi's great man-made river. In: Middle East eye. https://www.middleeasteye.net/opinion/trouble-ahead-gaddafis-great-man-made-river. Accessed 9 June 2020

Cramer W, Guiot J, Fader M, Garrabou J, Gattuso JP, Iglesias A, Lange MA, Lionello P, Llasat MC, Paz S, Penuelas J (2018) Climate change and interconnected risks to sustainable development in the Mediterranean. Nat Clim Change 8(11):972–980

Cudennec C, Leduc C, Koutsoyiannis D (2007) Dryland hydrology in Mediterranean regions—a review. Hydrol Sci J/Journal des Sciences Hydrologiques 52(6):1077–1087

De Sherbinin A (2014) Climate change hotspots mapping: what have we learned? Clim Change 123(1):23–37

Diffenbaugh NS, Giorgi F (2012) Climate change hotspots in the CMIP5 global climate model ensemble. Clim Change 114(3–4):813–822

Diffenbaugh NS, Giorgi F, Pal JS (2008) Climate change hotspots in the United States. Geophys Res Lett 35(16):L16709

Douglass AE (1929) The secrets of the Southwest solved by talkative tree rings. Natl Geogr Mag 56(6):737–770

Galbraith K (2015) Panhandling for water. In: Texas tribune, 17 June 2015. https://www.texastribune.org/2010/06/17/how-bad-is-the-ogallala-aquifers-decline-in-texas/. Accessed 9 June 2020

Garfin G, Franco G, Blanco H, Comrie A, Gonzalez P, Piechota T, Smyth R, Waskom R (2014) Southwest. In: Melillo JM, Richmond TC, Yohe GW (eds) Climate change impacts in the United States: the third national climate assessment. U.S. Global Change Research Program, Washington D.C., pp 462–486

Giorgi F (2006) Climate change hot-spots. Geophys Res Lett 33(8):L08707

Giorgi F, Lionello P (2008) Climate change projections for the Mediterranean region. Glob Planet Change 63(2–3):90–104

Gonzalez P, Garfin GM, Breshears DD, Brooks KM, Brown HE, Elias EH, Gunasekara A, Huntly N, Maldonado JK, Mantua NJ, Margolis HG, McAfee S, Middleton BR, Udall BH (2018) Southwest. In: Reidmiller DR, Avery CW, Easterling DR, Kunkel KE, Lewis KLM,

Maycock TK, Stewart BC (eds) Impacts, risks, and adaptation in the United States: fourth national climate assessment, vol II. U.S. Global Change Research Program, Washington, D.C., pp 1101–1184

Harris A (2020) Feds consider a plan to protect Miami-Dade from storm surge: 10 to 13-foot walls by the coast. In: Miami Herald 07 Feb 2020. https://www.miamiherald.com/news/local/environment/article239967808.html. Accessed 9 June 2020

Heimlich BN, Bloetscher F, Meeroff DM, Murley J (2009) Southeast Florida's resilient water resources: Adaptation to sea level rise and other impacts of climate change. Jupiter: Center for Environmental Studies at Florida Atlantic University. www.ces.fau.edu/files/projects/climate_change/SE_Florida_Resilient_Water_Resources.pdf. Accessed 11 Nov 2018

Howard G, Pedley S, Barrett M, Nalubega M, Johal K (2003) Risk factors contributing to microbiological contamination of shallow ground-water in Kampala, Uganda. Water Res 37:3421–3429

Iglesias A, Garrote L, Flores F, Moneo M (2007) Challenges to manage the risk of water scarcity and climate change in the Mediterranean. Water Resour Manage 21(5):775–788

IPCC (2013a) Climate change 2013: The physical science basis. In: Stocker TF, Qin D, Plattner G-K, Tignor M, Allen SK, Boschung J, Nauels A, Xia Y, Bex V, Midgley PM (eds) Contribution of working group I to the fifth assessment report of the Intergovernmental Panel on Climate Change. Cambridge University Press, Cambridge UK

IPCC (2013b) Van Oldenborgh GJ, Collins M, Arblaster J, Christensen JH, Marotzke J, Power SB, Rummukainen M, Zhou T (eds) Annex I: atlas of global and regional climate projections supplementary material RCP2.6. In: Stocker TF, Qin D, Plattner G-K, Tignor M, Allen SK, Boschung J, Nauels A, Xia Y, Bex V, Midgley PM (eds) Climate change 2013: the physical science basis. Contribution of working group I to the fifth assessment report of the Intergovernmental Panel on Climate Change. Cambridge University Press, Cambridge UK

IPCC (2013c) Van Oldenborgh GJ, Collins M, Arblaster J, Christensen JH, Marotzke J, Power SB, Rummukainen M, Zhou T (eds) Annex I: atlas of global and regional climate projections supplementary material RCP4.5. In: Stocker TF, Qin D, Plattner G-K, Tignor M, Allen SK, Boschung J, Nauels A, Xia Y, Bex V, Midgley PM (eds) Climate change 2013: the physical science basis. Contribution of working group I to the fifth assessment report of the Intergovernmental Panel on Climate Change. Cambridge University Press, Cambridge UK

IPCC (2013d) Van Oldenborgh GJ, Collins M, Arblaster J, Christensen JH, Marotzke J, Power SB, Rummukainen M, Zhou T (eds) Annex I: atlas of global and regional climate projections supplementary material RCP6.0. In: Stocker TF, Qin D, Plattner G-K, Tignor M, Allen SK, Boschung J, Nauels A, Xia Y, Bex V, Midgley PM (eds) Climate change 2013: the physical science basis. Contribution of working group I to the fifth assessment report of the Intergovernmental Panel on Climate Change. Cambridge University Press, Cambridge UK

IPCC (2013e) Van Oldenborgh GJ, Collins M, Arblaster J, Christensen JH, Marotzke J, Power SB, Rummukainen M, Zhou T (eds) Annex I: atlas of global and regional climate projections supplementary material RCP8.5. In: Stocker TF, Qin D, Plattner G-K, Tignor M, Allen SK, Boschung J, Nauels A, Xia Y, Bex V, Midgley PM (eds) Climate change 2013: the physical science basis. Contribution of working group I to the fifth assessment report of the Intergovernmental Panel on Climate Change. Cambridge University Press, Cambridge UK

IPCC (2014) Climate change 2014: synthesis report. In: Pachauri RK, Meyer LA (eds) Contribution of working groups I, II and III to the fifth assessment report of the Intergovernmental Panel on Climate Change. IPCC, Geneva

Joffe G (2017) A worsening water crisis in North Africa and the Middle East. In: The conversation, 31 April 2017. https://theconversation.com/a-worsening-water-crisis-in-north-africa-and-the-middle-east-83197. Accessed 9 June 2020

Kerr RA (2008) Climate change hot spots mapped across the United States. Science 321:909

Klarić Z (2017) Challenges of sustainable tourism development in the Mediterranean. In: IEMed. Mediterranean yearbook 2017. European Institute of the Mediterranean (IEMed), Barcelona, pp 261–269

Laessing U, Elumami A (2019) In battle for Libya's oil, water becomes a casualty. In: Reuters, 2 July 2019. https://www.reuters.com/article/us-libya-security-water-insight/in-battle-for-libyas-oil-water-becomes-a-casualty-idUSKCN1TX0KQ. Accessed 9 June 2020

Lapworth DJ, Nkhuwa DCW, Okotto-Okotto J, Pedley S, Stuart ME, Tijani MN, Wright J (2017) Urban groundwater quality in sub-Saharan Africa: current status and implications for water security and public health. Hydrogeol J 25(4):1093–1116

Lauffenburger ZH, Gurdak JJ, Hobza C, Woodward D, Wolf C (2018) Irrigated agriculture and future climate change effects on groundwater recharge, northern High Plains aquifer, USA. Agric Water Manag 204:69–80

Little JB (2009) The Ogallala Aquifer: saving a vital U.S. water source. In: Scientific American, 1 March 2009. https://www.scientificamerican.com/article/the-ogallala-aquifer/. Accessed 9 June 2020

London S (2017) Florida county aims for full usage of reclaimed water. In: Water World, 1 March 2017. https://www.waterworld.com/wastewater/reuse-recycling/article/16191861/florida-county-aims-for-full-usage-of-reclaimed-water. Accessed 9 June 2020

MacDonald AM, Calow RC, MacDonald DM, Darling WG, Dochartaigh BE (2009) What impact will climate change have on rural groundwater supplies in Africa? Hydrol Sci J 54(4):690–703

Mackun PJ (2019) About 14.6 million people live in 40 counties in five states. In: U.S. Census Bureau. https://www.census.gov/library/stories/2019/02/fast-growth-in-desert-southwest-continues.html. Accessed 9 June 2020

McGuire VL (2017) Water-level and recoverable water in storage changes, High Plains aquifer, predevelopment to 2015 and 2013–15. U.S. Geological Survey Scientific Investigations Report 2017–5040

Mileham L, Taylor RG, Todd M, Tindimugaya C, Thompson J (2009) The impact of climate change on groundwater recharge and runoff in a humid, equatorial catchment: Sensitivity of projections to rainfall intensity. Hydrol Sci J 54(4):727–738

Miller C (2018) Farmers are drawing groundwater from the giant Ogallala Aquifer faster than nature replaces it. In: The conversation, 7 Aug 2018. https://theconversation.com/farmers-are-drawing-groundwater-from-the-giant-ogallala-aquifer-faster-than-nature-replaces-it-100735. Accessed 9 June 2020

Miller JA, Appel CL (1997) Ground water atlas of the United States: Kansas, Missouri, and Nebraska, HA 730-D. U.S. Geological Survey, Reston

Missimer TM, Maliva RG (2018) Environmental issues in seawater reverse osmosis desalination: intakes and outfalls. Desalination 434:198–215

Moutahir H, Bellot P, Monjo R, Bellot J, Garcia M, Touhami I (2017) Likely effects of climate change on groundwater availability in a Mediterranean region of Southeastern Spain. Hydrol Process 31(1):161–176

NIC (2009) North Africa: the impacts of climate change to 2030 (selected countries). National Intelligence Council, Washington, D.C.

Obeysekera J, Barnes J, Nungesser M (2015) Climate sensitivity runs and regional hydrologic modeling for predicting the response of the greater Florida Everglades ecosystem to climate change. Environ Manage 55(4):749–762

Obeysekera J, Graham W, Sukop MC, Asefa T, Wang D, Ghebremichael K, Mwashote B (2017) Implications of climate change on Florida's water resources. In: Chassignet EP, Jones JW, Misra V, Obeysekera J (eds) Florida's climate: changes, variations, & impacts. Florida Climate Institute, Gainesville, pp 83–124

Popper DE, Popper F (1987) The Great Plains: From dust to dust, a daring proposal for dealing with an inevitable disaster. Planning 53:12–18

Pulido-Velazquez D, García-Aróstegui JL, Molina J-L, Pulido-Velazquez M (2015) Assessment of future groundwater recharge in semi-arid regions under climate change scenarios (Serral-Salinas aquifer, SE Spain). Could increased rainfall variability increase the recharge rate? Hydrol Process 29:828–844

Qi SL (2010) Digital map of the aquifer boundary of the High Plains aquifer in parts of Colorado, Kansas, Nebraska, New Mexico, Oklahoma, South Dakota, Texas, and Wyoming. U.S. Geological Survey Data Series 543

Quinn F, Edstrom J (2000) Great lakes diversions and other removals. Can Water Resour J 25(2): 125–151

Roque J, Woodcock A, Kinslow J, Foulkes B, Akhimie V (2017) A potable reuse demonstration scale program in east coast Florida. In: AMTA/AWWA Membrane technology conference, Long Beach, CA. 02/14/2017

Rosenberg NJ, Epstein DJ, Wang D, Vail L, Srinivasan R, Arnold JG (1999) Possible impacts of global warming on the hydrology of the Ogallala aquifer region. Clim Change 42(4):677–692

Scanlon BR, Faunt CC, Longuevergne L, Reedy RC, Alley WM, McGuire VL, McMahon PB (2012) Groundwater depletion and sustainability of irrigation in the US High Plains and Central Valley. Proc Natl Acad Sci 109(24):9320–9325

Scott CA, Megdal S, Oroz LA, Callegary J, Vandervoet P (2012) Effects of climate change and population growth on the transboundary Santa Cruz aquifer. Clim Res 51(2):159–170

SEFRCCC (n.d.) Southeast Florida regional climate change compact. https://southeastfloridaclimatecompact.org/. Accessed 28 Jan 2020

Senatore A, Mendicino G, Smiatek G, Kunstmann H (2011) Regional climate change projections and hydrological impact analysis for a Mediterranean basin in Southern Italy. J Hydrol 399(1–2): 70–92

Southeast Florida Regional Climate Change Compact Sea Level Rise Ad Hoc Work Group (2019). Unified sea level rise projection for Southeast Florida. 2019 Update, document prepared for the Southeast Florida Regional Climate Change Compact Steering Committee. https://southeastfloridaclimatecompact.org/unified-sea-level-rise-projections/#gallery-1. Accessed 9 June 2020

Stigter TY, Nunes JP, Pisani B, Fakir Y, Hugman R, Li Y, Tomé S, Ribeiro L, Samper J, Oliveira R, Monteiro JP (2014) Comparative assessment of climate change and its impacts on three coastal aquifers in the Mediterranean. Reg Environ Change 14(1):41–56

Stover S, Buchanan R (2017) The High Plains aquifer: can we make it last? GSA Today 27(6):44–45

Taylor RG, Koussis AD, Tindimugaya C (2009) Groundwater and climate in Africa—a review. Hydrol Sci J 54(4):655–664

Union of Concerned Scientists (2019) Florida: ground zero in the climate crisis. https://www.ucsusa.org/sites/default/files/attach/2019/05/Florida-Gound-Zero-in-the-Climate-Crisis-newer.pdf. Accessed 9 June 2020

USDA (n.d.) Ogallala aquifer initiative. United States Department of Agriculture Natural Resources Conservation Service. https://www.nrcs.usda.gov/wps/portal/nrcs/detailfull/ut/home/?cid=stelprdb1048809. Accessed 9 June 2020

Vose RS, Easterling DR, Kunkel KE, LeGrande AN, Wehner MF (2017) Temperature changes in the United States. In: Wuebbles DJ, Fahey DW, Hibbard KA, Dokken DJ, Stewart BC, Maycock TK (eds) Climate science special report: fourth national climate assessment, vol I. U.S. Global Change Research Program, Washington D.C., pp. 185–206.

Wehner MF, Arnold JR, Knutson T, Kunkel KE, LeGrande AN, (2017) Droughts, floods, and wildfires. In: Wuebbles DJ, Fahey DW, Hibbard KA, Dokken DJ, Stewart BC, Maycock TK (eds) Climate science special report: fourth national climate assessment, vol I. U.S. Global Change Research Program, Washington D.C., pp 231–256

Williams AP, Cook ER, Smerdon JE, Cook BI, Abatzoglou JT, Bolles K, Baek SH, Badger AM, Livneh B (2020) Large contribution from anthropogenic warming to an emerging North American megadrought. Science 368(6488):314–318

Woodhouse CA, Meko DM, MacDonald GM, Stahle DW, Cook ER (2010) A 1200-year perspective of 21st century drought in southwestern North America. Proc Natl Acad Sci 107 (50):21283–21288

Yoder D (2017) Southeastern Florida. Ground zero for sea-level rise. In: Mulroy P (ed) The water problem—climate change and water in the United States. Brookings Institution Press, Washington, D.C., pp 145–165

Chapter 14
Applied Climate Change Assessment and Adaptation

14.1 Introduction

"Do what you can, with what you have, where you are." —Theodore Roosevelt.

"If you fail to plan, you are planning to fail."—Benjamin Franklin.

The Methods in Water Resources Evaluation book series has a strong applied focus, concentrating on what is practicable for hydrogeologists, engineers, and water users, utilities, and suppliers, who normally operate under budget and time constraints. There is now no serious doubt in the scientific community that continued anthropogenic climate change is now inevitable. Even if widespread and effective mitigation measures were implemented and greenhouse gas emissions were drastically reduced today, the climate will continue to warm.

Virtually all the world will experience warmer conditions, whereas there will be geographic differences in the direction and magnitude of precipitation changes. With respect to water availability, there will be winners and losers. Climate change may benefit some areas and sectors, such as the northern parts of the northern hemisphere, where agricultural conditions may improve as a result of higher temperatures, a longer growing season, and greater rainfall. Other areas will experience decreased annual precipitation and more intra-annual and interannual variability in precipitation. Of particular concern are areas that are already experiencing water scarcity and increasing demands associated with population growth (e.g., Mediterranean region and southwestern North American) and are projected to experience even drier conditions. Over the next few decades the "inexorable march of population growth" and its ever-increasing demands on water resources will be a much more serious threat for most less developed countries than climate change (Carter and Parker 2009).

Groundwater resources in some areas with arid and semiarid climates will face increased pressure from climate-related increases in water demands and decreases in precipitation and, in turn, aquifer recharge. Areas endowed with more abundant

R. Maliva, *Climate Change and Groundwater: Planning and Adaptations for a Changing and Uncertain Future*, Springer Hydrogeology,
https://doi.org/10.1007/978-3-030-66813-6_14

mean annual rainfalls may experience more frequent and severe droughts, although there is less confidence in local drought predictions. Drier conditions will result in reduced surface water flows and thus potentially increased demands on groundwater as an alternative water source. Higher temperatures are expected to contribute to decreased soil moisture through higher evapotranspiration rates, which will increase irrigation demands. Decreased precipitation and soil moisture will tend to reduce aquifer recharge, although the timing and intensity of rainfall events can have a greater impact on recharge rates than changes in mean annual rainfall.

Climate change impacts land use and land cover either directly through changes in the degree and type of vegetation cover and indirectly through human responses to changing conditions. Land use and land cover changes can impact infiltration rates, runoff, and evapotranspiration rates and thus groundwater recharge and quality (Scanlon et al. 2005). Changes in land use and land cover can also impact radiative forcing and climate responses (Mahmood et al. 2010; Pielke et al. 2011).

Adaptation to climate change is essentially a local process that is carried out at multiple levels of societies. Water supply decisions are usually made by water users (where self-supplied), water utilities, irrigation districts, and regional water suppliers. In the case of water utilities, irrigation districts, and regional suppliers, in-house technical staff, often supported by external consultants, identify potential water sources and determine the source or combination of sources that can most reliably and least expensively provide their required water. State and national governments influence water supply decisions through permitting processes (which can limit the use of some sources), controlling surface water supplies from regional supply systems, performing large-scale planning, supporting and conducting scientific research (including climate modeling), construction of water infrastructure (e.g., dams and aqueducts), and various forms of subsidies.

For water-related information on climate change to find use in the applied realm, it needs to be available to decision makers in a form that is congruent with their existing decision-making process or is of such apparent value to warrant adoption of new procedures. An example of the latter might be the employment of advanced decision support systems that facilitate decision making under conditions of future climate uncertainty. However, it is important to recognize that water users and utilities vary greatly in their technical and financial resources to conduct climate change studies and employ advanced planning technologies. Hence, there is the need for simpler workflows that facilitate evaluation of the impacts of climate change on local water resources and are within the technical capabilities and financial resources of a wide range of potential users (i.e., decision makers). It must be stressed that climate change alone in most cases will not drive water planning, but it is rather another variable that needs to be mainstreamed into the planning process.

14.2 Prediction of Local Climate Changes

The fundamental challenge in planning for, and adapting to, climate change is that there is a high degree of uncertainty in future global and local conditions other than that it will become generally warmer. Climate change projections are now based on general circulation model (GCM) simulations for which a critical input is future emissions scenarios. The state of the art for predicting future hydrological conditions, referred to as the "top-down" approach, involves the downscaling of the GCM simulation results to increase their spatial resolution and then inputting the downscaled projections of precipitation and temperature into hydrological models that partition precipitation into evapotranspiration, runoff, and groundwater recharge. More sophisticated hydrological models may explicitly simulate various vadose zone properties. Groundwater recharge is next inputted into groundwater flow models.

As a number of workers have noted, there is a chain of uncertainty in the top-down approach which results in large uncertainties in the final hydrological projections. The main uncertainties are:

- which future emissions scenario will come to pass
- the response of the atmosphere to future GHG concentrations (i.e., the ability of GCMs to predict future climate)
- uncertainties in the downscaling of coarse geographic-scale GCM projections to smaller local areas of interest
- hydrological modeling—the ability of models to accurately simulate recharge rates from precipitation projections
- the human response to climate change and how that might affect water use, and land use and land cover and thus aquifer recharge.

Future GHG emissions are inherently unknowable as they will be based on decisions yet to be made and implemented. Comparative studies have shown that projected local changes in precipitation can vary greatly depending upon the GCM used. In a study of the effects of climate change on aquifer recharge in the transnational Abbotsford-Sumas aquifer in coastal British Columbia, Canada and Washington State, USA, the projected change in recharge rates obtained for the 2080s varied depending on the GCM used from −10.5 to +23.2% relative to the historical recharge rate (Allen et al. 2010).

A critical part of the climate change adaptation process is developing a firmer understanding of the effects of current climate conditions, including variability, on water demands and water supplies. Historical experiences can provide some insights into how climate variations can impact local water supplies and demands. For example, conditions during past droughts can provide some guidance on future drier conditions. However, the great threat posed by anthropogenic climate change is that future climate conditions may be more extreme than anything in the historical climate record, leaving water users and suppliers unprepared.

In many aquifers, estimated recharge rates under current climate conditions are poorly constrained because aquifer recharge is small component of aquifer water budgets. Water available for recharge and runoff is the difference between precipitation and actual evapotranspiration rates with the latter value very commonly having a high measurement error and spatial variability. The actual recharge process may be complex and challenging to accurately model. Aquifer recharge may be either direct (diffuse), whereby water enters the groundwater system by vertical infiltration and percolation at the site of precipitation or application, or indirect (focused), whereby rainfall runs off from the area of precipitation and infiltrates and percolates to the water table at another location, usually stream channels, depressions, and other low-lying areas where water accumulates. Infiltration and percolation may occur either through the sediment matrix or flow may be concentrated in secondary porosity features. Infiltrated water may be lost to ET before it percolates past the root zone and can become recharge. In practice, recharge rates, as a poorly measurable quantity, are often estimated as a "fudge factor" that is adjusted to calibrate numerical groundwater flow models.

A proposed solution to the chain of uncertainties is to evaluate the envelope of possible future hydrological and groundwater conditions by performing multiple simulations using different emissions scenarios, GCMs, downscaling methods, and/or hydrological models (e.g., Chen et al. 2011; Crosbie et al. 2011). By performing a large number of simulations using a suite of different emissions scenarios and models, the full range of plausible future recharge and runoff rates (and other parameters of interest) can theoretically be captured. The multi-scenario and multi-model approach is technically sound, but from an applied water management perspective it is often impractical. Running large numbers of simulations using different complex models and downscaling methods is technically demanding and very time consuming and thus expensive. Such effort is beyond the resources of all but perhaps the largest water utilities and regional water providers. A second and perhaps more important limitation of the multi-scenario and multi-model approach is that it inherently tends to give a wide range of projected future conditions, which makes the results poorly actionable. For example, a conclusion that recharge rates may change from −40 to +20% is too broad to justify the cost of the investigation and is akin to acknowledging that the future impacts of climate change on local groundwater recharge are largely unknown.

A study by the Water Utility Climate Alliance to investigate options for improving climate modeling to assist water utility planning for climate change noted that water utilities are looking for climate science to be actionable and that there is a need to substantially reduce the range of projections at the geographic scale at which utility planning is conducted so that such projections could be used by utilities to help make decisions on expensive or long-lived investments on infrastructure (Barsugli et al. 2009). Specifically,

> Those interested in adapting to climate change have expressed a sense of frustration that the projections on regional climate change are not precise enough to support incorporating climate change into many regional and local decision-making, particularly those involving large financial investments. The perception is that ranges of projected changes—whether in

> sea level rise, temperature, precipitation, or other variables—either cover too wide a range to be useful for policymaking or, as is often the case with precipitation, do not agree on whether there will be an increase or decrease. (Barsugli et al. 2009, pp. 1–2)

Multi-scenario and multi-model approaches have value for developing a general understanding of the range of potential impacts of climate change on groundwater resources but are not a practicable approach in the applied water supply realm. However, the alternative "bottom-up" approach is more practical from an applied perspective. There are compelling arguments that measures for adapting to climate change should reduce their reliance on specific climate projections and instead focus on a range of plausible futures (Carter and Parker 2009; Goulden et al. 2009). A suite of potential climate-induced changes to groundwater and water resources is considered to simulate the sensitivity of the water supply system of concern to climate changes.

For groundwater-sourced water supplies, simulations could be performed using existing or newly developed groundwater flow models. An important advantage of the bottom-up approach for assessing climate change impacts to groundwater is that in many groundwater-dependent areas in developed countries, groundwater flow models have already been developed for the management of resources under current climate conditions. For example, in the state of Florida multiple generations of calibrated groundwater flow models have been developed and subsequently refined by the state water management districts and U.S. Geological Survey that could be used to evaluate the impacts of changes in recharge rates and groundwater withdrawals on groundwater levels. Some of the models have been, or are in the process of being, expanded to included density-dependent solute transport to allow for the simulation of the impacts of sea level rise on saline-water intrusion. The impacts of a drying climate could be readily simulated using existing models by, for example, decreasing recharge rates by 10, 20%, etc., and/or increasing simulated groundwater pumping.

More complex integrated surface water and groundwater models and decision support systems are required for large water supply systems involving both surface water and groundwater. A state-of-the-art example, is the CALifornia Value Integrated Network (CALVIN) model, which is a state-wide economic-hydrologic-engineering water model that analyzes all of the state's water supply and delivery systems and allows for simulation of the economic demands for water and various operational options for the complex water storage and distribution system (Lund et al. 2009; Chou et al. 2012; Davis 2020).

Climate change scenarios to be considered can be selected based on professional judgment or stochastically generated. In the former case, the potential direction and possible magnitudes of changes can be evaluated from existing climate change modeling results, such as the CMIP5 ensemble summaries in the IPCC AR5 (IPCC 2013) or existing regional RCMs. An important uncertainty in the scenario-based approach to assessing vulnerability to climate change is assigning an accurate probability to a given scenario. If a tested scenario is found to have a significant impact on water supplies (e.g., projected demands would not be met), then more detailed investigations need to be performed to evaluate the likelihood of the climate change scenario occurring and when in the future it might occur. For example, if decreases in recharge of 20% or greater can compromise the groundwater supply of an area, then the questions need to be investigated as to how likely that change is and when it might occur, and thus if and when investments in alternative water sources and other adaptations (e.g., managed aquifer recharge) should be made.

Stochastic methods involve the running of a large number of randomly selected scenarios to allow for the evaluation of a greater space of possible future conditions. Rather than trying to predict and plan for a given climate future, robust (no or low regrets) strategies are developed that meet performance criteria under the widest range of possible future climate conditions (Lempert and Schlesinger 2000). In other words, societies should pursue robust strategies that will work reasonably well no matter what the future brings (Lempert and Schlesinger 2000). Stochastic methods that involve that performance of very large number of simulations have become more viable with the progressive increase in available computation power (Lempert and Schlesinger 2000).

The critical question as far as the selection of methodologies in the applied water management realm is whether they will in fact provide decision makers with actionable information that they did not already have. The benefits of investigations as far as providing new information that could be employed for water planning should exceed their costs.

14.3 Prediction of Sea Level Rise Impacts

Sea level rise (SLR) is the climate change impact that receives the greatest attention because of the enormous populations and economic assets that are in harm's way from both permanent direct and groundwater inundation and an increased likelihood, frequency, and magnitude of flooding during storm events. Vulnerable areas include coastal cities across the world, coastal (deltaic) agricultural areas, and low-lying islands. Coastal ecosystems are also under threat from SLR. For example, mangroves, which serve a vital role in coastal tropical ecosystems, may not be able to survive if relative SLR rates exceed 7 mm/yr (Saintilan et al. 2020). Barrier islands and beaches are dynamic features that undergo local erosion and accretion under existing climate conditions, and accelerating sea level rise is expected to increase the instability (rate of migration) of these features (Moore et al. 2007; Miselis and Lorenzo-Trueba 2017; Nienhuis and Lorenzo-Trueba 2019).

The Organisation for Economic Co-operation and Development (OCED) ranked the cities of the world with respect to their vulnerability to coastal flooding in 2070 in terms of both exposed populations and asset values (Table 14.1; Nicholls et al. 2007). The analysis considered projected increases in population and socioeconomic changes. Of particular concern is that large population increases are projected in many of the exposed cities. A recent study estimated that under a high emissions scenario up to 630 million (M) people will live on land below projected annual flood levels for 2100 and up to 340 M for mid-century, compared to roughly 250 M at present (Kulp and Strauss 2019). Societies vary greatly in their adaptive capacities, but the extent and magnitude of potential impacts of SLR rise at rates projected for high emissions scenarios are beyond the capability of even the wealthiest regions to fully protect against.

An annual SLR of 3 or 4 mm would be barely perceptible to the casual observer and even a rise of 3–4 cm over a decade may not have noticeable impacts. There may not be a strong impetus for immediate action, particularly in areas where resources are inadequate to deal with current societal problems. But yet continued SLR at current, and more ominously at accelerating rates, could eventually have devastating impacts on low-lying coastal areas, particularly on small islands where landward retreat may not possible. If often takes a major climate event to serve as a wakeup call to the threat posed by climate change. In New York City, the extensive damage caused (and vulnerability revealed) by the storm surge from Hurricane Sandy in October 2012 was the turning point for climate resiliency work in the city (deMause 2019).

14.3.1 Prediction of SLR Impacts

The impacts of sea level rise on local groundwater resources depend upon the rate of local (relative) SLR (RSLR), land surface elevation, aquifer hydrogeology, and water use. RSLR rates are a function of changes in global mean sea level (GMSL) and changes in land surface elevation, ocean circulation, and weather (e.g., wind direction and energy). NOAA (n.d) mapped global historical rates of RLSR from tidal gauge data (Fig. 14.1). Hot spots for RSLR include deltaic areas undergoing land subsidence. Some northern areas are experiencing negative RSLR (i.e., net

Table 14.1 Top ten cities exposed to coastal flooding by 2070 by population and asset value considering both climate and socioeconomic changes

Ranking	Population	Asset value
1	Calcutta (India)	Miami (USA)
2	Mumbai (India)	Guangzhou (China)
3	Dhaka (Bangladesh)	New York-Newark (USA)
4	Guangzhou (China)	Calcutta (India)
5	Ho Chi Minh City (Vietnam)	Shanghai (China)
6	Shanghai (China)	Mumbai (India)
7	Bangkok (Thailand)	Tianjin (China)
8	Rangoon (Myanmar)	Tokyo (Japan)
9	Miami (USA)	Hong Kong (China)
10	Hai Phong (Vietnam)	Bangkok (Thailand)

Source Nicholls et al. (2007)

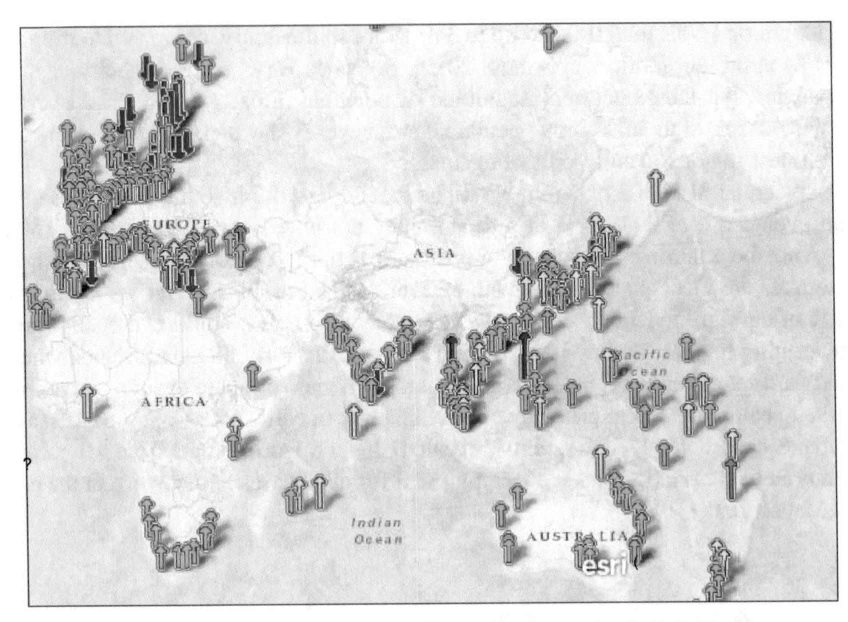

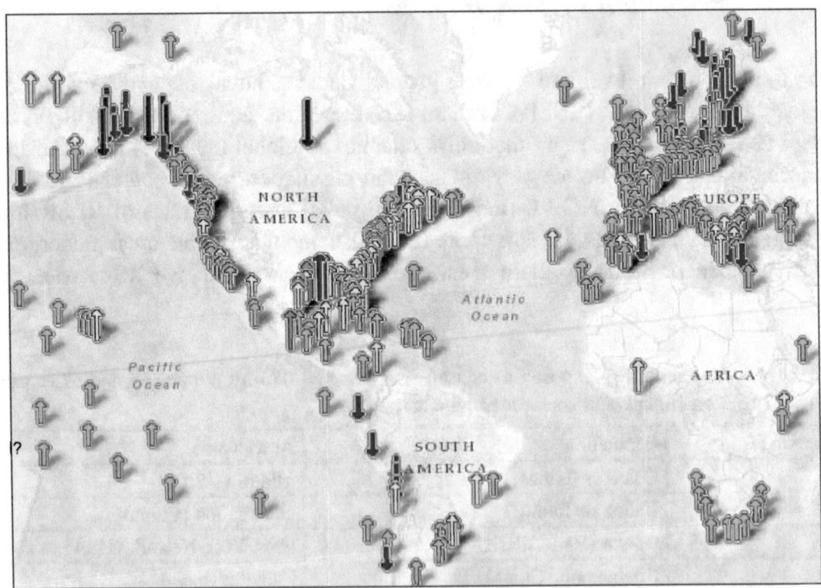

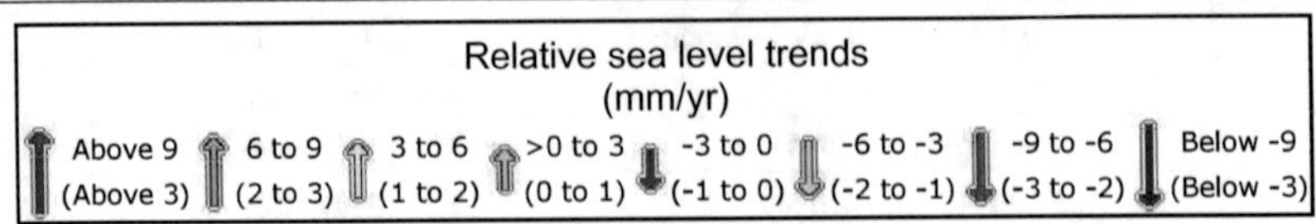

Fig. 14.1 Map of historical relative sea level rise trends. *Source* NOAA (n.d.)

lower sea levels) due to land masses rising from isostatic rebound (i.e., post-glacial rise in land surface due to the removal of the huge weight of ice sheets).

Sea level rise impacts will vary between changes in mean sea level, high tide (mean higher high water; MHHW), extreme tidal events (e.g. king tides), and storm surges and maximum wave heights from storms. MHHW is the average height of the highest tide recorded at a tide station each day during the recording period (19 years). King tides are colloquially taken as the highest tide in a year. As a first order approximation, changes in local sea level can be estimated by applying the historical difference between the rates of RSLR and GMSL to projected rates of GMSL rise. The rates of land surface elevation change and the differences between mean sea level and tidal levels are assumed as an approximation to not significantly change over the timeframe of interest.

According to NOAA, GMSL has risen by 0.14 inches (3.6 mm) per year during 2006–2015 (Lindsey 2019). The rate of GMSL rise has accelerated over the past several decades and all projections indicate continued acceleration. The greatest uncertainty is associated with the rate of melting of the Antarctic and Greenland ice sheets with recent estimates tending toward higher rates of melting and GMSL rise. A survey of 106 experts on climate change and sea level rise, chosen based on their (co)authorship of publications on the subject of sea level, projected under RCP2.6 a likely (central 66% probability) GMSL rise relative to 1986–2005 of 0.30–0.65 m by 2100, and 0.54–2.15 m by 2300 (Horton et al. 2020). Under RCP8.5, the experts projected a likely GMSL rise of 0.63–1.32 m by 2100, and 1.67–5.61 m by 2300; Horton et al (2020).

Areas vulnerable to direct and groundwater inundation can be estimated with GIS using the hydrostatic assumption (i.e., the "bathtub approach") by subtracting the surface elevations (obtained from a digital elevation model) from projected local sea level (e.g., MHHW). Alternatively, rising sea levels can be applied as a boundary condition for groundwater flow models of coastal areas. Vulnerability assessments are typically performed by mapping the areas that would be inundated for either arbitrary SLR increments (e.g., 0.25, 0.5, and 1.0 m), or using projected SLR values for future years. The bathtub approach tends to underestimate inundated areas by not considering storm and wave impacts.

Areas experiencing high local rates of RLSR rise due to subsidence are particularly vulnerable. For example, agricultural lands and fresh groundwater resources in the Nile River Delta are under threat from a combination of increased land subsidence, global sea level rise, and rising groundwater levels and associated soil salinization (Kotb et al. 2000; Hereher 2010; Hassaan and Abdrabo 2013; Stanley and Clemente 2017).

The groundwater quality impacts of sea level rise can be evaluated through density-dependent solute transport modeling. In coastal area with sufficiently high land elevations to avoid permanent inundation, the impacts of SLR on horizontal saline water intrusion into coastal aquifers over a short to mid-term (≤ 20 years) planning horizon are likely to be small, and usually much less than the potential impacts from population growth-induced increases in pumping. There is a greater vulnerability to salinization of surficial aquifers from storm inundation, particularly if coastal storms increase in frequency and intensity. Projected higher sea levels on the order of 1 m or more by the end of the century would profoundly impact many coastal communities and ecosystems.

As a broad generalization, impacts to groundwater supplies will likely be a small element of the overall impacts of sea level rise on coastal communities. Municipal water supply wells tend to now be located inland of areas subject to inundation from sea level rise because of vulnerability to saline intrusion under current conditions. Groundwater inundation in lower lying areas will likely be a greater threat from sea level rise than salinization of groundwater supplies. Household and community shallow wells in coastal area, including small islands, are vulnerable to contamination from storm overwash events, the frequency of which may increase with both sea level rise and an intensification of storms. The timing of large ($\geq$0.3 m, 1 ft) rises in sea level is beyond the common planning horizon used for water supply. Nevertheless, potential long-term future SLR is an important consideration for the siting and construction of long-lived infrastructure.

14.3.2 Sea Level Rise Adaptation

Countries, cities, and smaller communities with low-lying coastal areas will face difficult decisions between protection, accommodation and retreat. It is far beyond the scope of this book to review all the adaptation options to SLR. Protection mainly involves the construction of walls, berms (dikes), and other barriers to hold back the sea. Accommodation is the hardening or elevation of buildings and other infrastructure (Fig. 14.2). Dry floodproofing is sealing buildings so that water cannot enter their interior during flooding events. The remaining option is retreat. Managed retreat has been defined as “the planned, purposeful, coordinated movement of people and assets away from risk” (Siders 2019). Unmanaged retreat is simply people leaving an area on their own accord in an uncoordinated manner.

The Netherlands is the textbook example of adaptation in response to rising RSLR where for centuries a system of dikes, barriers, and pumps has been constructed and expanded to protect the country, approximately 25% of which is now below sea level. The Dutch strategy for addressing the additional challenges of climate change is both building larger and more robust barricades and now local retreat (Zuidema 2019). It is important to recognize that the Dutch model of protecting its coastline is not viable for the United States (and many other countries) because of its much longer coastline, and associated prohibitive costs, as well as because it would cause unacceptable environmental impacts (Pilkey et al. 2016).

Retreat is the strategy that most effectively eliminates risk as resistance measures, such as seawalls and levees, can fail or be breached (Pilkey et al. 2016; Siders 2019; Siders et al. 2019). Protective measures can also encourage development in at-risk locations by creating a false sense of security (Siders 2019). Protection and accommodation are attractive options in that they allow for continued occupation of coastal areas and are the least disruptive. However, large-scale protection and

Fig. 14.2 Accommodations to sea level rise includes elevation of structures. Recently rebuilt rest area adjoining Tampa Bay, Florida, (I-275 south of the Sunshine Skyway Bridge) that is elevated on pilings

accommodation against sea level rise is extraordinarily expensive. An initial estimate of a project to protect parts of Miami-Dade County, Florida, is US$ 8 billion (Harris 2020) and cost estimates to increase the resilience of New York City against SLR are in the tens of billions of dollars (deMause 2019). Invariably, cost estimates for protection and accommodation schemes will turn out to be too low.

Societies face very difficult decisions related to SLR and coastal developments. Some key questions raised by the U.S. Congressional Research Service (Folger and Carter 2016) include who is responsible for the costs associated with adjusting to SLR and who will bear the risks associated with vulnerable coastal development and infrastructure. Folger and Carter (2016) noted that

> Some stakeholders are concerned that governments at all levels are paying insufficient attention to the risks posed by sea-level rise; others are concerned that overestimating the risk of sea-level rise could result in foregoing current uses of coastal areas and promoting overinvestment and overdesign of sea-level rise mitigation and adaptation.

Governmental policies can result in maladapted incentives that discourage retreat, such as subsidized flood insurance and land use and development codes that allow further development in at-risk areas (Pilkey et al. 2016; Siders 2019). "Ceasing to advance is the first step in retreat" (Siders 2019, p. 223).

Siders (2019, p. 223), who is a strong advocate of managed retreat as an adaptation to SLR, concluded with respect to the United States that

> Retreat at large scale in the US—whether managed or unmanaged—is unlikely to happen this year. It may not happen this decade. The urgency in managed retreat is to set plans in place: to have conversations about what Americans want their future coast to look like; to consider the role managed retreat can play in achieving those goals; to identify areas where retreat may need to occur; to prohibit, limit, or modify development in those areas (e.g., allow new buildings only under temporary use permits); to determine threshold conditions and tipping points at which retreat will be pursued; and to establish preliminary procedures for retreat.

In areas with high-value assets and populations that would be cost prohibitive to move, protection and accommodation will continue to be pursued. Climate change protection schemes will have to be justified in turns of their cost-benefit ratio; i.e., overall project benefits should exceed projects costs. As a generalization, protection of major urban areas with their large populations and high-value commercial and industrial assets is economically justified. However, much of the global coastline will be necessarily be left behind and residents forced to retreat as its protection cannot be economically justified. Even if protection does have a favorable cost-benefit ratio, the financial resources may not be available. Small communities face a great challenge funding resilience projects because their smaller populations and tax bases result in per capita costs being much greater than in more populous urban areas (Morrison 2019). Coastal areas of poorer developing countries will face the formidable challenge of large impacted populations and scarce resources upon which there are many demands. A key issue is that investments in protection do not generate new value or a positive cash flow to recoup costs but rather protect against the loss of existing value (Morrison 2019).

As further explained by the OCED (2019)

> Modelling demonstrates that coastal protection is economically robust for 13% of the world's coastline – which accounts for 90% of the global coastal population and 96% of global assets. An implication of this is that the world is likely to see bifurcating coastal futures. On the one hand, the large majority of coastal inhabitants live in densely populated urban coastal areas, and are likely to continue to protect themselves even under high end sea-level rise due to the high cost-benefit ratios of coastal protection in these areas. This means engineered coasts with higher and higher defences, and possible catastrophic consequences in the case of sea wall failure. On the other hand, rural and poorer areas will struggle to maintain safe human settlements and will likely be forced to retreat from the coast.

With respect to groundwater, SLR can contribute to saline water intrusion. Salinization of shallow aquifers is of particular concern on small islands where freshwater lenses are a critical element of the local water supply. In rural areas, SLR can cause salinization of soils and shallow groundwater used for irrigation. However, in most urban coastal areas, wellfields used for public supply are now located inland, away from the coast.

Saline water intrusion is a slow process, and the most important adaptation task supporting groundwater use near coasts is a rigorous monitoring program that allows for the assessment of the rate of movement of the fresh-saline groundwater interface. Such saline water intrusion monitoring programs have already long been implemented in urban and suburban areas of developed countries for potentially impacted aquifers used for water supply. For example in Florida, salt-water intrusion monitoring is a normal permit requirement for near coastal wellfields. Groundwater (solute-transport) modeling calibrated using historical groundwater pumping and water quality (salinity) data is a critical tool for predicting the position of the saline-water interface under future sea level and groundwater pumping scenarios. Adaptation options to saline water intrusion include reducing pumping, moving pumping inland, increasing landward recharge, and development of hydraulic and physical salinity barriers.

The greatest groundwater-related threat to coastal areas from SLR will likely be rising water tables (groundwater inundation), which can cause local flooding (i.e., rise of water table to above land surface) and adversely impact underground infrastructure (e.g., basements, underground parking areas, septic systems, sewer pipes and utility equipment). Vulnerability to groundwater inundation can be mapped using detailed topographic maps (e.g., remotely sensing-derived digital elevation models) and sea level rise projections. Vigilance in the form of water level monitoring is critical for early identification of local groundwater inundation problems before adverse impacts to structures, septic systems, and other infrastructure occurs. Adaptation options for groundwater inundation include hardening of infrastructure, drainage, and retreat. Barriers constructed to protect against direct seawater encroachment, unless they are anchored in impervious rock or clay, will not protect against inland groundwater inundation.

14.4 Water Supply Adaptation Options

Adaptation options to climate-change induced water scarcity are essentially the same as used to address water scarcity under current climate conditions and are categorized as either demand management, new supply, and optimization solutions. Managed aquifer recharge is considered an optimization strategy as it involves making better use of existing water resources rather than creating a new water source. Decision makers will need to identify some combination of adaptation options to meet their current and future water needs.

14.4.1 Water Demand Management and Reallocation

Demand-side solutions involve the reduction in the consumption of water, ideally through increased efficiency of water use. Efficiency is defined as the ratio of output to input. The objective of demand-side solutions is usually to maintain (or preferably to increase) output while decreasing water input. Increased water use efficiency can be achieved by eliminating wasteful uses and adopting more water efficient technologies. Usually reducing demands is less expensive than developing new alternative sources of water. There is also an ethical argument that wasteful use of any scarce resource is inherently wrong.

The initial element of water demand management is the implementation of technological measures to increase water use efficiency (i.e., doing more with what you have; Turton 1999). Water use efficiency is relatively easy to apply and is unlikely to cause disruptions. For example, water efficient appliances are designed with the intended purpose of providing the same consumer satisfaction while using less water. Irrigation practices can be implemented that more directly provide water to the root zone of crops (e.g., micro-irrigation). Technologies to increase water use efficiency can be imported if they are not locally available. Increases in water use efficiency may also involve behavioral changes, which can be brought about through legal compulsion (e.g., building code requirements and lawn watering restrictions), education, and economic incentives (water pricing).

Revenues from water sales finance the operation of water utilities, which include fixed costs. Sudden decreases in water sales due to conservation has forced some utilities to increase their rates, placing some consumers in the position that they are using less water and paying more for it. This dilemma is due in part to utility water prices being based on the average cost to produce and distribute water, not the marginal cost to produce additional water. Water rates would increase even more in the absence of conservation due to the high marginal costs of obtaining additional alternative water sources.

Economic efficiency is a measure of the economic output or value derived from given amounts of water use. The second element of water demand management is the reallocation of water between sectors, particularly from agriculture to the residential, industrial, and commercial sectors (i.e., doing other things with water; Turton 1999). Given a limited amount of available water, the issue arises as to how that water should be best used. In the dry southwestern United States, where agricultural interests control a large part of the groundwater and surface water resources, the voluntary reallocation of water from agriculture to municipal use may be the least expensive source of additional water to supply growing communities. A water utility or supplier may make free market purchases of farm or ranch lands with their associated water rights.

Changes in the use of water are necessary and desirable in a dynamic society (National Research Council 1992). The benefits of reallocation can be great, but the process can be stressful and controversial if jobs and livelihoods are lost in the agricultural sector (Turton 1999). The Committee on Western Water Management of the United States National Research Council conducted a study of water transfer issues in the western United States, focusing on third party effects (National Research Council 1992). A key observation was that the primary parties in a voluntary water transfer (i.e., the buyer and seller) negotiate their own best interests and exercise control over whether or not the transfer will occur, but third parties who stand to be affected by the transfer are not represented. Water is typically transferred from rural to urban communities, and the discontinuation of agricultural activities can result in a loss of farm jobs and farm incomes, less agricultural business activity, and less household spending in the rural community, which affects non-farm related businesses (Office of Technology Assessment 1983; National Research Council 1992; Henderson and Akers 2008; Clayton 2009; Glennon 2009). The importance of rural community preservation is a social, political and economic issue for which there is no universal answer as there is no consensus on the value of rural agricultural communities (National Research Council 1992).

Reallocation also ties into food security and virtual water issues. While it makes eminent good economic sense to reallocate water from agricultural to municipal use in water scarce regions, the reduction in local food production will need to be compensated for by the importation of food grown in areas more richly endowed with water resources. The loss of local agricultural production could impact food prices. On a national level, reliance on the importation of food has national security implications. Reallocation (transfer) of water from agriculture to urban and sub-urban municipal uses is not an all or nothing proposition. Purchase of agricultural land for its water rights (allocations) can be an economically attractive means for water utilities and regional suppliers to obtain additional water for municipal use provided that the local water governance system allows the practice.

14.4.2 New Water Supply Options

Three main options are available for new, large-scale local water supplies beyond local surface water and groundwater resources:

- seawater and brackish groundwater desalination
- wastewater reuse
- water transfers

Desalination of seawater and brackish groundwater is a viable new water supply option for coastal areas and inland areas that are underlain at depth by aquifers containing brackish groundwater. Reverse-osmosis desalination costs have dropped greatly over the past several decades with total water costs for some large-scale seawater desalination plants dropping well below US$ 1 per cubic meter (Ghaffour et al. 2013). The costs of brackish groundwater desalination are usually even less. Desalination is thus economically viable for potable and high-value industrial uses but it is generally too expensive for agricultural irrigation.

A key feasibility issue for desalination facilities is identifying an economical and environmentally sound means of the disposal of the concentrate (residual salts). For any sizeable facility, great amounts of residual salt (in the form of highly saline water) is generated. Offshore outfalls are commonly used for coastal facilities but are facing increasing regulatory scrutiny. Concentrate disposal is a much greater challenge at inland locations where disposal options are more limited. Deep injection is used in some areas of the United States, particularly Florida, where hydrogeological conditions in large parts of the state are favorable for the practice.

Zero liquid discharge options are available but tend to be very expensive and still generate large amounts of salt that requires environmentally safe disposal or perhaps processing into marketable products.

Wastewater reuse (water recycling) is already widely practiced in areas facing water scarcity. Reclaimed water is used primarily for non-potable purposes, such as lawn (e.g., park and golf course), landscaping, and agricultural irrigation and some industrial uses (e.g., cooling water). Wastewater reuse avoids the environmental impacts from the disposal of the treated wastewater and, more importantly, makes otherwise used better-quality fresh groundwater and surface water available for higher value uses (i.e., potable supply). Wastewater reuse is most economical for utilities when there are large users (e.g., farms, parks, and golf courses) that can accept large flows and new residential developments where dual-piping systems can be installed at the time of construction. Installing new reclaimed water piping in existing residential communities tends to be cost prohibitive although it has been done in some communities. A residential irrigation reuse system was installed in parts of Cape Coral, Florida, starting in the early 1990s, to reduce demands on the overburdened freshwater aquifer (Godman and Kuyk 1997). In the City of Cape Coral, demand for reuse water at times exceeds its supply and the reuse system is augmented with surface water. Indeed, in many areas treated wastewater has transitioned from a disposal problem to a valuable resource where during at least some times of the year demands outstrips supplies.

Potable reuse of wastewater has understandably lagged behind nonpotable reuse but is receiving increasing attention. Direct potable reuse is being increasingly considered for urban water supply because the technology now exists to produce high-purity water that is of better quality than existing supplies, it is a local source of water (Leverenz et al. 2011), and it has a lesser carbon footprint than desalination (Cornejo et al. 2014). Utilities and regional water suppliers considering a potable reuse scheme should start a public education and relations program early to gain support for the projects and preempt disinformation campaigns.

New water transfer schemes are becoming less likely because of a combination of very high costs, strong political opposition from donor areas, and environmental concerns. Transfer projects have historically been implemented by national and state governments, regional water suppliers, and very large urban water users.

14.4.3 Optimization—Conjunctive Used and Managed Aquifer Recharge

Optimization solutions involve various forms of conjunctive use and managed aquifer recharge (MAR) to take best advantage of available water resources. Key issues are matching water quality with the minimum requirements for each planned use. Fresh groundwater is a particularly valuable resources in areas now facing water scarcity and may be under even greater demand under future climate conditions because of its high reliability. As a number of workers have noted, the most important value of groundwater may be to serve as bridge across drought periods when surface water supplies are reduced or unavailable. Hence, the basic conjunctive use strategy is to use more ephemerally available surface water when available and reserve groundwater for periods of inadequate surface water availability. Similarly, treated wastewater should be the primary water source used where its quality is acceptable for its intended use with groundwater reserved as a supplemental source.

Conjugate use schemes are most commonly implemented for irrigation water supply where surface water and reclaimed water are of suitable quality for the locally grown crops. Conjugate use for potable supply may be economically unattractive for some communities where it requires construction of new treatment facilities or where surface water has limited temporal availability and thus construction of storage facilities (e.g., reservoirs and aquifer storage and recovery systems) is required. Nevertheless, surface water treatment and storage are usually less expensive than seawater desalination and, therefore, may turn out to be the least expensive large-scale alternative water source in some areas.

Managed aquifer recharge includes a broad range of technologies that can be used to increase aquifer recharge and thus the amount of water stored in aquifers (Maliva 2019). A key element of MAR is the availability of a source of water for recharge for which there is not an immediate demand and the water would thus not otherwise be put to beneficial use. Surface water flows that would otherwise be lost to tide are an obvious attractive water source for MAR provided that the decline in freshwater flows does not harm offshore ecosystems. Reclaimed water that would otherwise be disposed of is another attractive source of water for MAR although water quality issues will be a major concern if indirect potable reuse may occur. In areas facing increasing water scarcity due to climate change, MAR may be the best option for offsetting decreases in recharge and additional groundwater withdrawals.

MAR includes both active and passive technologies. The former includes systems that inject treated water in wells and recharge using actively managed rapid infiltration basins. Passive technologies, which include stormwater best management practices and low impact development methods, involve modifications of the land surface to decrease (or locally capture) runoff and promote local infiltration and recharge. In arid regions where recharge is indirect (focused), the opposite strategy may be employed whereby measures are taken to increase runoff from catchment areas and concentrate it in settings with high permeabilities (e.g., ephemeral stream channels) where recharge actually occurs (Lohse et al. 2010). It is important to be aware of the entire toolbox of MAR technologies so as to be able to select and design a program that is most suitable for local water supplies and demands, and hydrogeological conditions.

14.5 Decision-Making Under Climate Uncertainty

The degree to which climate change needs to be considered in local water supply planning depends upon the water sources currently used for supply and the projected direction and magnitude of climate changes over the planning horizon in question. The available historical data and modeling results suggest that for most areas, changes in precipitation over the common 20 yr planning horizon will likely fall with the range of historical variability, i.e., there will not be a clear anthropogenic climate change signal above the historical "noise" in the data. It is generally not possible to conclusively determine whether recent drier conditions in a region are a manifestation of anthropogenic climate change or are part of the normal climate variation with a return of wetter conditions expected. However, there is some indication that recent extreme drought conditions in southwestern North America likely have an anthropogenic component (Williams et al. 2020). There is an obvious risk, for example, in relying on the expectation that a current dry period will soon break, as similar ones have done in past, when it might actually be the beginning of a long-term transition to drier conditions. Similarly, sea level rise at projected rates over the next couple of decades will likely not significantly impair most groundwater resources used for water supply, with the exception of some small low-lying islands.

Over the near and intermediate term, increases in water demand from population growth and economic development will likely exceed climate change impacts on groundwater. This is not to say the climate change does not need to be a consideration in water planning. Although anthropogenic climate change may not immediately impair local water supplies, it is critical that decisions not be made that commit water users, utilities, and suppliers to maladapted pathways that will eventually not be sustainable. Therefore, it is important for water supply decision makers to have at least a general roadmap as to where future water supplies will be obtained, when they will eventually be needed, and how climate change could impact those supplies.

The options available for addressing the impacts of climate change on water resources are largely the same available for coping with water scarcity in general. There is now an overall consensus in the climate change literature that water supply planning (and climate change adaptation planning in general) should focus on robust, no or low regrets solutions, rather than on solutions that are optimal for a narrow range of projected future climate conditions. Robust solutions include demand reduction (e.g., conservation and increased water use efficiency), climate-insensitive new water sources (e.g., desalination and wastewater reuse), and conjunctive use and managed aquifer recharge. Adaptative management is also recognized as being a critical element of water supply planning in an uncertain and changing environment, in which plans are adjusted based on experiences (monitoring data). Water supply systems should be designed and constructed to retain flexibility to allow for changes in operations over time.

Scenario-based (bottom-up) vulnerability assessments are the most practical methodology to examine the potential impacts of climate change and to explore adaptation options and strategies to improve the resilience of water supply systems and other infrastructure that could be impacted by changes to local groundwater systems (and climate change in general). A key part of the decision-making process is the intellectual exercise of exploring potential impacts and responses, which for climate change involves examining how various climate change scenarios could impact a water supply option (and the operations of a water supply and distribution system as a whole) and then what adaptation options could be taken and what are their associated costs and benefits. In logical argument terms, it is examining "if A, then B, and then C" type relationships. An example of this type of analysis is that if precipitation decreases by 10%, then a corresponding decrease in available surface water might seasonally occur, which would then be addressed by pumping additional groundwater, storing surface water underground in an aquifer by MAR during wet periods for use during dry periods, or mandating decreases in water use. For coastal communities at risk from sea level rise, vulnerability assessment should systematically consider the impact to water, wastewater, and other infrastructure from incremental increases in local sea level and groundwater levels. Adaptation options are then evaluated for each identified adverse impact.

The similar threshold risk assessment approach involves a systematic evaluation of the threshold or performance criteria for each component of a water system, which are the range of conditions for normal operation (Freas et al. 2010). Climate conditions that could cause the threshold conditions of components to not be met are identified, and then compared to the range of anticipated local changes in climate variables. Adaptations strategies are developed for identified vulnerable components.

The general topic of decision-making under uncertainty is a subject of much interest including dedicated books (e.g., Kochenderfer 2015) and college courses. Normative decision theory focuses on the identification of optimal decisions, which, in economics terms, involves using expected value (net benefits) methods that are based on the net present value and probability of various contingencies (Boardman et al. 1996). In general terms, multiple options are identified that could provide the required water supply for a geographic area (e.g., utility service area or water district). The future is divided into a series of mutually exclusive contingencies, which could include various climate change scenarios. For each supply option, the net present value of each contingency is then calculated, and a probability is assigned to each contingency. The expected net benefits (ENB) for each supply option are calculated as

$$\mathrm{ENB} = \sum \mathrm{P_i}(\mathrm{B_i} - \mathrm{C_i})$$

where P_i = probability of contingency "i", and B_i and C_i and the present value of the benefits and costs of contingency "i".

One contingency might be, for example, surface water flows decreasing below a threshold value required for a supply option under more frequent or severe future drought conditions, in which case the economic costs of the water shortage needs to be quantified. Assigning probabilities is typically a matter of expert professional judgment, which with respect to climate change is constrained by the inherent uncertainties associated with climate change projections.

Hydroeconomic decision support systems (DDSs) have been developed with the goal of facilitating analysis of uncertainty in complex water supply systems (Sect. 12.4; Yates et al. 2005; Huber-Lee et al. 2006; Purkey and Huber-Lee 2006; Sieber and Purkey 2011). The basic modeling procedure of first developing a reference ("business as usual") scenario and then developing a series of "what if" scenarios is well-suited for examining how climate change might impact both water supplies and demands and to evaluate potential water management options to respond to the changes. Decision makers will need to determine whether the time and cost involved in developing a DSS of their local water system is commensurate with the value of the actionable information potentially provided. Many smaller water supply systems are not particularly complex and simpler methods may be sufficient to evaluate the impacts of potential climate changes.

14.6 Prognosis and Recommendations

There is growing recognition that water supply has become an ever more difficult global challenge that will be compounded by climate change in some areas, including arid and semiarid regions already facing water scarcity. Water has been described as "the defining crisis of the twenty-first century" (Pearce 2006). Solomon (2010, p. 367) proclaimed that

> water scarcity is cleaving an explosive fault line between freshwater Haves and Have-Nots across the political, economic, and social global landscapes of the twenty-first century

Water scarcity problems are of obvious national and economic security importance and are receiving great governmental and academic attention. Water scarcity issues are also entering the general public consciousness. Recently published mass-market books provide approachable discussions of current and potential future freshwater challenges (e.g., Postel 1992; Reisner 1999; De Villiers 1999; Glennon 2002, 2009; Pearce 2006; Solomon 2010; Barnett 2011; Barbier 2019).

Groundwater resources in many areas of world are already being exploited at close to, or in excess of, sustainable levels, especially arid and semiarid regions where, by definition, local renewable water resources are limited. Climate change modeling results are clear in that temperatures will increase across the world, which will tend to result in greater water demands. However, it is likely that in most areas increases in population growth and economic development will have a greater impact on water demand than climate change. The effects of climate change on local precipitation and aquifer recharge are less clear, but climate modeling results indicate that some already dry regions, such as the southwestern North America, parts of Central America, and the Mediterranean region will likely become drier, which would exacerbate their existing water scarcity. Climate modeling results suggest (albeit with some uncertainty) that droughts may become more frequent and intense. Williams et al. (2020, p. 318) observed that the "effects of future droughts on humans will be further dependent on sustainable resource use because buffering mechanisms such as ground water and reservoir storage are at risk of being depleted during dry times." Hence, preserving and ideally augmenting groundwater resources is important for preserving their ability to serve as buffers or bridges across droughts. However, in some developing countries groundwater governance in weak and aquifer overdraft continues to meet immediate water demands.

Water scarcity in arid and semiarid lands occurs on three main levels. First, and most critical, is the supply of water to meet basic human consumption and sanitary needs, which involves the smallest volume of water, but has the greatest human impacts when not available. Second, is the supply of water needed for local food and industrial production. Third, is water used for amenities that are now considered basic elements of civilized life in developed countries. Groundwater use is ultimately self-limiting and as wells literally start to dry or the water becomes unacceptably salty, users will have to make do with lesser amounts of available water. Indeed, climate change adaption is ultimately a matter of proactively identifying and early implementing less disruptive options, rather than being forced to later reactively implement more disruptive adaptations.

Vulnerability assessments are a critical element of climate change adaptation. Vulnerability assessments involve evaluations of possible future climate changes and how they might impact systems of concern. Once climate change risks are identified, then the next step is an evaluation of the probabilities of such climate changes and associated impacts occurring, when they might occur, and the magnitude of the impacts. Possible adaptation options are then explored for identified risks whose impacts are considered unacceptable.

Climate change projections invariably have a high degree of uncertainty. Hence, it is prudent that at least general strategies be developed to respond to a variety of possible climate change scenarios and flexibility is retained in water management systems to allow for change in course as monitoring data provide more refined guidance on local changes. Robust, no or low regrets strategies that are functional under a wide range of future climate conditions are preferred.

The prognosis for the future state of local groundwater resources depends to a large degree on the ability of groundwater governance systems to control extractions to what are considered sustainable levels. Sustainable use of groundwater is generally taken as controlling long-term extractions so that neither unacceptable declines in aquifer water levels nor other adverse impacts occur. Other unacceptable impacts may include lowering of water levels in wetlands and decreases in lake water levels and stream and spring flows. Where governance is weak and groundwater use is largely uncontrolled, aquifer overdraft may continue until it is no longer economically viable to drill new deeper wells or deepen existing wells, pumping costs become too great, or water quality deteriorates to the point where the groundwater is no longer usable.

Where groundwater governance is strong (i.e., capable of actually controlling groundwater use), limitations on additional groundwater pumping is forcing users to invest in alternative water sources or aquifer augmentation (managed aquifer recharge). For example, in the state of Florida, limitations on the permitting of additional fresh groundwater withdrawals has forced utilities to invest in brackish groundwater desalination, wastewater reuse, seasonal surface water use, and aquifer recharge, as well as promoting conservation. The move to less climate-sensitive alternative water supplies has had the added benefit of increasing the resilience of the water supply systems to climate change. Florida and other states and developed countries have the financial resources to develop alternative water supplies although costs to users will increase.

Although developed countries may have sufficient resources to develop alternative water supplies, fragmentation of their water supply systems can be a hindrance to climate change adaptation (Mullin 2020). Small utilities may not have the financial resources on their own to develop alternative water supplies, and small water supply augmentation projects can be prohibitively expensive due to the lack of economies of scale. The solution to fragmentation is for small utilities to cooperatively develop alternative water supply projects. System interconnections can also increase resilience under current and future climate conditions, particularly if the systems use different water supplies.

The prognosis for less developed countries, particularly rural areas, is guarded. It is widely recognized that poor developing countries are more vulnerable to climate change than wealthier nations (Mendelsohn et al. 2006; Fankhauser and McDermott 2014), and that "the phenomenon's negative effect on quantity and quality of fresh water resources in these nations is becoming ever more apparent" (United Nations 2019). Groundwater use may already be unsustainable, groundwater governance is poor, and overall adaptive capacity is low. With respect to rural water supplies in Africa, accelerating groundwater development for irrigation could increase food security but could also threaten domestic supplies and, in some areas, lead to groundwater depletion (Bonsor et al. 2010). Bonsor et al. (2010, p. 25) concluded that "Although climate change will undoubtedly be important in determining future water security, other drivers (such as population growth and rising food demands) are likely to provide greater pressure on rural water supplies." Other rural areas are facing similar water supply challenges.

Adaptation options to climate change in rural areas will involve mainly community-scale interventions such as rainwater harvesting and small-scale managed aquifer recharge, and the drilling of new, deeper wells. The design and construction of such projects will likely require external support from either national or international agencies or non-governmental organizations. Resilience to droughts in sub-Saharan Africa, for example, can be improved with hand-pumped and motorized wells accessing deep (>30 m) groundwater (MacAllister et al. 2020). Increases in the functionality of wells can be achieved through the initiation of an intensive program of preventative operation and maintenance (MacAllister et al. 2020). It has been estimated that 50,000 water supply points (i.e., wells) in Africa have fallen into disrepair because the donors, governments, and nongovernmental organizations that built the infrastructure ignored the need to maintain it (Skinner 2009). Where water supplies have failed in rural areas of developing countries, the response has often been the retreat of populations to already crowded urban areas or to other rural areas to the detriment of their water resources.

References

Allen DM, Cannon AJ, Toews MW, Scibek J (2010) Variability in simulated recharge using different GCMs. Water Resour Res 46(10)

Barbier E (2019) The water paradox: overcoming the global crisis in water management. Yale University Press, New Haven

Barnett C (2011) Blue revolution: unmaking America's water crisis. Beacon Press, Boston

Barsugli J, Anderson C, Smith JM, Vogel JM (2009) Options for improving climate modeling to assist water utility planning for climate change. Water Utility Climate Alliance, Las Vegas

Boardman A, Greenberg D, Vining A, Weimer D (1996) Cost-benefit analysis: concepts and practice. Prentice Hall, Upper Saddle River

Bonsor HC, MacDonald AM, Calow RC (2010) Potential impact of climate change on improved and unimproved water supplies in Africa. RSC Issues Environ Sci Technol 31:25–50

Carter RC, Parker A (2009) Climate change, population trends and groundwater in Africa. Hydrol Sci J 54(4):676–689

Chen J, Brissette FP, Leconte R (2011) Uncertainty of downscaling method in quantifying the impact of climate change on hydrology. J Hydrol 401(3–4):190–202

Chou H, Buck CR, Zikalala G, Medellín-Azuara J, Lund J (2012) Updating the CALVIN hydro-economic optimization model of California: Central Valley groundwater. In: World environmental and water resources congress 2012: crossing boundaries, pp 2482–2490

Clayton JA (2009) Market-driven solutions to economic, environmental, and social issues related to water management in the western USA. Water 1:19–30

Cornejo PK, Santana MV, Hokanson DR, Mihelcic JR, Zhang Q (2014) Carbon footprint of water reuse and desalination: A review of greenhouse gas emissions and estimation tools. J Water Reuse Desalin 4(4):238–252

Crosbie RS, Dawes W R, Charles SP, Mpelasoka FS, Aryal S, Barron O, Summerell GK (2011) Differences in future recharge estimates due to GCMs, downscaling methods and hydrological models. Geophys Res Lett 38(11)

deMause N (2019) As the sea rises, will resiliency—rather than retreat—be enough to save waterfront NYC? City Limits. https://citylimits.org/2019/04/17/nyc-sea-rise-resiliency-or-retreat/

De Villiers M (1999) Water: fate of our most precious resource. Stoddart, Toronto

Fankhauser S, McDermott TK (2014) Understanding the adaptation deficit: why are poor countries more vulnerable to climate events than rich countries? Glob Environ Change 27:9–18

Folger P, Carter NT (2016) Sea-level rise and U.S. coasts: science and policy considerations. Congressional Research Service, Washington, DC

Freas K, Bailey R, Muneavar A, Butler S (2010) Incorporating climate change in water planning. In: Howe C, Smith JB, Henderson J (eds), Climate change and water. International perspectives on mitigation and adaptation. American Water Works Association, Denver and IWA Publishing, London, pp 173–182

Ghaffour N, Missimer TM, Amy GL (2013) Technical review and evaluation of the economics of water desalination: current and future challenges for better water supply sustainability. Desalination 309:197–207

Glennon RJ (2002) Water follies: groundwater pumping and the fate of America's fresh waters. Island Press, Washington, D.C.

Glennon RJ (2009) Unquenchable. America's water crisis and what to do about it. Island Press, Washington, D.C.
Godman RR, Kuyk DD (1997) A dual water system for Cape Coral. J Am Water Works Assoc 89 (7):45–53
Goulden M, Conway D, Persechino A (2009) Adaptation to climate change in international river basins in Africa: a review. Hydrol Sci J, 54(5):805–828
Harris A (2020) Feds consider a plan to protect Miami-Dade from storm surge: 10 to 13-foot walls by the coast. Miami Herald (February 07, 2020). https://www.miamiherald.com/news/local/environment/article239967808.html. Accessed 9 June 2020
Hassaan MA, Abdrabo MA (2013) Vulnerability of the Nile Delta coastal areas to inundation by sea level rise. Environ Monit Assess 185(8):6607–6616
Henderson J, Akers M (2008) Can markets improve water allocation in rural America? Federal Reserve Bank Kansas Econ Rev Fourth Quart 2008:97–117
Hereher ME (2010) Vulnerability of the Nile Delta to sea level rise: an assessment using remote sensing. Geomat Nat Hazards Risk 1(4):315–321
Horton BP, Khan NS, Cahill N, Lee JSH, Shaw TA, Garner AJ, Kemp AC, Engelhart SE, Rahmstorf S (2020) Estimating global mean sea-level rise and its uncertainties by 2100 and 2300 from an expert survey. NPJ Clim Atmos Sci, 3, 18. https://doi.org/10.1038/s41612-020-0121-5
Huber-Lee A, Schwartz C, Sieber J, Goldstein J, Purkey D, Young C, Soderstrom E, Henderson J, Raucher R (2006) Decision support systems for sustainable water supply planning. AWWA Research Foundation, Denver
IPCC (2013) In: Stocker TF, Qin D, Plattner G-K, Tignor M, Allen SK, Boschung J, Nauels A, Xia Y, Bex V, Midgley PM (eds) Climate change 2013: the physical science basis. Contribution of working group I to the fifth assessment report of the Intergovernmental Panel on Climate Change. Cambridge University Press, Cambridge, UK
Kochenderfer MJ (2015) Decision making under uncertainty: theory and application. MIT Press, Cambridge, MA
Kotb TH, Watanabe T, Ogino Y, Tanji KK (2000) Soil salinization in the Nile Delta and related policy issues in Egypt. Agric Water Manag 43(2):239–261
Kulp SA, Strauss BH (2019) New elevation data triple estimates of global vulnerability to sea-level rise and coastal flooding. Nat Commun 10(1):1–12
Lempert RJ, Schlesinger ME (2000) Robust strategies for abating climate change. Clim Change 45 (3–4):387–401
Leverenz HL, Tchobanoglous G, Asano T (2011) Direct potable reuse: a future imperative. J Water Reuse Desalin 1(1):2–10
Lindsey R (2019) Climate change: global sea level (November 19, 2019). https://www.climate.gov/news-features/understanding-climate/climate-change-global-sea-level. Accessed 10 June 2020
Lohse KA, Gallo EL, Kennedy KR (2010) Possible tradeoffs from urbanization on groundwater recharge and water quality. Southwest Hydrol 18–32
Lund JR, Howitt RE, Medellín-Azuara J, Jenkins MW (2009) Water management lessons for California from statewide hydro-economic modeling using the CALVIN Model. University of California, Davis
MacAllister DJ, MacDonald AM, Kebede S, Godfrey S, Calow R (2020) Comparative performance of rural water supplies during drought. Nat Commun 11(1099)
Mahmood R, Pielke RA Sr, Hubbard KG, Niyogi D, Bonan G, Lawrence P, McNider R, McAlpine C, Etter A, Gameda S, Qian B (2010) Impacts of land use/land cover change on climate and future research priorities. Bull Am Meteor Soc 91(1):37–46
Maliva RG (2019) Anthropogenic aquifer recharge. Springer, Cham
Mendelsohn R, Dinar A, Williams L (2006) The distributional impact of climate change on rich and poor countries. Environ Dev Econ 11(2):159–178
Miselis JL, Lorenzo-Trueba J (2017) Natural and human-induced variability in Barrier-Island response to sea level rise. Geophys Res Lett 44(23):11922–11931

Moore LJ, List JH, Williams SJ, Stolper D (2007) Modeling barrier island response to sea-level rise in the Outer Banks, North Carolina. In: Coastal sediments' 07, pp 1153–1164

Morrison J (2019) Who will pay for the huge costs of holding back rising seas? Yale Environment 360. https://e360.yale.edu/features/who-will-pay-for-the-huge-costs-of-holding-back-rising-seas/. Accessed 10 June 2020

Mullin M (2020) The effects of drinking water service fragmentation on drought-related water security. Science 368(6488):274–277

National Research Council (1992) Water transfers in the west, efficiency, equity, and the environment. National Academy Press, Washington, D.C.

Nicholls RJ, Hanson S, Herweijer C, Patmore N, Hallegatte S, Corfee-Morlot J, Chateau J, Muir-Wood R (2007) Ranking of the world's cities most exposed to coastal flooding today and in the future. OCED (Organisation for Economic Co-operation and Development), Paris

Nienhuis JH, Lorenzo-Trueba J (2019) Simulating barrier island response to sea level rise with the barrier island and inlet environment (BRIE) model v1.0. Geosci Model Dev 12(9):4013–4030

NOAA (n.d). U.S. sea level trends map. https://tidesandcurrents.noaa.gov/sltrends/sltrends.html. Accessed 11 June 2020

OECD (2019) Responding to rising seas: OECD country approaches to tackling coastal risks. OECD Publishing, Paris. http://www.oecd.org/environment/cc/policy-highlights-responding-to-rising-seas.pdf. Accessed 10 June 2020

Office of Technology Assessment (1983) Institutions affecting western agricultural water use. In: Water-related technologies for sustainable agriculture in the U.S. arid/semiarid lands (OTA-F-212). U.S. Congress Office of Technology Assessment, Washington, D.C., pp 109–145)

Pearce F (2006) When rivers run dry. Beacon Press, Boston

Pielke RA Sr, Pitman A, Niyogi D, Mahmood R, McAlpine C, Hossain F, Goldewijk KK, Nair U, Betts R, Fall S, Reichstein M (2011) Land use/land cover changes and climate: Modeling analysis and observational evidence. Wiley Interdiscip Rev Clim Change 2(6):828–850

Pilkey OH, Linda Pilkey-Jarvis L, Pilkey KC (2016) Retreat from a rising sea: hard choices in an age of climate change. Columbia University Press, New York

Postel S (1992) Last oasis: facing water scarcity. WW Norton, New York

Purkey DR, Huber-Lee A (2006) A DSS for long-term water utility planning. Southwest Hydrol 5 (4):18–31

Reisner M (1999) Cadillac desert: the American West and its disappearing water, 2nd edn. Penguin, New York

Saintilan N, Khan NS, Ashe E, Kelleway JJ, Rogers K, Woodroffe CD, Hortin BP (2020) Thresholds of mangrove survival under rapid sea level rise. Science 368(6495):1118–1121

Scanlon BR, Reedy RC, Stonestrom DA, Prudic DE, Dennehy KF (2005) Impact of land use and land cover change on groundwater recharge and quality in the southwestern US. Glob Change Biol 11(10):1577–1593

Siders AR (2019) Managed retreat in the United States. One Earth 1(2):216–225

Siders AR, Hino M, Mach KJ (2019) The case for strategic and managed climate retreat. Science 365(6455):761–763

Sieber J, Purkey D (2011) WEAP: water evaluation and planning system. User guide. Stockholm Environment Institute, US Center, Somerville, MA. https://www.weap21.org/downloads/WEAP_User_Guide.pdf. Accessed June 12, 2020

Skinner J (2009) Where every drop counts: tackling rural Africa's water crisis. IIED Briefing. International Institute for Environment and Development. https://pubs.iied.org/pdfs/17055IIED.pdf. Accessed 12 June 2020

Solomon S (2010) Water. The epic struggle for wealth, power, and civilization. Harper Collins, New York

Stanley JD, Clemente PL (2017) Increased land subsidence and sea-level rise are submerging Egypt's Nile Delta coastal margin. GSA Today 27(5):4–11

Turton AR (1999) Water scarcity and social adaptive capacity: towards an understanding of the social dynamics of water demand management in developing countries. MEWRE Occasional Paper No. 9. School of Oriental and African Studies, University of London

UC Davis (2020) CALVIN Project Overview. University of California, Davis Center for Watershed Sciences. https://calvin.ucdavis.edu/node. Accessed 12 June 2020

United Nations (2019) Unprecedented impacts of climate change disproportionately burdening developing countries, delegate stresses, as Second Committee concludes general debate (press release 8 October 2019). https://www.un.org/press/en/2019/gaef3516.doc.htm. Accessed 12 June 2020

Williams AP, Cook ER, Smerdon JE, Cook BI, Abatzoglou JT, Bolles K, Baek SH, Badger AM, Livneh B (2020) Large contribution from anthropogenic warming to an emerging North American megadrought. Science 368(6488):314–318

Yates D, Sieber J, Purkey D, Huber-Lee A (2005) WEAP21—a demand-, priority-, and preference-driven water planning model. Water Int 30(4):487–500

Zuidema T (2019) The Dutch are building a barricade against climate change. PBS. https://www.pbs.org/wnet/peril-and-promise/2019/07/dutch-barricade-against-climate-change/. Accessed 11 June 2020

Printed in the United States
by Baker & Taylor Publisher Services